DEEP-SEA BIODIVERSITY

DEEP-SEA BIODIVERSITY

Pattern and Scale

MICHAEL A. REX

RON J. ETTER

HARVARD UNIVERSITY PRESS | Cambridge, Massachusetts | London, England 2010

Library of Congress Cataloging-in-Publication Data

Rex, Michael A., 1946

Deep-sea biodiversity : pattern and scale / Michael A. Rex, Ron J. Etter.

p.   cm.

Includes bibliographical references and index.

ISBN 978-0-674-03607-9 (alk. paper)

1. Deep-sea biology.   2. Marine biodiversity.   I. Etter, Ron J., 1955–

II. Title.

QH91.8.D44R49   2010

578.77′9—dc22   2009024415

Any opinions, findings, and conclusions or recommendations expressed in this material are those
of the author(s) and do not necessarily reflect the views of the National Science Foundation
(OCE-0726382).

For R. R. Hessler and H. L. Sanders

CONTENTS

Color illustrations follow page 50

This book is a synthesis of geographic patterns of biodiversity in the deep-sea benthos that inhabit soft sediments. Patterns of biodiversity in terrestrial environments have been a central focus of ecology for more than a century and are the primary inspiration for modern ecological theory. However, our knowledge of these patterns in the deep sea extends back only to the 1960s, when Robert R. Hessler and Howard L. Sanders of the Woods Hole Oceanographic Institution developed the first effective gear for collecting the small macrobenthos that dominate biodiversity and designed a program to systematically sample the benthos from the continental shelf to the abyssal plain. Their discovery that deep-sea species diversity exceeds that of coastal systems (Hessler and Sanders 1967) completely changed the way we look at life in the deep sea and added a whole new dimension to our understanding of global biodiversity. Sanders' (1968) stability-time hypothesis was the first attempt to place deep-sea diversity within a conceptual context, transforming what had been a purely descriptive field dating back to the H.M.S. *Challenger* expedition (1872–1876) into a theory-driven science of quantitative sampling and experimentation. Their revolutionary research and the rapid development of new sampling technology and deep-sea submersibles stimulated an unprecedented expansion of work on deep-sea ecology that has revealed not only a rich fauna uniquely adapted to this environment but also a complex seascape of unimagined habitats, including hydrothermal vents, cold seeps, soaring massive carbonate towers, frozen methane hydrates, asphalt and mud volcanoes, cold-water coral reefs, and a topography every bit as varied and spectacular as that of the terrestrial realm. Far from being the isolated, stable environment it was long assumed to be, the deep sea is now understood to be a dynamic and integral part of the global biosphere. Since most of the deep sea remains unexplored, we can hardly guess what other wonders exist there.

This book focuses on biodiversity in the immense expanse of soft sediments that cover most of the deep seafloor. Since Hessler and Sanders' pioneering studies, a large international literature has appeared on community structure in soft sediments, but there has been little effort to synthesize patterns and relate them to ecological and evolutionary processes that act at different scales of time and space. We do not imagine that our efforts to do this are in any way definitive or comprehensive. Quite the contrary— important gaps in existing information will be only too obvious to the reader, and the generality of our conclusions can only be tested by further research and exploration. However, we felt that given the huge proliferation of data it was time to attempt a synthesis and that the results would be useful in identifying new hypotheses and in planning and coordinating future research programs. We were also motivated by the fact that deep-sea ecology has not been incorporated into mainstream ecology, which is based on terrestrial, freshwater, and marine coastal systems. One can scarcely find the term "deep sea" in the indices of ecology textbooks and major reference works. This is partly because deep-sea ecology grew out of the tradition of oceanography and has persisted largely as a separate discipline with its own institutions, scientific journals, societies, and scientific meetings. We hope that this book will make ecological patterns in the deep sea more generally accessible. It is important to recognize that the deep sea covers fully two-thirds of the globe and is now known to harbor high biodiversity at all levels of organization: genetic, species, community, and seascape. Current ecological and evolutionary theory is based on only a subset of living phenomena —organisms found outside of the deep sea.

The deep sea is an energy-poor environment, and many features of community structure and evolutionary opportunity seem linked to food supply. The most reliable indication of nutrient input, especially at larger scales, is benthic standing stock, which is the raw material that is ultimately shaped into biodiversity. Hence, our first task was to compile a global database on abundance and biomass of the four principal size categories of deep-sea organisms: the bacteria, meiofauna, macrofauna, and megafauna. Chapter 1 describes patterns of standing stock among the four groups with depth and among ocean basins and reviews the process of pelagic-benthic coupling that regulates them. Subsequent chapters cover patterns of alpha species diversity at local, regional, and among-basin scales; changes in diversity through geological time; and depth-related beta diversity. Most of our analyses of diversity represent the North Atlantic and adjacent seas, which are the most intensively sampled deep basins.

While much has been learned about patterns of diversity and their potential ecological causes, little is known about the evolutionary-historical processes that have generated this rich and highly endemic fauna. How and where did all of these species originate? Much of our own research during the past 15 years has been devoted to developing molecular genetic approaches to assess geographic variation in deep-sea mollusks in order to find answers. We summarize this and related research to propose a model of population differentiation and speciation in the deep-sea benthos. Finally, we attempt to integrate patterns of deep-sea biodiversity across scales and relate them to the ecological and evolutionary processes that potentially shape them.

Writing this book has been a humbling experience. We were overwhelmingly impressed by the sheer volume of outstanding research, most of it logistically difficult, that has been accomplished during just a few decades of intense effort by deep-sea ecologists. The constant flow of new and exciting discoveries and the generation of new ideas have been exhilarating to everyone privileged to be involved.

We are very grateful for the help that colleagues have provided. Foremost, we thank Robert R. Hessler and Howard L. Sanders, to whom this book is dedicated. Their innovative science, encouragement, and generosity inspired our own pursuit of deep-sea ecology. It is hard to think of any topic in this book that does not stem in some way from their research interests or from many hours of pleasant and stimulating conversation. Ann Downer-Hazell and Michael Fisher, our editors at Harvard University Press, invited us to write this book and provided much-needed encouragement and advice. Peter Strupp of Princeton Editorial Associates managed the entire production process. Support for the book came from the Joseph P. Healey Grant Program at the University of Massachusetts–Boston; from CeDAMar (Census of the Diversity of Abyssal Marine Life), a division of Census of Marine Life; and from Harvard University Press.

Many of the ideas expressed here and the published articles cited originated from the Deep-Sea Biodiversity Working Group, chaired by M.A.R., at the National Center for Ecological Analysis and Synthesis in Santa Barbara, California. Jack Cook of the Woods Hole Oceanographic Institution's graphic services department expertly prepared many of the figures. Maria Mahoney did a superb job of typing and helping to organize the entire manuscript and handling all of the requests to authors and publishers to use figures and illustrations. She was untiringly cheerful and supportive throughout, despite the chaotic state of the material she received from the authors. We are very grateful to Andrea Rex for patiently reading, editing, and improving the text.

The staff of the Ernst Mayr Library of the Museum of Comparative Zoology, Harvard University, was extremely helpful in locating the literature used to compile the databases on standing stock and species diversity. The Ocean Sciences Division of the National Science Foundation has generously funded our deep-sea research programs for many years.

For fruitful discussions, data, photographs, and illustrations we especially thank Carol Stuart, Craig Smith, Lisa Levin, Gilbert Rowe, Gordon Paterson, John Lambshead, Craig McClain, Ingrid Krönke, Thomas Soltwedel, Hjalmar Thiel, Cindy Lee Van Dover, John Zardus, Moriaki Yasuhara, John Allen, Thomas Cronin, Richard Haedrich, Krista Baker, Paul Snelgrove, Peter Auster, Stacey Doner, Frederick Grassle, Kenneth Smith, Jr., James Blake, Elizabeth Boyle, Selina Våge, George Wilson, Philippe Bouchet, Frank Müller-Karger, Remy Luerssen, James Yoder, Maureen Kennelly, Anders Warén, Joëlle Galéron, Stefanie Keller, George Hampson, James Barry, Linda Kuhnz, Nicholas Johnson, Scott France, Jesús Pineda, Paul Tyler, John Gage, David Billett, Joan Bernhard, Ann Vanreusel, Maarten Raes, Robert Carney, Chih-Lin Wei, Ian MacDonald, Myriam Sibuet, John Gray, Angelika Brandt, Brigitte Ebbe, Dieter Fiege, Jody Deming, Pedro Martinez, Nancy Maciolek, Brian Bett, Andrew Gooday, Anastasios Tselepides, Dorte Janussen, Julian Gutt, David Thistle, Linda Sedlacek, Barbara Hecker, Bruce Corliss, and Elva Escobar-Briones. Any errors, misinterpretations, and other shortcomings are entirely our own.

DEEP-SEA BIODIVERSITY

1

SETTING THE STAGE: PATTERNS OF BENTHIC STANDING STOCK

Without bold, regular patterns in nature, ecologists do not have anything very interesting to explain.

John H. Lawton (1996)

You can't make progress on processes without understanding the patterns.

Anthony J. Underwood et al. (2000)

The story of deep-sea ecology begins, paradoxically, in the sunlit waters of the ocean's surface. Except for isolated chemosynthetic systems (Van Dover 2000, Tunnicliffe et al. 2003, Levin 2005), which are the only sources of in situ primary production, energy supply to the deep-sea benthos originates as surface production that sinks through the water column. Regulation of photosynthesis, climatic forcing, and trophic dynamics of surface production are highly complex and are not the subject of this book, but an excellent account of them can be found in Falkowski and Raven (2006).

Most production of organic carbon is recycled within the euphotic zone by a highly efficient network of microbial action and zooplankton grazing. Carbon flux downward through the aphotic water column varies with depth, distance from productive coastal waters, and seasonality. However, on average, only about 1% of surface production reaches the deep-sea floor (Lampitt and Antia 1997, Fischer et al. 2000), making the deep sea an extremely energy-poor environment. In this chapter we first discuss the spatial and temporal patterns of surface production, the export and gravitational settling of particulate organic matter, and the benthic response to nutrient input. These processes, together referred to as pelagic-benthic coupling, determine geographic and temporal variation of standing stock in the deep-sea benthos. We then present a global-scale synthesis of bathymetric and among-basin patterns of biomass and abundance and relate these to food supply. As we show in subsequent chapters, patterns of standing stock are essential to understanding both the community structure of the deep-sea benthos and the evolutionary potential of the deep-sea ecosystem.

SURFACE PRODUCTION

Satellite imagery of surface chlorophyll concentration has made it possible to assess patterns of primary productivity at regional and global scales and over daily to decadal time scales (Campbell and Aarup 1992, Antoine et al. 1996, Behrenfeld and Falkowski 1997, Yoder et al. 2001, Longhurst 2007). Satellite-derived chlorophyll distribution data have revolutionized plankton ecology, greatly improved our understanding of the global carbon cycle and climate change, and now make it possible to consider the large-scale distribution of the potential food supply in the deep sea.

Spatiotemporal variation in production on an oceanwide scale is shown in Figure 1.1 for the North Atlantic (Sun et al. 2006). Annual production shows strong geographic variation (Figure 1.1A). Production is generally highest in the coastal regions, particularly in areas of upwelled nutrient-rich waters, such as off West Africa, or river discharge, as from the Orinoco and the Amazon; it is lowest in subtropical central gyres. It also tends to increase with latitude, although this pattern is modulated by current patterns and regional productive hotspots. Productivity is not only higher, but also more seasonal at higher latitudes (Figure 1.1B), and this has important implications for export flux to the deep ocean: when primary production is highly pulsed, population growth rates of phytoplankton and zooplankton can become decoupled, allowing more phytoplankton to sink before it is consumed. Seasonal plankton blooms also promote the growth of larger, denser phytoplankton cells such as diatoms, which sink faster (Ducklow et al. 2001).

On regional scales, surface production is spatially and temporally very dynamic. Figure 1.2 shows sea surface temperature and production in the western North Atlantic on a single spring day (the complete seasonal cycle of production can be found in Yoder et al. 2002). The temperature image clearly reveals the meandering warm core of the Gulf Stream and the complex eddy systems that it spawns. The image of surface production taken simultaneously shows three eddies being formed northwest of the Gulf Stream (anticyclonic warm-core rings) that convey warm nutrient-poor water from

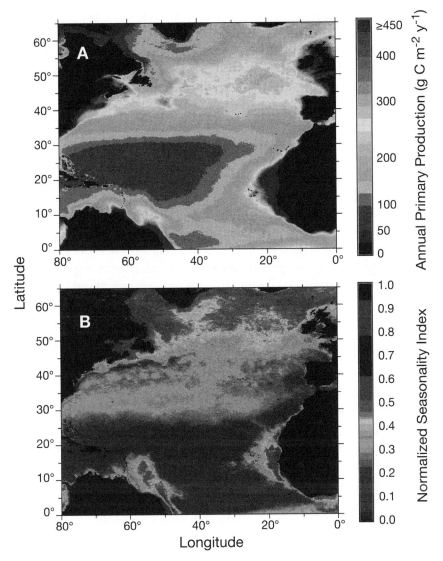

Figure 1.1. (A) Annual surface productivity (g C m^{-2} y^{-1}) in the North Atlantic estimated from SeaWiFS satellite imagery. (B) Seasonality of surface production measured as a normalized Berger and Wefer (1990) seasonality index. High values of the index represent a more seasonal pattern of production. Higher and more seasonally pulsed production is generally found in coastal waters and at high latitudes. A color version of this figure is included in the insert following p. 50. Modified from Sun et al. (2006), courtesy of Bruce Corliss, and reproduced with permission of the authors and Elsevier.

the Sargasso Sea toward the continental shelf. Eddies can also advect productive coastal waters seaward (Crawford et al. 2005). Nutrient upwelling
caused by eddies promotes localized surface production (McGillicuddy et al.
1998, Oschlies and Garçon 1998). Oceanographic frontal systems created by
colliding water masses enhance production and result in heavy export flux
and sedimentation of organic carbon to the bottom (Ryan et al. 1999, Nodder et al. 2003, Martin and Sayles 2004, Grove et al. 2006). A similar phenomenon is observed in narrow zones, even far out at sea, induced by
upwelling along major equatorial currents (C. R. Smith et al. 1996). Patterns
of surface production are translated down through the water column to the
seabed as particulate organic carbon (POC) flux. Thus, at very large scales,
surface production is a template of food supply to the benthos, although it
is a template that becomes distorted and fainter with increasing depth. The
distribution of total organic carbon in surface sediments of the deep sea rec-
ognizably mirrors surface production at global scales, but at regional scales
there are marked differences owing to offshore transport, fluvial input, and
lateral advection (Seiter et al. 2004, 2005).

ORGANIC CARBON FLUX TO THE BENTHOS

Much has been learned from sediment trap studies (Honjo et al. 2008) concerning the nature of sinking material and its rate and pattern of descent.
Sediment traps moored at specific depths above the bottom intercept settling POC with a funnel that collects and concentrates the material in a
sequence of rotating compartments, where it is preserved. On recovery, the
amount of material can be used to estimate the rate of flux integrated over
the time that individual compartments are exposed, as well as its variation
during deployment, which can last for a year or more. Sediment traps provide reasonably consistent and reproducible results, especially when moored
several hundred meters above the seafloor, where the vertical component
of the flux is less compromised by the lateral advection of POC in near-
bottom currents.

Most of the downward flux of POC from the euphotic zone to the deep
sea is in the form of phytodetrital aggregates (Beaulieu 2002). These consist
of intact and degraded phytoplankton cells, particularly diatoms and cocco-
lithophores (which are relatively large and contain dense mineral support
structures), foraminiferans, zooplankton exoskeletons, fecal pellets, protozoans,
and bacteria, all packaged in a gelatinous matrix measuring millimeters to

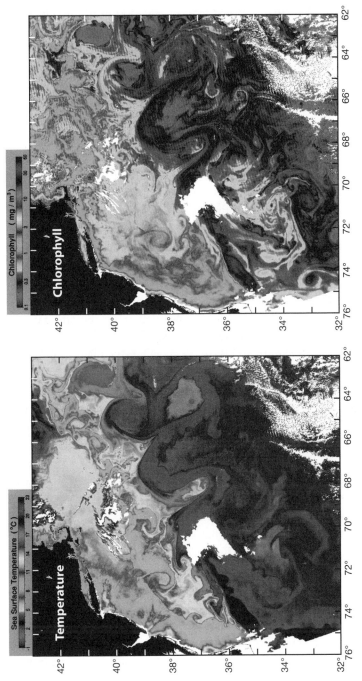

Figure 1.2. MODIS (Moderate Resolution Imaging Spectroradiometer) images from the National Aeronautics and Space Administration's Aqua Satellite of sea surface temperature and surface chlorophyll concentration in the western North Atlantic on April 18, 2005. Three anticyclonic Gulf Stream eddies (warm core rings) can be observed northwest of the Gulf Stream. The images illustrate the extraordinary spatial complexity of oceanographic features and surface production at regional scales. A color version of this figure is included in the insert following p. 50. Images courtesy of NASA and GeoEye.

centimeters in size. Aggregates sink at rates of 100–300 m d^{-1}, accelerating with increased depth (Berelson 2001, 2002). Consequently, biogenic material produced at the surface can reach bathyal depths on the continental margins in days to weeks and the abyssal plains in weeks to about a month, much faster than had previously been imagined. Most of the organic matter exported from the euphotic zone is respired in the upper 2000 m of the water column by surface-derived bacteria that gradually become inhibited by the physical conditions in the ocean's interior. At greater depths the aggregates are colonized by barophilic bacteria and flagellates, which continue to actively degrade organic carbon (Lochte and Turley 1988).

Carbon flux is increased by around 10% through the vertical diel migration of zooplankton, which consume phytoplankton near the surface and transport it actively in their guts to mesopelagic depths (Ducklow et al. 2001). In addition to the rain of phytodetrital aggregates, there is a component of free-sinking zooplankton fecal pellets (Fowler and Knauer 1986, Graf 1989, Pfannkuche and Lochte 1993, Ramaswamy et al. 2005) and discarded suspension-feeding structures of larvaceans (Robison et al. 2005). Superimposed on this background of small sinking debris are sporadic rapidly sinking parcels such as carcasses of dead fish (Stockton and DeLaca 1982), marine mammals (C. R. Smith and Baco 2003), jellyfish (Billett et al. 2006), macroalgae (Harrold et al. 1998), and terrestrial plant material (Wolff 1979). We have more to say about these patchy food sources in our discussion on how local diversity is maintained. In addition to vertical flux through the water column, there can be significant river discharge (Müller-Karger et al. 1988, 1991), lateral advection in currents, and downslope transport of organic material on continental margins (Hecker 1990a, Van Weering et al. 2001, Belicka et al. 2002), particularly in submarine canyons that cut the slope face (Vetter and Dayton 1998, Okey 2003).

Vertical particulate flux is seasonal and closely linked to annual cycles of surface production, particularly at higher latitudes (Deuser and Ross 1980). Flux can also occur episodically during periods of normally low surface production from transient phytoplankton blooms induced by wandering mesoscale eddies (Conte et al. 2003). After spring blooms, sedimentation can be so heavy that phytodetritus accumulates as an amorphous flocculant green fluff on the seabed, which lasts for weeks before it is remineralized by bacteria or consumed by foraminiferans and metazoans (Billett et al. 1983). Figure 1.3 illustrates a striking example of this in the abyssal eastern North Atlantic, where green phytodetritus has collected in depressions after being

Figure 1.3. Fresh phytodetritus accumulated in depressions on the seafloor of the Porcupine Abyssal Plain, eastern North Atlantic (4850 m). A color version of this figure is included in the insert following p. 50. Photograph courtesy of Andrew Gooday, National Oceanography Centre, Southampton, UK.

redistributed on the bottom by internal tidal currents, creating a patchy distribution of food.

Given the huge spatiotemporal variability of surface production and export flux, as well as the potential for horizontal dispersion of sinking particles by mesoscale eddies and deep currents, the organic carbon input to a particular site on the bottom must correspond to a large volume of ocean. Deuser et al. (1988) proposed the useful expression "statistical funnel" for the spatial domain supplying particles to a point in the deep-water column. The sea surface area contributing to the funnel is its catchment area. Comparisons of long-term flux measured at sediment traps with overhead production estimated from satellite imagery (Deuser et al. 1990) and modeling efforts (Siegel and Deuser 1997, Siegel and Armstrong 2002, Siegel et al. 2008) suggest that catchment areas have dimensions approximating hundreds of kilometers in diameter. Catchment areas appear to vary in size, shape, and position interannually, depending on changing oceanographic conditions (Waniek et al. 2005). There is also considerable geographic variation in the efficiency of export flux through the upper water column to the ocean's interior (Boyd and Trull 2007, Buesseler et al. 2007, Buesseler and Lampitt 2008).

Given the logistical difficulties and expense of deep-sea research, not to mention the vicissitudes of research funding, very few studies have been able to integrate the entire process of pelagic-benthic coupling from surface production, to flux through the water column, to the benthic response. The most comprehensive and long-term program to date was conducted by K. L. Smith and colleagues over a 15-year period from 1989 to 2004 at Station M (4100 m) in the northeast Pacific (Smith et al. 1994, Smith and Druffel 1998, Smith and Kaufmann 1999, Smith et al. 2001, 2002, 2006). For the last 8 years of this ongoing study it was possible to correlate POC flux measured by sediment traps with concurrent satellite-derived surface chlorophyll *a* distributions. Export flux from the surface was estimated using the models of Behrenfeld and Falkowski (1997) and Laws (2004) and was based on two circular catchment areas of 50 and 100 km radius and an oval measuring 200 × 600 km oriented parallel to the current.

Figure 1.4 shows the sea-surface temperature, the rate of surface chlorophyll production, the calculated net primary productivity, the export flux originating from the catchment areas, and the POC flux measured at sediment traps moored 50 and 600 m above the bottom. POC flux shows strong seasonal and interannual variation, as widely observed in sediment trap studies. Cross-correlation analyses showed that POC flux was most significantly correlated with export flux at a time lag of 1–3 months. Catchment area was relatively scale invariant, but the smallest area was associated with the most rapid response of POC flux, suggesting that flux has a strong vertical component.

POC flux was also correlated with two ocean-scale climate indices reflecting the El Niño–Southern Oscillation, the North Pacific Oscillation, and a regional upwelling index, which showed time lags of 6–10 months and 3 months, respectively. K. L. Smith et al. (2006) showed that an empirical multiple regression model, using POC flux as a response variable and export flux and climate indices with appropriate time lags as explanatory variables, predicts actual POC flux with a high degree of statistical significance (Figure 1.5A). The same basic multiple regression model was applied to data from another important long-term sediment trap study in the eastern North Atlantic (Lampitt et al. 2001), using regional measures of the explanatory variables. This also provided convincing and statistically significant ($P < 0.001$) agreement between actual and predicted POC flux (Figure 1.5B). From the

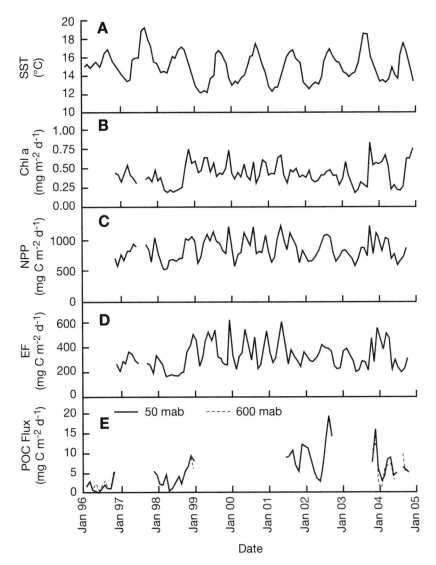

Figure 1.4. A 10-year time series of: (A) sea surface temperature (SST), (B) satellite-derived rate of production of chlorophyll *a* (Chl a) at the sea surface, (C) estimated net primary production (NPP), (D) estimated downward export flux (EF) from the euphotic zone, (E) particulate organic carbon (POC) flux measured at deep moored sediment traps at Station M in the eastern North Atlantic off California. The lines in panels A–D represent a surface catchment area of 50 km centered over the sediment traps. Catchment areas of a 100 km radius circle and an oval 200 km wide and 600 km long give similar results (K. L. Smith et al. 2006). In panel E the two lines indicate POC flux at sediment traps moored 50 and 600 m above the bottom (mab), 4050 and 3500 m in depth, respectively. From K. L. Smith et al. (2006), with permission of the authors and the American Society of Limnology and Oceanography.

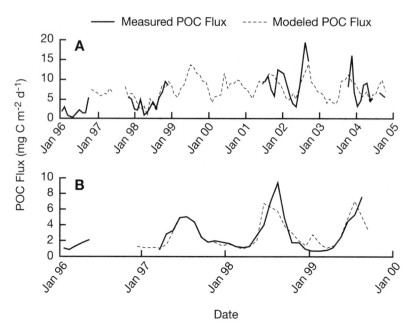

Figure 1.5. Time series of POC flux measured at sediment traps and modeled POC flux in the eastern North Pacific (A) and eastern North Atlantic (B). The predicted POC flux is from a multiple regression model that uses POC flux as a response variable and regional climatic indices and export flux estimated from satellite-derived surface chlorophyll concentration as explanatory variables. Agreement between measured and predicted POC flux is highly significant ($P < 0.001$). From K. L. Smith et al. (2006), with permission of the authors and the American Society of Limnology and Oceanography.

very consistent results at Atlantic and Pacific sites, Smith et al. (2006) suggested that it might now be possible to model long-term, large-scale, deep-sea processes by using satellite sensing and climate indices.

THE BENTHIC COMMUNITY

Before discussing the biological response to nutrient input at the seafloor, it is first necessary to introduce the cast of characters that make up benthic communities and describe how they are collected. As there are now extensive, accessible accounts of the taxonomic diversity and natural history of deep-sea organisms (Gage and Tyler 1991, Herring 2002, Koslow 2007, Nouvian 2007), we relegate most of this information to Appendix A, which readers can peruse according to their interest. In Appendix B, we show examples of deep-sea benthic sampling gear, concentrating on the main workhorses used during the past several decades to assess community structure. For more comprehensive accounts of sampling gear, see Rowe (1983), Gage and Tyler (1991), and Gage and Bett (2005).

Deep-sea remote sampling is fundamentally different from sampling more accessible and familiar environments. Until the recent development of high-resolution sonar imagery and computerized reconstructions of topography (see Figure 6.3), it was not possible to get a precise visual impression of the deep seascape on scales of 1–1000 km. As the deep sea is totally dark, photographs have narrow horizons at around 5 m, similar to that seen in Figure 1.3. Apart from limited benthic sampling from manned submersibles and remotely operated vehicles, most deep-sea sampling is still done blind by lowering gear from ships on cables to what is thought to be an appropriate habitat.

To help conceptualize this procedure, imagine yourself to be in a helicopter flying at around 5000 m and looking out the window during the day at the landscape below. To sample the landscape, you dangle a collecting device to the ground on a very long cable and then winch it back up. Suppose your terrestrial sampling gear collects organisms over an area of about 0.25 m^2, as with quantitative deep-sea sampling gear, or perhaps even 1000 m^2, as with larger qualitative trawls. How representative of the landscape do you expect your sample to be? Now imagine doing this at night without being able to observe the landscape! If you extend this analogy to the whole deep-sea ecosystem, which covers two-thirds of the planet and has been sampled quantitatively only a few thousand times, mostly in the Atlantic (see Figure 1.11), you will have an idea of just how meager is our knowledge of this environment and how much uncertainty and potential error attend the data that we do have.

The deep-sea benthos has traditionally been divided into four main categories according to organism size: bacteria, meiofauna, macrofauna, and megafauna. The categories are fairly coherent taxonomic assemblages that exhibit different spatial patterns of standing stock and may represent four peaks in a polymodal biomass-size spectrum for deep-sea organisms (Schwinghamer 1985, Lampitt et al. 1986). They also require different sampling methods. The persistent use of size-class distinctions in deep-sea research probably devolves, as much as anything, from the very different kinds of taxonomic expertise and technology required to study each group. Very few studies include all size categories assessed simultaneously at the same sites. (Examples of more integrated sampling programs are described in Sibuet et al. 1989, C. R. Smith et al. 1997, Galéron et al. 2000, Heip et al. 2001, Aller et al. 2002, Rowe et al. 2003, and Hughes and Gage 2004.)

The group containing the largest organisms is the megafauna, defined loosely as "animals readily visible in photographs" of the seafloor (Grassle et al. 1975, p. 458) or collected by large trawls with mesh sizes of 1–3 cm (Haedrich and Rowe 1977). The megafauna includes epibenthic mobile foragers (seastars, brittle stars, sea urchins, sea cucumbers, decapod crustaceans, sea spiders, cephalopods, and benthopelagic fishes), sessile epibenthic suspension feeders (crinoids, sponges, sea anemones, and corals), infaunal burrowing forms (heart urchins, some sea cucumbers, decapod crustaceans, sipunculids, echiurans, and enteropneusts), as well as giant scavenging amphipods and isopods. The latter are known from baited time-lapse cameras or traps (Isaacs 1969, Schulenberger and Hessler 1974), but their whereabouts while they are not actively scavenging remain a mystery. Most megafaunal elements range in size from centimeters to decimeters, although some fishes, cephalopods, stalked crinoids, gorgonian corals, and deep-burrowing forms attain sizes on the order of 1 m in length.

The next largest size category is the macrofauna. This group includes representatives of all major marine phyla, but it is dominated by polychaete worms, peracarid crustaceans, and mollusks, in that order of relative importance. These organisms are much smaller than the megafauna, ranging in size from less than 1 mm to around 1 cm. They are primarily infaunal, living in the top 1–5 cm of sediment or at the sediment-water interface. The macrofauna is defined by the mesh size of sieves used to separate animals from sediments. The current standard for American programs is 300 μm and for European Union programs 250 μm. These mesh sizes provide nearly identical results for estimating macrofaunal abundance and biomass (Gage et al. 2002).

The meiofauna consists of two different groups of minute organisms, the metazoan meiofauna and protozoan foraminiferans, which share the upper 1–5 cm of sediment with the macrofauna. The metazoan meiofauna is heavily dominated by nematodes (usually more than 75% and often as high as 90% or more of individuals), followed by harpacticoid copepods (3–12%) and rarer miniaturized species belonging to a variety of other invertebrate groups, including polychaetes, mollusks, turbellarians, ostracods, and pseudocoels (Vincx et al. 1994). This group is partly defined by sieve mesh size, the currently accepted lower limit being 32 μm (Soltwedel 2000). Most metazoan meiofaunal species range in size from 32 to 1000 μm. Since some meiofaunal individuals are retained on 300 μm sieves, and larvae and juveniles of some macrofaunal taxa pass through 300 μm sieves and are captured on 32 μm sieves, most contemporary studies define the metazoan meiofauna taxonomically to include nematodes and other typical meiofaunal groups, rather than by size alone. Deep-sea foraminiferans include calcareous, agglutinated, and soft-shelled taxa (Gooday 2003, Gooday et al. 2004). Despite their abundance and ecological importance (Bernstein and Meador 1979, Gooday 1986, Bernhard et al. 2008), surprisingly little is known about the geographic patterns of their density and especially of their biomass (Vincx et al. 1994, Soltwedel 2000). This is partly because it is difficult to unambiguously discriminate living from dead specimens and to determine the actual biomass of living protoplasm (Bernhard 1992, Gooday et al. 1992). Although many foraminiferans are of meiofaunal size, they are often as large as macrofaunal taxa, and some agglutinated species actually attain megafaunal size of around 10 cm (Gooday et al. 2001). If the even larger agglutinated protozoans called xenophyophores are actually foraminiferans, as some recent molecular genetic studies suggest (Pawloski et al. 2003), then members of this group reach 25 cm (Tendal 1972). Hence, they do not fit easily into the conventional scheme of size categories.

Bacterial standing stock is assessed by taking small subcores from box-core or multicore samples. Density is then estimated by DNA staining and epifluorescence microscopy (Deming and Colwell 1982), and biomass is determined by applying a conversion factor to cell count or mean biovolume (Deming and Yager 1992, Boetius et al. 1996) or by biochemical quantification of phospholipid concentration (Boetius and Lochte 2000). Although research on quantifying standing stock in deep-sea bacteria has developed more recently than work on the mega-, macro-, and meiofauna and there are still far fewer studies available, a fairly clear picture of bathymetric and geographic variation is beginning to emerge.

There also exists a deep-sea nanobiota consisting of protozoans, large prokaryotes, and yeastlike cells that span a size range of 2–50 μm, but are not as small as bacteria (Burnett 1973, 1981, Snider et al. 1984, Cunningham and Ustach 1992). Too little is presently known of this potentially important component of the benthos to evaluate geographic patterns of standing stock.

RESPONSE OF THE BENTHOS
TO FOOD SUPPLY

The composite response of the benthos to POC flux is estimated by sediment community oxygen consumption (SCOC), the rate of organic matter mineralization. SCOC can be measured by in situ respirometers deployed by submersibles (K. L. Smith et al. 1998) or by using free vehicles (Smith 1987). The latter are autonomous experimental packages that are emplaced on the seafloor and recovered at the surface by flotation devices after they acoustically release their ballast. The best long-time-series record for SCOC and its relationship to synchronously measured POC flux comes from Station M in the eastern North Pacific (Smith and Kaufmann 1999). Figure 1.6 shows a 7-year record of POC flux and SCOC. As noted earlier, POC flux varies on seasonal and interannual time scales; it peaks in summer and fall and is lowest during the winter. SCOC shows a similar, but more muted, pattern of variation; its fluctuations are an order of magnitude less than for POC. Average POC trends significantly downward over the 7-year period because of a long-term trend of decreasing surface production. Average SCOC, however, remains relatively constant. A plot of the ratio of POC to SCOC (Figure 1.7) shows a steady decline over the 7-year period, indicating a growing deficit in POC. At the end of the period in 1996, food supply was only about half of the metabolic demand of the benthos. Later measurements at the same site showed that the decline in POC flux was actually part of a decadal-scale cycle (Smith et al. 2006).

Other attempts to detect seasonal fluctuations in SCOC that correspond to variations in POC flux have yielded mixed results. Graf (1989) observed a significant increase in sediment oxygen consumption following a pulse of chlorophyll deposition at bathyal depths, but other studies have failed to find seasonal variation in SCOC that tracks changes in POC flux (Sayles et al. 1994, Lampitt et al. 1995, Witbaard et al. 2000). This lack of response may be related to differences in the quality of food, but these studies were also of shorter duration and had less continuity of sampling. The variation in SCOC shown for Station M (Figure 1.6B) makes it seem likely that detecting a convincing seasonal signal in SCOC requires regular sampling and a time series approaching a decade. Results of all of these studies are consistent with a uniform average rate of SCOC irrespective of variation in POC flux on annual to interannual time scales. If this is correct, the benthos can sustain prolonged periods of diminished food supply. However, it is important to point

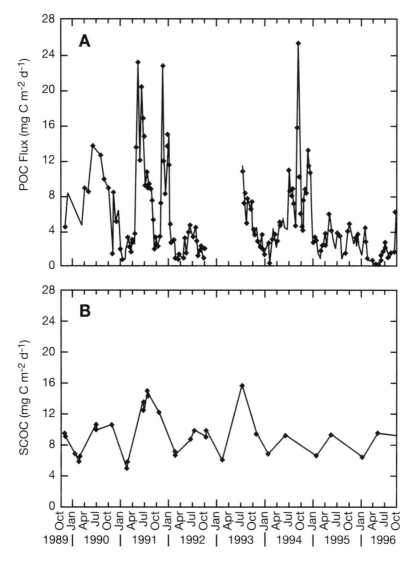

Figure 1.6. Synchronous time series of POC flux (A) and SCOC (B) at Station M in the eastern North Pacific. POC flux trends downward over the 7-year period, but SCOC remains relatively stable. From K. L. Smith and Kaufmann (1999), with permission of the authors and the American Association for the Advancement of Science.

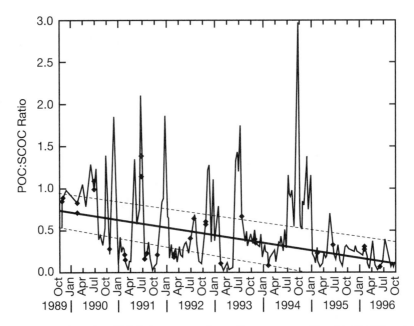

Figure 1.7. The ratio of POC flux measured at sediment traps and benthic SCOC at Station M in the eastern North Pacific. The decreasing trend in the ratio indicates a progressive deficit in food supply to the benthos over the 7-year period. From K. L. Smith and Kaufmann (1999), with permission of the authors and the American Association for the Advancement of Science.

out that SCOC is measured on very small spatial scales and largely reflects the metabolic demand of bacteria and meiofauna, rather than that of macrofauna and megafauna, which are more sparsely distributed.

Not all benthic organisms respond equally to phytodetrital pulses. From mid-bathyal to abyssal depths, benthic biomass and the process of organic matter mineralization are dominated by bacteria (Rowe et al. 1991). Tracer experiments using artificial enrichment of sediments show that bacteria begin to take up and assimilate labile material within hours to days (Moodley et al. 2002, Witte et al. 2003a, Bühring et al. 2006) and can exhibit a significant increase in biomass in days to weeks (Boetius and Lochte 1996), but deposition may have to be quite heavy for metabolic activity to be translated into population growth (Pfannkuche et al. 1999). Foraminifera also show rapid uptake of fresh organic matter (Moodley et al. 2002) and population increase in weeks to months, although this response appears to involve mainly a minority of opportunistic species that colonize the phytodetritus but are rare in the background sediment assemblage (Gooday 1988, Drazen et al. 1998, Heinz et al. 2002, Ernst and van der Zwaan 2004). Foraminifera appear to have at least two adaptive feeding types (Linke 1992). Some species feed continuously and have a fairly constant low metabolism. Others maintain a high metabolic rate, which allows them to exploit phytodetritus falls rapidly, but at the expense of metabolizing their own protoplasm during lean times. The elevation of SCOC following phytodetritus deposition at abyssal depths is largely attributable to bacteria and foraminifera, which can very efficiently exploit fresh organic material and reproduce asexually.

The response of metazoans is not as immediate and is more varied, probably owing to their diversity of lifestyles, lower rates of somatic growth, and the energetic cost of sexual reproduction (Gooday et al. 1996). Pfannkuche (1993) and Gooday et al. (1996) found no evidence of a meiofaunal population response to phytodetrital deposition in the eastern North Atlantic, although nematodes do show significant seasonal shifts of body-size spectra suggestive of an annual cycle of reproduction and mortality (Soltwedel et al. 1996). Over a 2-year period at Station M, Drazen et al. (1998) found no seasonal change in metazoa as a whole (meiofaunal and macrofauna combined), but did observe significant seasonal shifts in the density of two meiofaunal taxa, nematodes and harpacticoid copepods, with peak densities being reached 9 months after phytodetrital settlement events.

Owing to their much lower density and higher mobility, it is much more difficult to detect seasonal variation in the macro- and megafauna or to distinguish temporal from spatial changes owing to migration. In situ labeling experiments indicate that the macrofauna consume phytodetritus within days and can play a role in initial carbon degradation (Witte et al. 2003b), although their response varies among taxa and geographically for the macrofauna as a whole (Sweetman and Witte 2008a,b). At Station M, Drazen et al. (1998) observed seasonal density shifts in polychaetes, tanaids, and isopods, with peak densities attained 9 months after phytodetritus deposition, although trends among years were inconsistent. Pfannkuche (1993) found no significant change in macrofaunal biomass following depositional events in the eastern North Atlantic. A small number of species in the major deep-sea macrofaunal taxa—polychaetes, peracarid crustaceans, and mollusks—do show seasonal periodicity in reproductive cycles, especially at bathyal depths, which presumably relates to seasonal variation in food availability (reviewed in Young 2003).

An accumulation of phytodetritus in the Porcupine Abyssal Plain in the summer of 1996 was not followed by an increase in the abundance of either meiofauna or macrofauna as a whole (Galéron et al. 2001). The pulse did co-incide with a heavy recruitment event in one opportunistic meiofaunal polychaete and was followed by a weakly significant increase in macro-faunal polychaetes over the next 10–18 months (Galéron et al. 2001, Van-reusel et al. 2001). No phytodetrital cover was observed during the follow-ing 2 years, evidently because there was a dramatic increase in the abundance and activity of megafaunal grazers (holothurians and ophiuroids), which im-mediately consumed the phytodetritus as it reached the seafloor (Bett et al. 2001, Ginger et al. 2001, Iken et al. 2001). The increase in megafaunal abun-dance had occurred over a 10-year period from 1989 to 1999 (Billett et al. 2001). It was not accompanied by an increase in megafaunal biomass, and several species actually showed a concomitant decrease in body size. Al-though a number of megafaunal species participated, the boom was domi-nated by just one species, *Amperima rosea,* a previously rare small holothurian with early sexual maturity, high fecundity, and a feeding preference for phyto-detritus (Wigham et al. 2003a,b). The precise causes of this megafaunal pop-ulation explosion and whether it was part of a cycle or a stochastic event remain unclear.

At Station M in the eastern North Pacific, changes in the epibenthic megafauna were monitored over a 14-year period from 1989 to 2002. Ini-tial short-term analyses on annual time scales in the mid-1990s revealed no consistent changes in abundance or distribution (Lauerman and Kaufmann 1998). Eight large echinoderms had fairly steady abundance levels for a decade, from 1989 until 1999, but six of these experienced major changes following the 1997–1999 El Niño (Ruhl and Smith 2004). One abundant holothurian disappeared completely and the other species either increased or decreased by one to two orders of magnitude. The abundance shifts lagged El Niño indices by 1 to 2 years and the POC flux by 6 months to 1 year. The period from 1998 to 2002 was one of enhanced upwelling and POC flux, suggesting that different megafaunal species may have a competitive edge under circumstances of either high or low food availability. Studies at both Station M and in the eastern North Atlantic indicate that major pop-ulation level changes in the megabenthos can occur on interannual to decadal time scales. The Station M study strongly suggests that variation in invertebrate megafaunal abundance is linked to climatically driven changes in food supply and recruitment success (Ruhl and Smith 2004, Ruhl 2007, 2008). Megafaunal benthopelagic fishes also show strong shifts in abundance on interannual and decadal time scales, though this has been more difficult to associate with climate change and POC flux, possibly because of even larger lag effects (Bailey et al. 2006).

Integrated long-term studies of surface production, POC flux, SCOC, and benthic community structure have not been carried out in the tropical deep sea, where the lower variation in surface production (Figure 1.1B) would predict more stable nutrient input to the benthos. Short-term benthic sampling programs spanning 2 years, including the summer, fall, and winter seasons, detected no change in benthic abundances at abyssal depths in the tropical Atlantic (Cosson et al. 1997, Galéron et al. 2000).

Although there are very promising advances in determining the diets and resource partitioning within and among some deep-sea species (Miller et al. 2000, Wigham et al. 2003a, Madurell and Cartes 2006, Cartes et al. 2007) and the basic carbon budget of the benthic community (Pfannkuche 1992, Rowe et al. 2003), deciphering the precise trophic structure and dynamics of the benthic food web remains a major challenge. However, there is very compelling correlative evidence linking carbon flux to benthic standing stock. POC flux is a highly accurate predictor of bacterial biomass (Deming and Yager 1992, Deming and Baross 1993), which is a significant positive function of chloroplastic pigment equivalents (CPE) in sediments that reflect surface-derived phytodetritus. The correlation with CPE is weaker than that for POC flux, possibly because bacteria respond so rapidly to deposition events, but CPE, which becomes mixed into the sediments by bioturbation, represents food potential averaged over a longer time period (Pfannkuche and Soltwedel 1998). Meiofaunal biomass and abundance are also positive functions of POC flux (Tietjen et al. 1989, Danovaro et al. 1999) and CPE (Pfannkuche 1985, Soltwedel 2000; Figure 1.8). Similarly, macrofaunal biomass and abundance are both positively related to POC flux (Rowe et al. 1991, DeMaster et al. 1994, Cosson et al. 1997, Smith and Demopoulos 2003) and CPE (Tselepides et al. 2000a). There are also strong positive correlations between bacterial and meiofaunal biomass (Rowe et al. 1991) and densities (Vanreusel et al. 1995), meiofaunal and macrofaunal abundances (Sibuet et al. 1989, Tietjen 1992), and biomass (Sibuet et al. 1989), and even meiofaunal and megafaunal abundance especially for the deposit-feeding megafauna (Sibuet et al. 1989).

Remarkably, despite the complexity of pelagic-benthic coupling, SeaWiFS satellite imagery of annual surface production in the North Atlantic provides a statistically significant prediction of Holocene foraminiferan abundance (Sun et al. 2006) and the abundance and biomass of the macro-

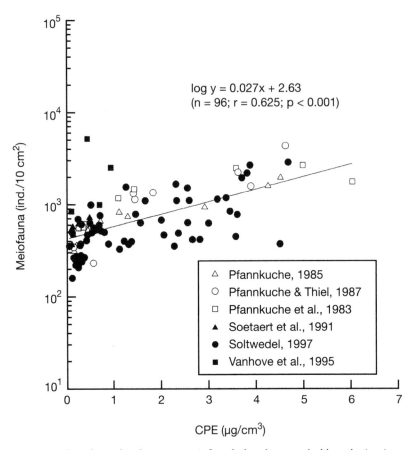

Figure 1.8. The relationship between meiofaunal abundance and chloroplastic pigment equivalents in deep-sea sediments. Data are from the Arctic Ocean, the temperate and tropical east Atlantic Ocean, the Mediterranean Sea, and the Southern Ocean. From Soltwedel (2000), with permission of Thomas Soltwedel and Elsevier.

fauna (Johnson et al. 2007). Abyssal megafaunal abundance is also a positive significant function of surface production (Thurston et al. 1994).

Increased surface production does not always result in increased benthic standing stock. For example, in the eastern Mediterranean, climatically forced changes in deep-water formation from 1992 to 1999 caused cooling of bottom water and an upward displacement of nutrient-rich water that tripled surface production. The increased surface production and export flux led to a 90% reduction in bacterial biomass and to a steep decline in their meiofaunal consumers, evidently because the abundance and activity of bacteria were inhibited by bottom-water cooling (Danovaro et al. 2001a).

THE THIN VENEER OF BENTHIC LIFE

Within around 100 m of the seafloor, sinking POC that fuels the benthic community enters the benthic boundary layer (BBL). The BBL consists of those parts of the lower water column and upper sediment that are influenced directly by the sediment-water interface (Boudreau and Jørgensen 2001). It is a zone of intensified biological activity, chemical exchange between the water and sediment habitats, and redistribution of organic material by biogenic processes and deep currents. Most of the standing stock of the BBL appears to reside in the infaunal and epifaunal sediment community (K. L. Smith 1992). Compared to terrestrial and other marine environments, the layer of benthic animal life in the deep sea is extraordinarily thin. Figure 1.9 shows the vertical distribution within the sediment of the macrofauna, metazoan meiofauna, foraminifera, and bacteria at abyssal sites. Some 90% of the macrofauna and meiofauna occupy only the top few centimeters of sediment. In the more organically rich sediments of the continental margins, the meiofauna and macrofauna can extend down to around 10 cm or more, but their standing stock is still concentrated in the top 3 cm (Jumars and Eckman 1983, Vincx et al. 1994). Bacteria penetrate deeper and decrease more gradually in the sediment (Figure 1.9). Indeed, prokaryotes remain abundant in subseafloor sediments hundreds of meters below the surface, although most appear to be either dormant or have extremely low metabolism (Parkes et al. 2005).

As noted earlier, the burrowing megafauna is relatively poorly known because it is so difficult to sample. Threadlike vertical burrows (0.5 mm wide) of sipunculids reaching down 2 m are known from deep-sea sediment cores (Weaver and Schultheiss 1983, Romero-Wetzel 1987). The extensible feeding proboscis of infaunal echiurans has been captured in deep-sea bottom photographs and measures an impressive 30–60 cm, suggesting that the subsurface burrow containing the body and retracted proboscis must be fairly spacious (Ohta 1984, Bett and Rice 1993). However, most burrows of larger infaunal elements such as polychaetes and echinoderms appear to be restricted to the top 10–30 cm of sediment (Aller and Aller 1986, Gage and Tyler 1991, Hughes et al. 2005). The epifauna of the deep-sea soft sediment environ-

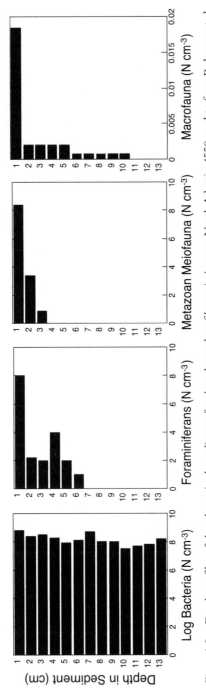

Figure 1.9. Depth profiles of abundance in the sediment for abyssal samples of bacteria (eastern North Atlantic, 4550 m, data from Relexans et al. 1996), foraminifera (central North Pacific, 5874 m, data from Snider et al. 1984), metazoan meiofauna (central North Pacific, 5874 m, data from Snider et al. 1984), and macrofauna (Arabian Sea, 4424 m, data from Witte 2000). For the macrofauna the cores were sectioned in three parts: 0–1, 1–5, and 5–10 cm. Most meiofaunal and macrofaunal life is concentrated in the top few centimeters of sediment. Bacteria penetrate deeper, but the proportion of the microbiota that is active or dormant is unclear.

ment, especially at great depths, is also sparse and relatively compressed. Most of the mobile megafaunal foragers and sessile suspension feeders probably reach no more than 20–30 cm above the surface, with occasional slender gorgorian corals and crinoids approaching 1 m in height.

BIOTURBATION IN DEEP-SEA SEDIMENTS

Deposit-feeding animals mix POC down into the sediments by their feeding and burrowing behavior. This process, called bioturbation, occurs in several modes. Diffusive bioturbation refers to a gradual subduction of POC by the cumulative activity of small burrowing forms (R. C. Aller 1982). Nondiffusive bioturbation is the direct transport of POC to a specific sediment horizon by larger vertical burrow dwellers that feed at the surface and deposit fecal material deep in their burrows or transport unconsumed surface POC by ventilating their burrows (J. N. Smith and Schafer 1984, Levin et al. 1997). Downward transport can also be caused by fresh phytodetritus infiltrating vacant burrows and creating a minihotspot for bacteria and meiofaunal feeding (Aller and Aller 1986).

The effects of bioturbation can be determined by measuring sediment profiles of excess naturally occurring radiotracers (^{210}Pb or ^{234}Th) scavenged by sinking POC or by introducing labeled algae or glass beads onto the sediment surface and examining their sediment profiles at a later date. Nondiffusive subduction of material can be quite rapid. Levin et al. (1997) showed that tube-building maldanid polychaetes living on the upper continental slope transport ^{13}C-labeled diatoms down to 10 cm in 1–2 days. Not surprisingly, since many deposit feeders are selective, the process of bioturbation seems to be selective in terms of the size and quality of the transported particles (Wheatcroft and Jumars 1987, C. R. Smith 1992, C. R. Smith et al. 1993, Pope et al. 1996).

The depth in the sediment of bioturbation, called the mixed-layer depth, can be measured by radionuclide profiles. C. R. Smith (1992) theorized that bioturbation intensity in the deep sea is largely a function of POC flux. High POC flux promotes higher density and larger size of macro- and megabenthos, and hence greater potential for bioturbation. Later in our global synthesis of benthic standing stock, we show that the decrease in POC flux with depth is very convincingly associated with reduced density and size of deep-sea organisms. Smith and Rabouille (2002) reviewed data on the relationship of the mixed-layer sediment depth to POC flux and depth in the

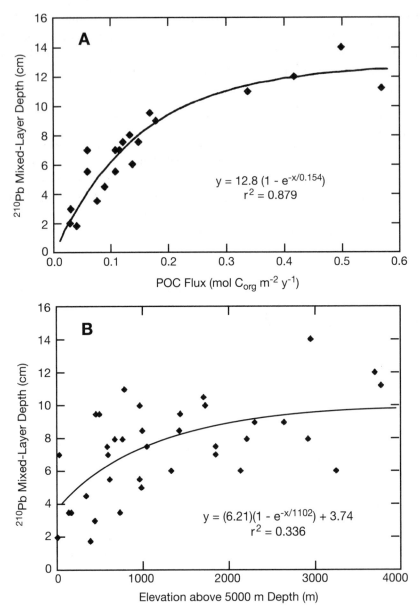

Figure 1.10. The relationship of the mixed-layer depth in the sediment to POC flux (A), and the mixed-layer depth in sediments from the abyss to the upper bathyal zone (B). From Smith and Rabouille (2002), with permission of the authors and the American Society of Limnology and Oceanography.

deep sea (Figure 1.10). The mixed-layer depth increases with POC flux and reaches an asymptote caused by a shift from food limitation to other physical factors that prevent deeper burrowing. Since POC flux decreases with depth, the mixed-layer depth decreases from bathyal to abyssal depths (Figure 1.10) and, as might be expected, it corresponds well with the occupancy depth of macrofaunal and megafaunal organisms discussed earlier (centimeter to decimeter scales). The veneer of seafloor life becomes thinner with increasing depth. Smith and Rabouille (2002) note that the relationship between the mixed-layer depth and POC flux trends downward at flux levels high enough to create oxygen minimum zones (OMZs), where organisms are generally very small and the mixed-layer depth is reduced; in other words the overall relationship is unimodal (Smith et al. 2000).

GEOGRAPHIC PATTERNS OF STANDING STOCK

The seabed is the ultimate sediment trap (Herman et al. 2001, Beaulieu 2002), and benthic standing stock represents the gradual incorporation through assimilation and growth of the 1% of surface production that eventually reaches the bottom minus the cost of respiration (80–90%; Pfannkuche 1992, Rowe et al. 2003) and the loss to long-term burial in sediments as refractory material (a few percentage of input; Ståhl et al. 2004). Standing stock is the culmination of pelagic-benthic coupling, whatever form it takes, and is the most directly relevant measure of ecological and evolutionary opportunity in the deep sea. It is the raw material that is shaped into biodiversity.

We compiled data on spatial patterns of standing stock from 132 studies that reported estimates from a total of 2702 samples taken between 15 and 8376 m. A map of the sampling stations (Figure 1.11) reflects the current status of deep-sea exploration. All of the samples were collected during and after the 1960s, when reasonably accurate quantitative sampling gear was first deployed. Earlier voyages of discovery—such as the British *Challenger*, the American *Blake* and *Albatross*, and the Danish *Galathea* expeditions—generally used qualitative sampling gear, which was ineffective in capturing the small organisms that dominate the deep-sea benthos. Modern quantitative samples are heavily concentrated in the Atlantic and Arctic oceans, the Mediterranean Sea, and the Gulf of Mexico because these regions are within closer ship-steaming distance from American and European oceanographic centers. Much of the Indo-Pacific, by far the largest deep-sea realm, has yet

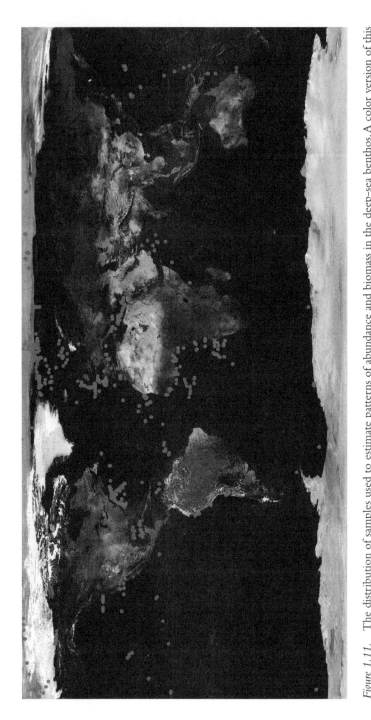

Figure 1.11. The distribution of samples used to estimate patterns of abundance and biomass in the deep-sea benthos. A color version of this figure is included in the insert following p. 50. The map was created by using iMap 3.1 software (http://www.biovolution.com).

to be explored, so the available data have a strong geographic bias and span only five decades.

The database includes both abundance and biomass for bacteria, metazoan meiofauna, macrofauna, and megafauna. We examine bathymetric patterns within and among size categories using linear regression and the analysis of covariance (ANCOVA). For the two largest datasets, metazoan meiofauna and macrofauna, we also look at the horizontal among-basin variation in standing stock associated with differences in nutrient input from overhead production, topographic focusing, lateral advection of sediments, and proximity to OMZs. We plot regressions or clusters of points for these cases and show whether they deviate from the overall standing stock–depth trends in a predictable way. A preliminary analysis of only bathymetric trends within the deep sea restricted to bathyal and abyssal depths (200–6000 m) was presented by Rex et al. (2006).

We attempted to standardize the database by selecting studies that met certain sampling criteria and then used partial regression to control for potential effects of sampling techniques and geographic distribution. For bacteria, we used studies that integrated (or normalized) standing stock over a 2- to 17-cm depth in the sediment. Bacteria penetrate deeper within the sediment and show a slower rate of decline than do the meiobenthos and macrobenthos (Levin et al. 1991b, Relexans et al. 1996, Soltwedel et al. 2000). Their distribution with depth in sediments can also be irregular (Aller et al. 2002). Consequently, integrating standing stock over different sediment depths is a potential source of error that could be especially important for this group. We tested this by using partial regression to statistically remove the effects of the sediment depth over which standing stock was assessed. In studies of deep-sea bacteria, density and biomass are typically provided in terms of direct counts and organic carbon content, respectively. We converted all data to number of cells per square meter and grams of carbon per square meter.

For meiofauna, we use studies that report the entire metazoan meiofaunal community, rather than individual taxa such as nematodes or harpacticoid copepods. Samples were hand sorted or strained out on 20 to 74 μm sieves (most were 32–45 μm). Densities were converted to number per square meter. Biomass was reported as wet weight, dry weight, or ash-free dry weight. We converted biomass to grams of carbon per square meter using conversion factors from Soltwedel (2000). Data were not available to establish a comparable database on standing stock in foraminiferans, especially for biomass, but we do show later how known patterns of abundance fit the general trend observed in the four basic size categories.

For macrofauna, again, we used only studies that reported the entire macrofaunal community. Sieve size ranged from 250 to 520 μm (most were 300–500 μm). Densities were converted to number per square meter. Biomass, if not reported in units of organic carbon (e.g., wet weight, dry weight, ash-free dry weight), was converted to grams of carbon per square meter using conversion factors from Rowe (1983). For the megafauna, we used studies that reported both invertebrates and demersal fishes. Density was converted to number per square meter and biomass converted to grams of carbon per square meter using the conversion factors of Rowe (1983).

BATHYMETRIC PATTERNS OF STANDING STOCK

Patterns of abundance with depth are shown in Figure 1.12. The vertical axis indicates the residuals of the log of density (number of individuals per square meter) with the effects of longitude and latitude removed by partial regression. One must control for the effects of geographic position in order to examine bathymetric patterns independently because, as we show later, there is considerable among-basin variation in abundance associated with other factors mentioned earlier, such as overhead production, topography, surface ice, currents, and OMZs. We add the y-intercept to the partial residuals to recover the appropriate scale of the dependent variable. The vertical axis can be read directly as standing stock with the effects of longitude and latitude held constant. Figure 1.12 shows how density varies solely as a function of depth.

The regression statistics for each size category and the ANCOVA are given in Table 1.1. The ANCOVA allows us to compare the slopes and elevations of the regressions among the four size groups. Elevations of regression lines can be compared with ANCOVA only when the slopes do not differ statistically (Zar 1984). In this analysis, since the four slopes are heterogeneous (Table 1.1), we performed Tukey multiple comparison tests (Zar 1984) to determine which slopes differed significantly. Where there were significant differences, we used the Johnson-Neyman test (Huitema 1980) to calculate the depth range over which the elevations are not significantly different ($P > 0.05$). In other words, the analysis tells us whether standing stock–depth relationships differ among size groups and exactly what the dif-

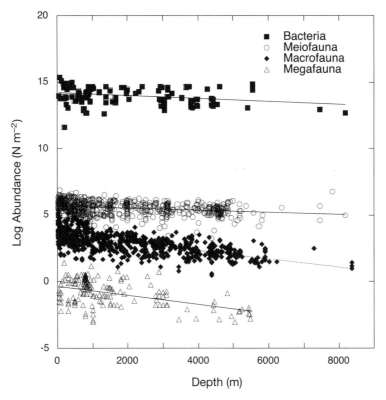

Figure 1.12. Relationships of abundance to depth in four size classes of the deep-sea benthos: bacteria, metazoan meiofauna, macrofauna, and megafauna. The influence of longitude and latitude are removed by partial regression. References for the four groups are given in Appendix C. Regression equations and statistics are provided in Table 1.1. Larger-size classes decrease more rapidly with depth than do smaller-size classes.

ferences are in terms of the levels of standing stock and the rates at which standing stock changes with depth.

Relationships of density to depth are significant and negative for all four groups (Table 1.1). Regressions are very highly significant for the meio-, macro-, and megafauna. The weakest relationship is for bacteria, which have relatively high densities even at abyssal (>4000 m) and trench (>6000 m) depths. To assess the effect of the difference in the depth of sediment over which bacterial inventories are integrated, we performed partial regressions that controlled for depth, latitude, longitude, and depth in the sediment. Depth in the sediment does have a significant effect on the density of bac-

Table 1.1. Regressions, ANCOVAs, and Multiple Comparison Tests for the Relationships of Abundance and Biomass to Depth in Bacteria, Metazoan Meiofauna, Macrofauna, and Megafauna

| Standing Stock | Regressions | | | | ANCOVA | | | Multiple Comparisons | | | |
	Group	Equation	N	F		d.f.	F	Contrast	Slope	Elevation	Johnson–Neyman Test
Abundance	Bacteria (B)	$Y = 14.195 - 0.00011X$	120	10.254**	Slope	3, 1755	61.604***	B–Me	n.s.	B > Me	
	Meiofauna (Me)	$Y = 5.710 - 0.00008X$	672	52.455****	Elevation	3, 1758	15,475.56***	B–Ma	B < Ma		0 m
	Macrofauna (Ma)	$Y = 3.588 - 0.00031X$	812	667.813****				B–Mg	B < Mg		0 m
	Megafauna (Mg)	$Y = -0.245 - 0.00035X$	159	47.748****				Me–Ma	Me < Ma		0 m
								Me–Mg	Me < Mg		0 m
								Ma–Mg	n.s.	Ma > Mg	
Biomass	Bacteria (B)	$Y = -0.423 - 0.00006X$	111	6.349*	Slope	3, 931	52.209***	B–Me	B < Me		0–516 m
	Meiofauna (Me)	$Y = -0.559 - 0.00018X$	232	63.977****	Elevation	3, 934	44.471***	B–Ma	B < Ma		1306–2262 m
	Macrofauna (Ma)	$Y = 0.281 - 0.00045X$	542	632.876****				B–Mg	B < Mg		0 m
	Megafauna (Mg)	$Y = -0.843 - 0.00036X$	54	43.251****				Me–Ma	Me < Ma		2509–3931 m
								Me–Mg	Me < Mg		0–292 m
								Ma–Mg	n.s.	Ma > Mg	

Note: Under multiple comparisons, the inequality signs indicate significant differences ($P < 0.05$) and the direction of the differences. For comparisons for which slopes differ, the Johnson–Neyman test provides the depth range over which regression lines do not differ significantly in elevation. The regressions are plotted in Figures 1.12 and 1.13.
*$P < 0.05$, **$P < 0.01$, ***$P < 0.001$, ****$P < 0.0001$.

teria ($F = 6.275$; d.f. $= 4,115$; $P = 0.0136$) when the influence of depth, latitude, and longitude are held constant. Controlling for depth in the sediment has no effect on the slope of the residuals in Figure 1.12, but it does elevate the regression line (the intercept increases from 14.20 to 14.80). Including the effect of depth in the sediment does not alter the overall patterns among the groups discussed later, but does caution that depth in the sediment over which bacterial abundance is integrated can matter in estimating standing stock per unit surface area of sediment. Since we cannot adjust for this variable in a consistent way for the other three size categories, comparisons among groups are made by using the regression for bacteria shown in Figure 1.12. As we have shown elsewhere (Rex et al. 2006), the relationship of bacterial density to depth between the shelf-slope transition and abyssal depths (200–6000 m) is not significant; that is, given the available data, there is no evidence that bacterial densities decline from the upper slope to the abyssal plain.

The multiple comparisons reveal that slopes do not differ in the bacteria-meiofauna and macrofauna-megafauna combinations, and that in these cases the elevation of the bacteria is higher than that of the meiofauna and that of the macrofauna is higher than that of the megafauna (Table 1.1). For the combinations where slopes do differ significantly, the Johnson–Neyman tests show that the elevations are significantly different throughout the depth range sampled (i.e., the regression lines would converge and become statistically indistinguishable in terms of elevation only at some point lower than 0 m). The sequence of significant differences in elevation is: bacteria > meiofauna > macrofauna > megafauna. This is so obvious from Figure 1.12 that it scarcely needs statistical validation, but the analysis provides a good introduction to how the ANCOVA and multiple comparison tests help sort out the more complicated relationships between biomass and depth presented later.

Perhaps the most interesting result is that the slope of the macrofauna regression is significantly steeper than that of the meiofauna regression (see also Rex et al. 2006 and McClain et al. 2009). Among all the combinations of slopes the macrofauna-meiofauna difference is the most pronounced ($Q = 17.78$ versus 7.28–10.26 for other significant differences). This supports, on a global scale, Thiel's (1975, 1979b) contention that average metazoan size decreases with depth in the deep sea. Thiel first showed this by regressing the densities of meiofauna and macrofauna against depth just as we do here. An ANCOVA revealed that the meiofaunal regression had a higher elevation and lower slope than the macrofaunal regression. Since meiofauna are smaller, the implication is that average size must decrease with depth. Thiel compared 48 observations of meiofaunal density off Portugal and West Africa to the macrofaunal density-depth relationship reported by Rowe (1971a) for samples collected in the western Atlantic, Gulf of Mexico, and eastern Pacific. Subsequent attempts to measure size-depth trends yielded conflicting results (cf. Shirayama 1983, Sibuet et al. 1989, Pfannkuche and Soltwedel 1998, Hughes and Gage 2004), partly perhaps because of methodological differences or limited geographic scope. On the very large scales examined here, Thiel's prediction that average metazoan body size decreases with depth is strongly supported. The decrease in average organism size with depth is perhaps the single most basic macroecological trend in the deep sea. The association between body-size categories and abundance in the deep-sea benthos also generally agrees with the negative relationship of population density to body size in terrestrial systems (Brown 1995).

Biomass presents a more complicated picture, but one that largely corroborates the abundance patterns (Figure 1.13). All four size groups show significant decreases in biomass with depth (Table 1.1), but the relationship for bacteria again is especially weak. Depth in the sediment has no significant effect ($F = 2.0698$; d.f. = 4,106; $P = 0.1532$) in explaining bacterial biomass once the influences of depth, latitude, and longitude are removed. Including sediment depth in calculating the residuals of biomass does not change the slope, but does elevate the regression line (the intercept increases from -0.423 to -0.246). The relative positions of the lines representing the different size classes are unaffected.

The only case in which slopes do not differ is the one between the macrofauna and the megafauna, and here the elevation for the macrofauna is significantly higher. For most of the depth range, the biomass of animals declines faster than that of bacteria and the macro- and megafauna decrease faster than the meiofauna. Macrofaunal biomass exceeds bacterial biomass above 1306 m, but bacteria predominate below 2262 m (the regression lines cross and are not significantly different in elevation within this depth range; Table 1.1). Similarly, macrofaunal biomass is higher than meiofaunal biomass above 2509 m, but meiofaunal biomass is higher below 3931 m. Meiofaunal biomass exceeds megafaunal biomass throughout most of the depth range (>292 m). All of these trends indicate that with increasing depth, community biomass becomes progressively more dominated by bacteria and meiofaunal elements.

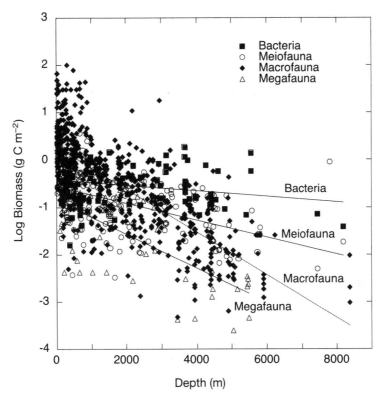

Figure 1.13. Relationships of biomass to depth in four size classes of the deep-sea benthos: bacteria, metazoan meiofauna, macrofauna, and megafauna. The influence of longitude and latitude is removed by partial regression. References for the four groups are given in Appendix D. Regression equations and statistics are provided in Table 1.1. Smaller-size classes replace larger-size classes with increasing depth.

Some features of this set of relationships have been observed before on regional scales, such as the steeper descent of larger organisms with depth (Thiel 1975, Heip et al. 2001, Flach et al. 2002) and the dominance of bacteria at abyssal depths (Rowe et al. 1991, 2003). However, the present analysis is the first of all four major size groups at very large scales. The exponential decrease of standing stock of the community and within size categories (Figures 1.12, 1.13) is a general trend in the deep sea, at least for the areas sampled (Figure 1.11).

Bacterial standing stock remains high and decreases only slightly from the continental shelf to trench depths. Within the bathyal and abyssal zones

(200–6000 m), which constitute most of the deep-sea ecosystem, bacterial abundance and biomass show no statistically detectable decrease with depth (Rex et al. 2006). However, the standing stock of bacteria may not reflect either its trophic role in the benthos or POC flux in a direct and immediate way. There is an indigenous barophilic microbial community that is adapted to rapidly degrade POC under the tremendous pressures and frigid temperatures found at great depths (Deming 1985, Lochte and Turley 1988). However, part of the community appears to be composed of surface-derived bacteria that were entrained by sinking phytodetritus and made dormant by the extreme physical conditions in the deep sea (Deming and Baross 2000). The proportion of active to inactive bacteria in deep-sea sediments and the reason for the accumulation of such high concentrations of bacteria remain unknown.

Danovaro et al. (2008a) have shown that viral production is high in deep-sea sediments. They proposed that a viral infection of bacteria causes lysis and that the release of labile organic material, in turn, nourishes uninfected cells. This viral shunt could both promote microbial production and limit energy transfer to higher consumer levels.

Among animals, the rate of decrease in abundance and biomass with depth is highest in larger-size classes. In terms of biomass, there is a shift from communities dominated by macrofauna at shelf and upper bathyal depths to those dominated by bacteria and meiofauna at abyssal depths. Larger organisms appear to be more vulnerable to the decrease in food supply with depth, and Rex (1973) and Thiel (1975) proposed that this is caused by an Allee effect. Larger animals require more energy, and the extreme energy constraints at great depths may depress population densities for some large species below the levels needed to be reproductively viable.

Although some energy must flow from bacteria to consumer levels, the size categories do not represent discrete trophic levels in an Eltonian food chain (Haedrich and Rowe 1977). This is readily apparent in Figure 1.13, which shows broad overlap of biomass values for the four size classes. Trophic levels in a food pyramid are normally separated by order-of-magnitude differences in biomass. It seems much more likely that the deep-sea benthic food web is complex, with predators and detritivores in all size categories. For example, in our earlier discussion of the benthic response to organic carbon flux, we pointed out that phytodetrital aggregates settling on the bottom are quickly exploited by bacteria and foraminifera at one end of the size spectrum and by epibenthic holothurians at the other end. The remaining labile organic matter available to the meiofauna and macrofauna must be very restricted.

INTERREGIONAL COMPARISONS OF
STANDING STOCK

Four subsets of the database—abundance and biomass for both macro- and meiofauna—are sufficiently large and represent a broad enough range of habitats to permit interregional comparisons. Our approach is to regress standing stock against depth and then compare specific sites to the overall regression line to observe if they depart from the average trend in a predictable direction, that is, whether their levels of standing stock are higher or lower than average. Though the choice of sites to compare is necessarily ad hoc and differences in sampling intensity do not permit rigorous statistical tests, the results are quite consistent and support the relationship of food supply to standing stock evinced in the bathymetric analyses. The most obvious and demonstrable difference among sites is the level of productivity. Within these two size categories we are able to control for sieve size and gear deployed, as well as for geographic location. Both sieve mesh size and gear are known to influence sampling efficiency in the meiofauna and macrofauna (Gage 1975, Bett et al. 1994, Gage et al. 2002, Hughes and Gage 2004, Gage and Bett 2005).

Table 1.2 shows the statistically independent influence of depth, latitude, longitude, sieve mesh size, and sampling gear on abundance and biomass in both the meiofauna and macrofauna. Geographic position and sampling methods both have statistically significant effects in at least some cases, although their predictive value (*F*-statistic) is very subordinate to depth in all cases. For consistency we remove geographic and methodological variables from all the standing stock–depth comparisons. In our discussion of how deep-sea standing stock corresponds to regional scale surface production, we adopt Nixon's (1995) classification of eutrophic, mesotrophic, and oligotrophic areas as having productivities of 300–500, 100–300, and <100 g C m^{-2} y^{-1}, respectively.

Regions of High Productivity

Upper to mid-slope depths off Cape Hatteras in the western North Atlantic experience atypically high organic carbon flux. This is caused by a unique

Table 1.2.   Effect Tests for Geographic Position and Sampling Methods

	Meiofauna		Macrofauna	
Source	Abundance	Biomass	Abundance	Biomass
Depth	72.29****	70.19****	779.79****	583.52****
Latitude	1.09	0.36	1.32	116.72****
Longitude	0.82	0.08	13.03***	13.84***
Sieve size	27.82****	9.27**	127.51****	0.42
Sampling gear	7.80***	3.62*	5.93**	10.71****

Note: Values in the table are the *F*-statistics and their significance levels for the independent effects on standing stock of each source variable with the effects of all other source variables removed by multiple regression. For meiofauna, sieve mesh size ranged from 20 to 74 μm, and sampling gear included grabs, coring devices, and multicorers. For macrofauna, sieve mesh size ranged from 250 to 520 μm, and gear included grabs, coring devices, and anchor dredges. Geographic position and sampling methods all have significant independent effects in some cases, but all are subordinate in importance to depth.
$^*P < 0.05$, $^{**}P < 0.01$, $^{***}P < 0.001$, $^{****}P < 0.0001$.

combination of enhanced surface production associated with Gulf Stream–induced upwelling, outwelling of nutrients from nearby coastal bays, and rapid topographic funneling across the narrow shelf and down the steep incised slope (Schaff et al. 1992, Blake and Diaz 1994, Rhoads and Hecker 1994). The organic carbon deposition rate is estimated to be 170 g m^{-2} y^{-1} or less, as high as is found in productive estuaries and major river deltas and much higher than in normal slope habitats (DeMaster et al. 1994). Macrofaunal abundance (Figure 1.14) and biomass (Figure 1.15) reach some of the highest values recorded at continental shelf and upper bathyal depths. Meiofaunal abundance (Figure 1.16) and especially biomass (Figure 1.17) are similarly quite high at these depths, although abundance trails off to low values at middle to lower bathyal and abyssal levels.

Oxygen minimum zones are formed where strong upwelling promotes unusually high surface production. Decomposition of sinking phytodetritus depletes the oxygen in the water column, and if circulation is weak a persistent regional hypoxic layer can develop. Vast areas of the eastern Pacific Ocean, northern Indian Ocean, and eastern South Atlantic contain midwater (200–1000 m) OMZs with concentrations of less than 0.5 ml l^{-1} of dissolved oxygen (Levin 2003). Where these zones impinge on the continental margin, they create an extreme benthic environment of low oxygen and very high sediment organic carbon. The core of benthic OMZs supports

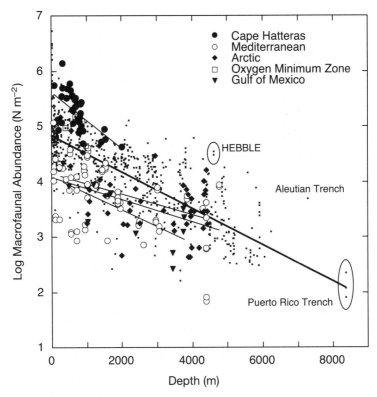

Figure 1.14. Interregional variation in macrofaunal abundance. The effects of longitude, latitude, gear type, and sieve mesh size are removed by partial regression. The thick line is the overall regression including all data. Regions or habitats with relatively high food supply tend to be above the average trend and those with low food supply tend to be below it. See text for detailed explanation. Only significant regressions ($P < 0.01$) are plotted (Cape Hatteras above the heavy line; Arctic, Mediterranean, and Gulf of Mexico below). See Appendix C for references.

a highly adapted dense community of bacteria, protozoa, and metazoan meiofauna that can tolerate low levels of oxygen. At the upper and lower boundaries, where sediments are still organically enriched and more oxic, macrofaunal and megafaunal densities are elevated. Helly and Levin (2004) estimate that OMZs impact more than a million square kilometers of the seafloor, which makes them a very significant feature of the deep seascape at the upper bathyal depths. Figures 1.14 and 1.15 show the elevated macrofaunal abundance and biomass found in the OMZ (400–1000 m) of the northwest Arabian Sea (Levin et al. 2000).

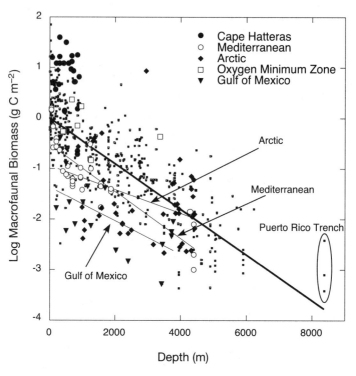

Figure 1.15. Interregional variation in macrofaunal biomass. The effects of longitude, latitude, gear type, and sieve mesh size are removed by partial regression. The thick line is the overall regression including all data. Regions or habitats with relatively high food supply tend to be above the average trend, and those with low food supply tend to be below it. See text for detailed explanation. Only significant regressions (P < 0.01) are plotted. See Appendix D for references.

At the high-energy benthic boundary layer experiment (HEBBLE) site (4800–4900 m on the Nova Scotia Rise in the western North Atlantic), the Deep Western Boundary Current can accelerate to speeds of 20–40 cm s⁻¹ (the highest recording is 73 cm s⁻¹), which is fast enough to resuspend sediments and cause benthic storms that last days to weeks (Richardson et al. 1981, Hollister and McCave 1984). Abyssal currents are typically much more sluggish, about 3–4 cm s⁻¹. Episodes of strong flow erode and transport sediments, and intervening weaker flows (<10 cm s⁻¹) permit sediment redeposition. These cycles of erosion and deposition stimulate bacterial production in sediments, resulting in microbial biomass levels 6–10 times higher than at similar depths in more tranquil regions (Baird et al. 1985, Aller

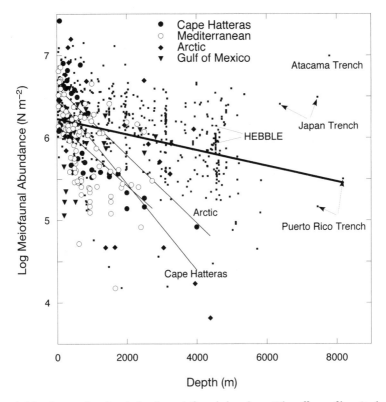

Figure 1.16. Interregional variation in meiofaunal abundance. The effects of longitude, latitude, gear type, and sieve mesh size are removed by partial regression. The thick line is the overall regression including all data. Regions or habitats with relatively high food supply tend to be above the average trend and those with low food supply tend to be below it. See text for detailed explanation. Only significant regressions ($P < 0.01$) are plotted. See Appendix C for references.

1989). The enhanced food availability from bacterial growth and deposition of organic-rich sediment permits an unusually large abundance of macrofauna and meiofauna (Thistle et al. 1985, 1991), as shown in Figures 1.14 and 1.16. Conditions conducive to generating deep benthic storms are widespread, especially in the western Atlantic and southern reaches of the Indo-Pacific (Hollister and McCave 1984, Hollister et al. 1984).

Trenches represent one of the last frontiers of deep-sea exploration. Investigations of trench faunas began only in the 1950s with the Russian *Vityaz* and Danish *Galathea* expeditions (Wolff 1956, Belyaev 1966). There are, even now, very few quantitative samples from trenches. Formed by the

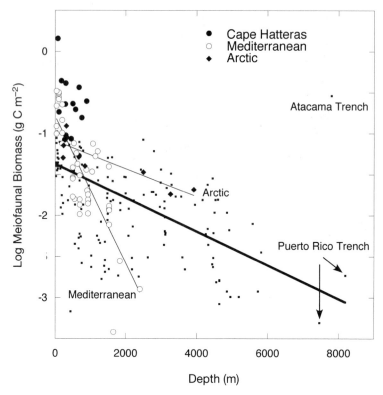

Figure 1.17.   Interregional variation in meiofaunal biomass. The effects of longitude, latitude, gear type, and sieve mesh size are removed by partial regression. The thick line is the
overall regression including all data. Regions or habitats with relatively high food supply tend
to be above the average trend and those with low food supply tend to be below it. See text
for detailed explanation. Only significant regressions (*P* < 0.01) are plotted. See Appendix D
for references.

geological forces of ocean crust subduction, trenches are enormous, very
deep narrow valleys that extend along the continental margins. Owing to
the global configuration of tectonic plates and spreading patterns, trenches
are much more common in the Indo-Pacific than in the Atlantic. Although
trenches are considerably deeper than abyssal plains, by as much as 5000 m,
they can support comparable or even higher benthic standing stock because
many of them lie directly under productive coastal waters and their steep-
walled topography concentrates sinking organic material at the narrow trench
floor. Figures 1.14 and 1.16 show, respectively, the unusually high macro-

faunal abundance in the Aleutian Trench (Jumars and Hessler 1976) and meiofaunal abundance in the Japan Trench (Shirayama and Kojima 1994). High benthic standing stock also occurs in the Ogasawara Trench south of Japan (Shirayama 1984). An extreme example of elevated meiofaunal standing stock is found in the Atacama Trench off Chile (Figures 1.16, 1.17), which is located in an area of very high surface productivity caused by strong upwelling (Danovaro et al. 2002, 2003). Meiofaunal abundance and biomass not only exceed abyssal levels, but are actually similar to those found in coastal systems.

However, not all trenches have high standing stock. The Puerto Rico Trench in the tropical western Atlantic lies in a region of relatively low surface productivity (Figure 1.1A). Meiofaunal (Tietjen et al. 1989) and macrofaunal (Richardson et al. 1995) levels of standing stock conform more closely to the general trend with depth (Figures 1.14–1.17). Conditions at the bottom of the Puerto Rico Trench, around 8000 m, appear to represent an even more impoverished extension of the abyssal environment. Thus, although limited sampling has been done at these extreme depths, there appear to be high-input and low-input trenches with corresponding levels of standing stock.

Other common topographic features such as canyons (Vetter 1994) and seamounts (Rogers 1994) can have high productivity and high benthic abundance. Until recently these habitats have been difficult to sample for benthos because of their steep rugged terrain. When comparable data become available, it will be very interesting to include them in the analysis of benthic standing stock.

Regions of Low Productivity

The Mediterranean Sea is an unusually oligotrophic marine system. Offshore surface productivity ranges from 94 g C m^{-2} y^{-1} in the western Mediterranean (Morel and André 1991) to 20–60 g C m^{-2} y^{-1} in the Aegean and Cretan Seas of the eastern Mediterranean (Dugdale and Wilkerson 1988, Psarra et al. 2000). These levels of production generally fit the definition of oligotrophic (<100 g C m^{-2} y^{-1}; Nixon 1995), especially in the Eastern Basin, and are comparable to or below the most oligotrophic regions of the central Atlantic (Sathyendranath et al. 1995, Behrenfeld and Falkowski 1997; Figure 1.1A). Nutrient input to the deep benthos of the Mediterranean is further diminished by warm water temperatures throughout the water column. Abyssal temperatures are around 13°C compared to 0°–2°C in the Atlantic. These relatively high temperatures intensify the remineralization of sinking organic material by bacterial degradation. Consequently, rates of nutrient input to the benthos and organic content of sediments are extremely low, especially in the Eastern Basin (Danovaro et al. 1999). Macrofaunal and meiofaunal abundance and biomass are predictably all depressed in the Mediterranean (Figures 1.14–1.17).

The Central Arctic Basin is extremely oligotrophic because of permanent ice cover and a surface freshwater lens (Clarke 2003). Estimated annual production is only 10–15 g C m^{-2} y^{-1} (Wheeler et al. 1996, Gosselin et al. 1997). In surrounding coastal regions seasonal production can be very high (Anderson 1989, Behrenfeld and Falkowski 1997), and some of this is evidently transported downslope. Lower bathyal and abyssal standing stock seaward from the shelf-slope transition is generally low, but there are some conspicuously high values for macrofaunal abundance and biomass and meiofaunal abundance on the continental slope and rise closer to the shelf edge (Figures 1.14–1.16; Clough et al. 1997, Kröncke 1998, Kröncke et al. 2000). A clear case of enrichment near the shelf is meiofaunal biomass just north of Svalbard (Figure 1.17), where there is only seasonal ice cover and an influx of organically rich Atlantic water (Soltwedel et al. 2000). Meiobenthic standing stock and sediment chloroplastic pigment equivalents here are actually similar to those found in the Porcupine Seabight of the eastern North Atlantic around 30° latitude to the south (Pfannkuche and Thiel 1987).

Early estimates of production in the central Gulf of Mexico were 20–30 g C m^{-2} y^{-1} (Walsh et al. 1989, Müller-Karger et al. 1991). Basinwide average production was 105–210 g C m^{-2} y^{-1}, primarily because of a more productive zone, 280–330 g C m^{-2} y^{-1}, near the Mississippi River discharge. More recent estimates of production from SeaWiFS satellite imagery are about 120–140 g C m^{-2} y^{-1} in the central Gulf and 116–340 g C m^{-2} y^{-1} in coastal areas (F. Müller-Karger and R. Luerrsen, personal communication). Thus most of the Gulf appears to be mesotrophic *sensu* Nixon (1995). It is also possible that widespread cold seeps export organic nutrients to the surrounding benthic environments (MacAvoy et al. 2002). The first quantitative studies of deep-sea benthos in the Gulf of Mexico indicated that macrofaunal biomass and abundance were about an order of magnitude less than in the western North Atlantic, where surface production is three to six times higher (Rowe and Menzel 1971, Rowe et al. 1974). As in the Mediterranean, biodegradation of sinking organic material may be accelerated by relatively warm temperatures and the temperature in the deepest Gulf is 4°C. Macrofaunal abundance and biomass show a clear decline with depth and are generally low (Figures 1.14, 1.15). Meiofaunal abundance, although appearing low, shows no significant decrease with depth (Figure 1.16). No data for meiofaunal biomass were available. More recent estimates of macrofaunal abundance in the northern Gulf of Mexico suggest that abundance is less than in the North Atlantic, but higher than in the Arctic and Mediterranean (Wei et al. in press; see Figure 3.25).

The recently discovered "bare zone" of the southern Pacific Basin may be a truly extreme region of low productivity and POC flux (Rea et al. 2006). This huge area of seafloor ($\sim 2 \times 10^6$ km^2), about the size of the Mediterranean Sea, has only a very thin layer of sediment deposited over the ocean crust, a circumstance owing to low surface production, remoteness from terrestrial influence, and a comparatively shallow dissolution of calcite. The deep-sea benthos of the bare zone has not been sampled. It would be very interesting to do so because the environment of this region suggests that it may represent the abyssal endpoint in terms of diminished standing stock.

Foraminiferans

Figure 1.18 shows the relationship of abundance to depth in deep-sea foraminiferans. As we noted earlier, there are very few estimates of biomass available, and the abundance database ($N = 73$) is considerably smaller than for the other size classes ($N = 120$–182, Table 1.1). Foraminiferan abundance does decrease significantly with depth. Despite the paucity of data, differences in abundance among regions correspond quite well to the patterns observed with meiofaunal and macrofaunal abundance (Figures 1.14, 1.16). For example, observations from the depocenter at Cape Hatteras, the OMZ in the Arabian Sea, the Arctic Yermak Plateau north of Svalbard, and the tropical upwelling region in the eastern Atlantic are consistently and predictably above the overall regression line. Values for the Gulf of Mexico are consistently lower and those representing the Eurasian Arctic tend to be lower, especially at great depths. However, estimates from the eastern North Atlantic (Porcupine Seabight; Goban Spur; and the Madeira, Porcupine, and Cape Verde abyssal plains) show wide variation, possibly because of local ecological conditions or methodological differences in measuring abundance in

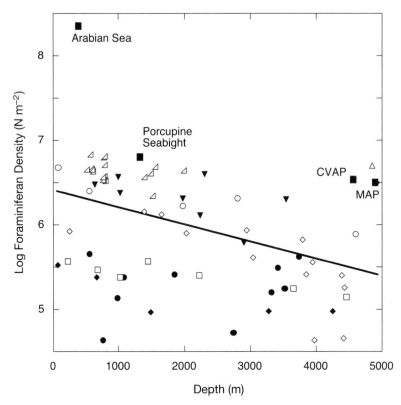

Figure 1.18. The relationship of foraminiferan abundance to depth. Though available data are sparse, the departure of regions or habitats from the overall general trend largely mirrors those for the macrofauna and meiofauna in Figures 1.12–1.17. Sites with relatively high food supply tend to be above the average trend and those with low food supply below it. See text for detailed explanation. Gulf Mexico (solid circles), tropical eastern Atlantic upwelling region (open circles), tropical eastern Atlantic region with no upwelling (solid diamonds), eastern North Atlantic (open boxes), Arctic, Yermak Plateau (inverted solid equilateral triangles), Cape Hatteras (open right triangles), Eurasian Arctic (open diamonds), Porcupine Abyssal Plain (open equilateral triangles), other individual sites indicated by name and solid squares (CVAP is the Cape Verde Abyssal Plain and MAP is the Madeira Abyssal Plain). Data are from Bernhard et al. 2008 (Gulf of Mexico), Cutter et al. 1994 (Cape Hatteras, western North Atlantic), Gooday 1996 (eastern North Atlantic), Gooday et al. 2000 (Arabian Sea), Heip et al. 2001 (eastern North Atlantic), Kröncke et al. 2000 (Eurasian Arctic), Soltwedel 1997 (tropical eastern Atlantic), Soltwedel et al. 2000 (Arctic, Yermak Plateau). The overall regression is: $Y = 6.410 - 0.000204X$, $N = 73$, $R^2 = 0.175$, $F = 15.04$, $P < 0.0002$.

this difficult group. We found no comparable data for the HEBBLE, Mediterranean, or trench sites. Though there appear to be regional and depth differences in the relative abundance of foraminiferans versus metazoan meiofauna (cf., e.g., Tietjen 1971, Gooday 1986, 1996, Soltwedel 1997a, Heip et al. 2001, Bernhard et al. 2008), their abundance-depth trends are globally quite similar (Figures 1.12, 1.18). This reinforces the conclusion that average organism size decreases with depth.

THE EASTERN AND WESTERN NORTH ATLANTIC

One size class, the macrofauna, is widely collected enough to allow an ocean-wide comparison between the eastern and western corridors of the North Atlantic, the most well-sampled areas of the World Ocean (Figure 1.11). Using a much more limited database, Levin and Gooday (2003) found that macrofaunal density at bathyal depths was higher in the western North Atlantic, but that biomass was higher in the eastern region. Consistent with the latter, Lampitt et al. (1986) estimated that invertebrate megafaunal biomass may be an order of magnitude higher in the Porcupine Seabight off Ireland than in the western North Atlantic south of New England (Haedrich et al. 1980). Limited SCOC data suggest that oxygen consumption is similar in the eastern and western North Atlantic (Duineveld et al. 1997).

In Figures 1.19 and 1.20 we compare the macrofaunal abundance and biomass in the western and eastern North Atlantic, again controlling for longitude, latitude, type of sampling gear, and sieve size. Bathymetric patterns of abundance and biomass are both remarkably similar for the eastern and western corridors, although both measures of standing stock appear to be somewhat higher in the abyssal eastern North Atlantic. Thus, despite some clear regional variation (Figures 1.14, 1.15), the same general trend of exponentially decreasing standing stock with depth obtains across very large spatial scales, at least in the macrofauna.

CONCLUSIONS

Food supply to the benthos occupying the vast soft sediment environment of the deep sea is derived ultimately from surface production. There is a necessary ecological connection between the patterns of surface production and

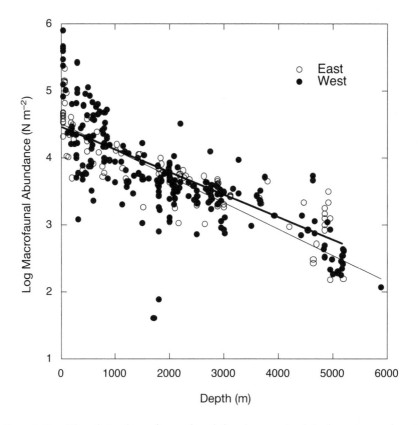

Figure 1.19. The relationships of macrofaunal abundance to depth in the eastern and western North Atlantic. The thick line is east. The regressions are not significantly different.

those of benthic standing stock, but how does the process work and on what scales of time and space does it operate?

Although a great deal remains to be learned, the basic features of pelagic-benthic coupling are beginning to emerge. Satellite imagery of surface chlorophyll concentration has made it possible to map global productivity on daily to decadal time scales. These patterns provide the initial template of potential energy supply to the benthos. Sediment trap studies have shown that POC flux decreases exponentially through the water column and tracks the seasonality of surface production. POC flux to a benthic site represents export flux downward from surface production, as well as its transformation through the water column and lateral advection over a large volume of ocean on variable time scales. Present estimates suggest surface catchment areas on

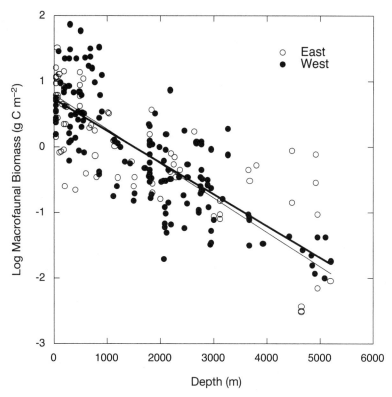

Figure 1.20. The relationships of macrofaunal biomass to depth in the eastern and western North Atlantic. The thick line is east. The regressions are not significantly different.

the order of 100 km in size and weeks to months of settling time for abyssal sites. Only about 1% of surface production reaches the abyssal plain, making most of the deep sea an extremely energy-poor system.

The relationships between the arrival of POC and its conversion into benthic standing stock are poorly characterized. SCOC, which represents primarily bacterial and protozoan metabolism, tracks POC flux in some studies but not others. Experiments and observations to assess the response of benthic organisms to phytodetrital settlement suggest that only bacteria and a minority of foraminiferans show immediate growth and reproduction. The response in meiofaunal and macrofaunal elements appears to be taxonomically selective and is expressed through seasonal periodicity of reproductive cycles or modest population growth with time lags of about a year.

Mobile megafaunal grazers, mainly echinoderms, appear to be very effective exploiters of phytodetritus accumulations and show marked changes in species composition and abundance on interannual time scales. The most conspicuous responses have been in highly opportunistic species that are un-common in background communities.

Based on present evidence, the overall impression is that while season-ality of nutrient input is certainly reflected in the population dynamics and reproductive cycles of deep-sea organisms, the response is neither dramatic nor widespread in the community. It seems largely confined to very small organisms and to some large epibenthic species. For the community as a whole, it seems likely that standing stock is fairly steady and represents food supply averaged over large spatial scales in terms of its origin and temporal scales on the order of years to decades.

There is good evidence that geographic variation in standing stock is closely linked to geographic variation in food supply. Standing stock is a pos-itive function of POC flux and chloroplastic pigment equivalents in sedi-ments. As we have shown here, all major size categories of the benthos decrease exponentially with depth on a global basis as an obvious conse-quence of surface production and POC flux decreasing seaward from pro-ductive coastal waters and being remineralized while settling through a deeper water column. Despite all of the complexities of surface-benthic coupling and uncertainty about scaling, POC flux at depth estimated from satellite-derived surface production provides a significant prediction of regional and oceanwide variation of standing stock. The association of bathymetric pat-terns with food supply is reinforced by the horizontal interregional com-parisons. Eutrophic, mesotrophic, and oligotrophic regions have predictable relative levels of deep-sea benthic standing stock. Similarly, special habitats with elevated food supply such as those close to oxygen minimum zones, sediment depocenters, and topographies that concentrate sinking phyto-detritus consistently have elevated standing stock.

Compared to terrestrial and coastal environments, metazoan life in the deep sea occupies a very thin layer, primarily the top few centimeters of sed-iment. Standing stock decreases exponentially down the continental margin and reaches extremely low levels in the abyssal plain. Energy constraints drive a decrease in average organism size with depth, so that the great abyssal plains of the World Ocean are heavily dominated by bacteria, foraminiferans, and the most minute metazoans.

Geographic patterns of biomass and abundance are vital to understanding deep-sea community structure. They have implications for energy use, species diversity, body size, geographic range size, population density, and evolutionary potential. However, the macroecological relationships among these variables are seldom explored using deep-sea data, largely because relevant information is not collected simultaneously and standing-stock measurements are usually reported for single size categories of the benthos. Hence, relationships between standing stock and species diversity, for example, are often indirect and based on data from completely different sampling programs. Understanding community structure and function in the deep sea will require a more integrated and comprehensive approach to sampling in which whole communities are collected contemporaneously and then analyzed at the species level. It would also be very useful if international standards were adopted for sampling gear and sample-processing protocols. Compiling the standing-stock database explored in this chapter presented a considerable challenge because the actual data are not readily accessible (for much of it we developed a computer program to read values from published plots) and because inconsistent methods, particularly in estimating biomass, meant converting data to the same variables and units. All of these data incompatibility problems are quite understandable and expected in a rapidly growing field of research with many independent initiatives. A remedy would be to take advantage of the relatively new opportunity to make raw data available as electronic attachments to published papers and to develop a central depository of data on deep-sea community structure to which investigators could contribute and which could be accessed for synthetic analyses to test new theories.

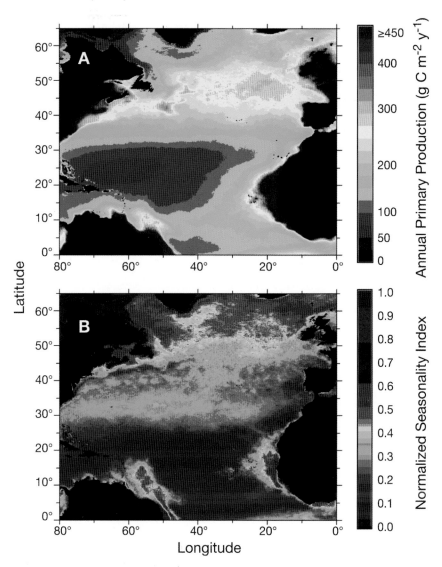

Figure 1.1. (A) Annual surface productivity (g C m^{-2} y^{-1}) in the North Atlantic estimated from SeaWiFS satellite imagery. (B) Seasonality of surface production measured as a normalized Berger and Wefer (1990) seasonality index. High values of the index represent a more seasonal pattern of production. Higher and more seasonally pulsed production is generally found in coastal waters and at high latitudes. Modified from Sun et al. (2006), courtesy of Bruce Corliss, and reproduced with permission of the authors and Elsevier.

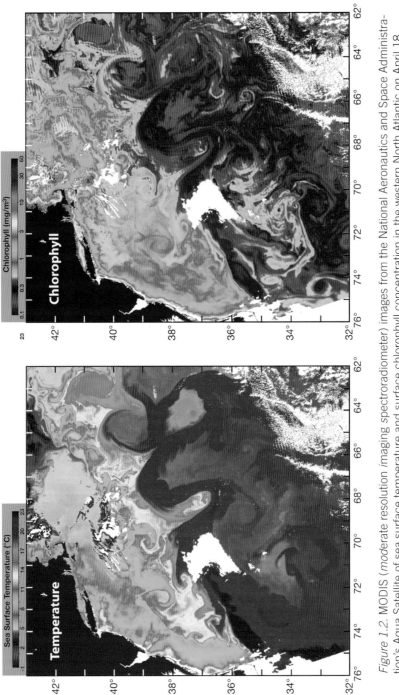

Figure 1.2. MODIS (*moderate resolution imaging spectroradiometer*) images from the National Aeronautics and Space Administration's Aqua Satellite of sea surface temperature and surface chlorophyll concentration in the western North Atlantic on April 18, 2005. Three anticyclonic Gulf Stream eddies (warm core rings) can be observed northwest of the Gulf Stream. The images illustrate the extraordinary spatial complexity of oceanographic features and surface production at regional scales. Images courtesy of NASA and GeoEye.

Figure 1.3. Fresh phytodetritus accumulated in depressions on the seafloor of the Porcupine Abyssal Plain, eastern North Atlantic (4850 m). Photograph courtesy of Andrew Gooday, National Oceanography Centre, Southampton, UK.

Figure 1.11. The distribution of samples used to estimate patterns of abundance and biomass in the deep-sea benthos. The map was created by using iMap 3.1 software (http://www.biovolution.com).

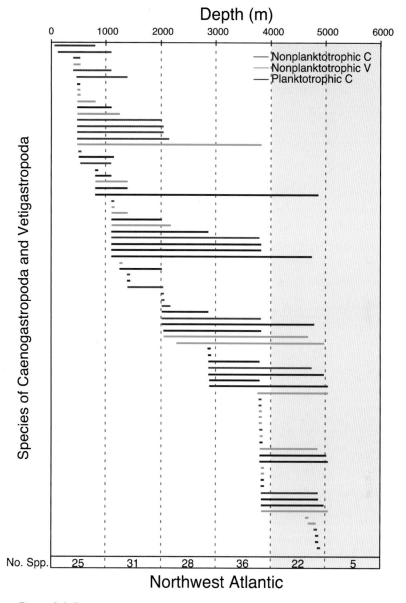

Figure 3.3. Bathymetric ranges for deep-sea gastropods south of New England in the western North Atlantic. Mode of development is indicated. Planktotrophic C (Caenogastropoda) and Nonplanktotrophic V (Vetigastropoda) are assumed to have higher larval dispersal ability than Nonplanktotrophic C. The abyssal zone is shaded. The number of species (bottom) is the number of coexisting ranges in 1000 m depth intervals. From Rex et al. (2005a), with permission of the authors and the University of Chicago Press. Species names and depth ranges are provided in the online edition of *American Naturalist* (2005), vol. 165, no. 2.

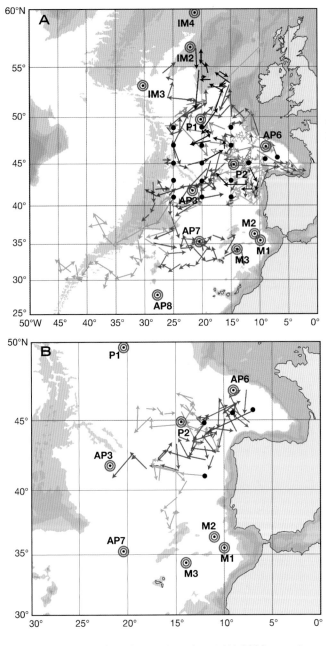

Figure 6.2. Subsurface float trajectories at (A) 1000 m and
(B) 1500–2000 m in the eastern North Atlantic from the Eurofloat program. The colors indicate individual floats. Each vector
represents the direction and distance moved by a subsurface
float at 3-month intervals from November 1996 to December
2002. Modified from Speer et al. (2003) and http://www.ifremer
.fr/lpo/eurofloat/disptraj.html, with permission of Dahlem University Press.

0325
0400
0475
0550
0625
0700
0775
0850
0925
1000
1075
1150
1225
1300
1375
1450
1525
1600
1675

Figure 6.3. Sea beam bathymetry depicting the rugged topography of a section of the upper slope off North Carolina, U.S., measuring approximately 7 × 8 km, looking south and color coded by depth (25-m intervals). From Mellor and Paull (1994), with permission of the authors and Elsevier.

Figure A.5. Macrofaunal polychaete. The epibenthic carnivore *Eunoe spica*, 25 mm in length, collected at 595 m in the Weddell Sea off Antarctica. Courtesy of Brigitte Ebbe, photograph by Dieter Fiege.

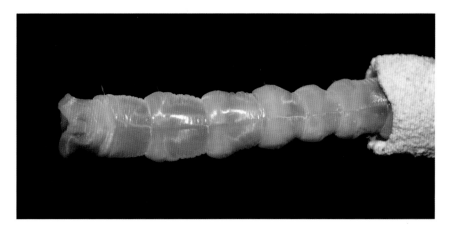

Figure A.7. Macrofaunal polychaete. A subsurface deposit feeding-tube dweller of the family Maldanidae, 10 mm in length, collected at 595 m in the Weddell Sea off Antarctica. Courtesy of Brigitte Ebbe, photograph by Dieter Fiege.

Figure A.10. Macrofaunal polychaete. A burrow-dwelling surface deposit feeder, *Aphelochaeta* sp., image of body 6 mm across, collected at 1000 m in Monterey Bay off California. See also Figure A.29 in the text for the quill worm *Hyalinoecia tubicola*. Courtesy of James Barry and Linda Kuhnz, 2003 MBARI.

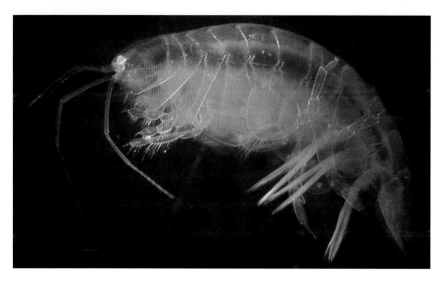

Figure A.11. Macrofaunal peracarid crustacean. The tube-dwelling filter-feeding amphipod *Ampelisca unsocalae*, length 4 mm, collected at 1040 m in Monterey Canyon off California. Courtesy of James Barry and Linda Kuhnz, 2003 MBARI.

Figure A.13. Macrofaunal peracarid crustacean. A juvenile deposit-feeding isopod, *Acanthaspidia* sp., 4 mm in length, collected at 2700 m in the Weddell Sea off Antarctica. Courtesy of Angelika Brandt.

Figure A.14. Macrofaunal peracarid crustacean. A deposit feeding cumacean, *Cyclaspis* sp., length 3 mm, collected at 3000 m in the Lazarev Sea off Antarctica. Courtesy of Angelika Brandt.

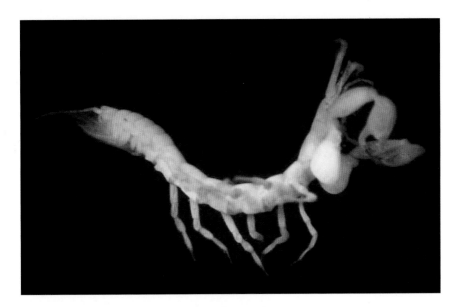

Figure A.15. Macrofaunal peracarid crustacean. A deposit-feeding tanaid, *Neotanais antarcticus*, length 10 mm, collected at 4000 m off Elephant Island (Antarctica) in the Southern Ocean. Courtesy of Angelika Brandt.

Figure A.16. Macrofaunal mollusk. The deposit-feeding protobranch bivalve *Deminucula atacellana*, 2.67 mm in length, collected at 2040 m in the western North Atlantic. Courtesy of Elizabeth Boyle and Selina Våge.

Figure A.17. Macrofaunal mollusk. The deposit-feeding gastropod *Benthonella tenella*, 3.84 mm in height, collected at 3800 m in the western North Atlantic. Photograph by Michael Rex, from Rex and Etter (1990), with permission of Elsevier.

Figure A.19. Megafaunal fish. The sit-and-wait predator *Bathypterois* sp. at 1600 m on the Yakutat Seamount in the Corner Rise Seamounts of the western North Atlantic. Courtesy of Peter Auster and the Deep Atlantic Stepping Stones Science team / NOAA/IFE/URI-IAO.

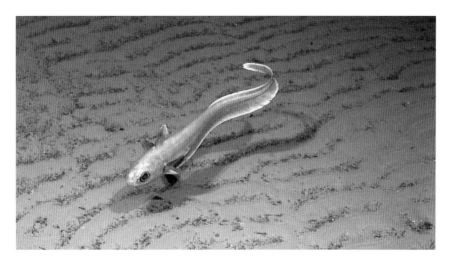

Figure A.20. Megafaunal fish. A generalist foraging both on the bottom and in the water column: *Coryphaenoides rupestris*, at 1340 m on the Manning Seamount of the New England Seamounts of the western North Atlantic. Courtesy of Peter Auster and the Deep Atlantic Stepping Stones Science team / NOAA/IFE/URI-IAO.

Figure A.24. Megafaunal fish. A forager on bottom invertebrates: *Hydrolagus affinis*, at 2300 m off Newfoundland. Courtesy of Richard Haedrich, Krista Baker, Chevron Canada Ltd., and the SERPENT project.

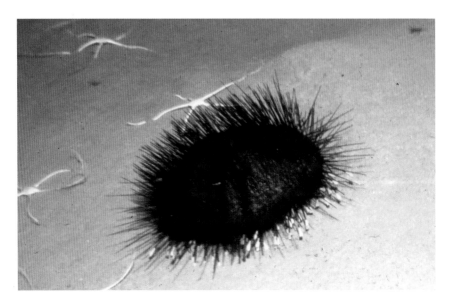

Figure A.25. Megafaunal invertebrate. The deposit-feeding sea urchin *Hygrosoma petersii* and the omnivorous ophiuroid *Ophiomusium lymani* at 1800 m in the western North Atlantic south of New England. Courtesy of Ruth Turner, from Rex et al. (2006), with permission of Inter-Research.

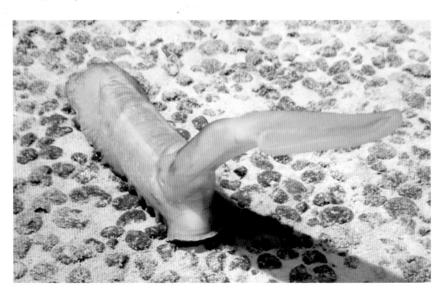

Figure A.26. Megafaunal invertebrate. The deposit feeding holothurian *Psychropotes longicauda* at 5000 m in a manganese nodule field in the tropical East Pacific. The animal is 35 cm in length. Copyright Ifremer-Nautile/Cruise Nodinaut 2004 to study biodiversity of the French Claim Nodule Area. Courtesy of Joëlle Galéron and Stefanie Keller.

Figure A.27. Megafaunal invertebrate. A filter-feeding brisingid seastar at 5600 m in the central North Pacific, perched on a manganese nodule. Manganese nodule beds cover much of the abyssal seafloor, particularly in the Indo-Pacific. Courtesy of Robert Hessler.

Figure A.28. Megafaunal invertebrate. The suspension-feeding sea pens *Pennatula aculeata* at 432 m in Lydonia Canyon on the U.S. continental margin of the western North Atlantic. Courtesy of Barbara Hecker, ACSAR project.

Figure A.30. Megafaunal invertebrate. The omnivorous anemone *Actinoscyphia aurelia* at 1947 m in Atwater Canyon in the eastern Gulf of Mexico. Courtesy of Ian R. MacDonald, photo by Will Sager.

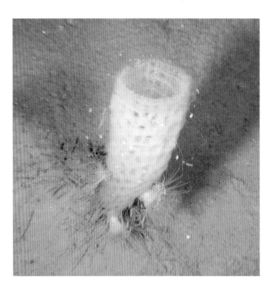

Figure A.31. Megafaunal invertebrate. A suspension-feeding glass sponge with anemones at 2876 m in the Gulf of Mexico off Florida. Courtesy of Ian R. MacDonald.

2

LOCAL SPECIES DIVERSITY

The species richness in deep-sea assemblages is already known to be extraordinarily high.

J. Frederick Grassle and Judith P. Grassle (1994)

Is deep-sea species diversity really so high?

John S. Gray (1994)

Deep-sea biodiversity—hyperdiverse or hype?

P. John D. Lambshead and Guy Boucher (2003)

In this chapter we begin our discussion of deep-sea biodiversity by examining species diversity and its potential causes at local spatial scales and on relatively short ecological time scales. What does local scale mean in the deep-sea benthic environment? As a useful generality, for our purposes, we can define it as an area of relatively uniform ecology and faunal composition. This typically occurs only at restricted depths because community makeup, the sedimentary regime, food availability, and physical factors can change rapidly with depth, especially in the bathyal zone, as we will see in Chapter 5. With remote sampling, the only indication of local scale is a sample or a set of replicate samples taken as closely as navigation permits at a single station. Permanent bottom stations where submersibles or free vehicles are used to take samples, carry out photographic observations, and conduct experiments in close proximity (e.g., Grassle and Morse-Porteous 1987, K. L. Smith et al. 2006) might also be considered to represent local scale. Thus in terms of spatial dimensions, local means roughly a range of centimeters to tens or possibly hundreds of meters at bathyal depths. However, as we will see in the discussion of spatial dispersion, significant ecological heterogeneity can occur anywhere within this range. At abyssal depths, local scale may effectively increase as the environment becomes more spatially homogeneous. Patterns and causes of diversity at local scales, as loosely defined here, do appear to be distinguishable from patterns and causes at larger spatial and longer time scales, as we discuss in subsequent chapters.

BACKGROUND

Views on deep-sea biodiversity have changed dramatically since Edward Forbes (1844) predicted that the great depths were nearly devoid of life. Late-nineteenth to mid-twentieth century voyages of exploration found evidence of life throughout the oceans at all depths. In 1967, Hessler and Sanders discovered that the deep-sea fauna is actually more diverse than temperate subtidal faunas. Early misconceptions stemmed mainly from the use of ineffective trawling nets with large mesh sizes more suitable for collecting coastal organisms. Since deep-sea species are typically much smaller than their shallow-water counterparts, most individuals simply passed through the nets and diversity was grossly underestimated. Recognizing this problem, Hessler and Sanders (1967) designed the epibenthic sled (see Appendix B for a recent version) specifically to collect large samples of the minute sparsely distributed animals that inhabit deep-sea sediments and the sediment-water interface. To illustrate how successful sleds are at collecting deep-sea material, Sanders (1977) proposed measuring their yield in "Challenger units." The *Challenger* expedition (1872–1876), circumnavigating the globe, collected a total of 6667 deep-sea benthic specimens. Single sled samples taken in the deep sea during an afternoon of trawling contain 0.5–40 Challenger units (i.e., up to about one-quarter million individuals). Sled samples revealed a much richer macrofauna than ever expected and raised the still challenging theoretical dilemma of how so many species could coexist in what appeared to be an energy-starved and ecologically uniform environment.

However, sled samples presented a number of difficulties for comparing and interpreting species diversity. Sleds are not quantitative samplers. They are towed for about 1 km on the seafloor with the distance varying depending on depth, bottom conditions, anticipated abundance of the benthos, weather, and so on. Since diversity is sample-size dependent, that is, larger samples accumulate more species, the number of species acquired in sled samples cannot be compared directly. To address this difficulty, Sanders (1968) developed rarefaction methodology, an inferential numerical method to resample the relative abundance distribution of species in order to predict how many species would have been encountered at progressively smaller sample sizes. Estimated diversities among samples could then be compared at a common normalized sample size. Hurlbert (1971) presented an analytical solution to rarefaction including its statistical properties. The measure has since been referred to as the Sanders-Hurlbert expected number of species $E(S_n)$ and remains the most widely used diversity index in deep-sea ecology. It is calculated as

$$E(S_n) = \sum_{i=1}^{S} \left[1 - \frac{\binom{N-N_i}{n}}{\binom{N}{n}} \right]$$

where $E(S_n)$ is the expected number of species in a sample normalized (i.e., rarefied) to n individuals, N is the total number of individuals in the sample, S is the number of species in the total sample, and N_i is the number of individuals in the ith species.

Rarefaction has the advantage of enabling us to portray diversity as an interpolated curve representing the relationship between the number of species and sample size, rather than just the dimensionless point of most diversity measures. The height of the curve represents species richness and its rate of ascent the evenness of the relative abundance distribution of species (see, e.g., Figures 2.1 and 2.2). A further advantage is that the curve tells us something about the adequacy of the sampling. If it becomes asymptotic, the local biota has been thoroughly collected. If it continues to increase rapidly at N, diversity has been incompletely sampled. There are a number of problems involved in interpreting $E(S_n)$, particularly when applied to quantitative samples, which we discuss as we go along. For an overview of measuring species diversity, see Magurran (2004).

THE LEVEL OF LOCAL SPECIES DIVERSITY

The deployment of box corers (see Appendix B) provided greatly improved estimates of local macrofaunal species diversity. Small quantitative samples also avoid the potential problem that epibenthic sleds might traverse a number of habitats, each with its distinctive fauna, thus giving inflated estimates of local diversity—a difficulty that cannot be overcome by rarefaction, which assumes a random distribution of species. The most intensive box-core sampling program remains the Atlantic Continental Slope and Rise (ACSAR)

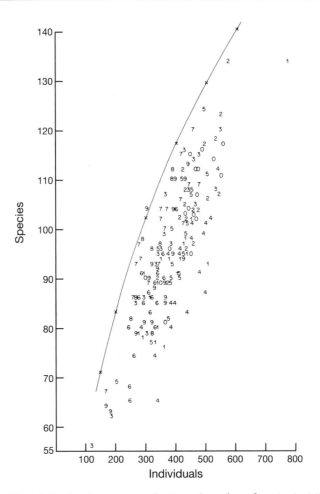

Figure 2.1. The relationship between sample size and number of species in 168 box-core samples (900 cm²) taken at 10 sites distributed over a 180-km transect along the 2100-m iso-bath off New Jersey and Delaware (United States) in the western North Atlantic. The rare-faction curve is for the combined sample over the range of sample size and diversity in individual samples. From Grassle and Maciolek (1992), with permission of the authors and the University of Chicago Press.

study, a baseline biotic survey for petroleum exploration in the bathyal zone of the U.S. eastern continental margin (Maciolek et al. 1987a,b, Diaz et al. 1994). Figure 2.1 shows the most geographically controlled assessment of species diversity in this study: 168 samples from the inner nine subcores (0.09 m²) of 0.25 m² box cores collected at 10 sites along a 176 km seg-ment of the 2100 m isobath off New Jersey and Delaware (Grassle and

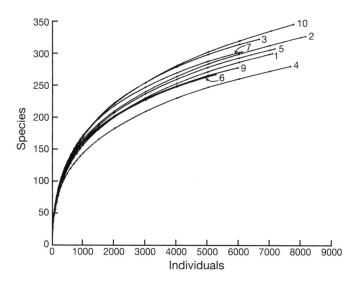

Figure 2.2. Rarefaction curves for the combined fauna (lumped replicates) from nine samp-ling sites along the 2100-m isobath off New Jersey and Delaware (United States) in the west-ern North Atlantic. Modified from Grassle and Maciolek (1992), with permission of the au-thors and the University of Chicago Press.

Maciolek 1992). The average number of species per sample is about 100, which seems impressively high for such a small area of sediment and a sample of only around 400 individuals. Variation in species richness is also quite high (55–135 species) and is clearly a positive function of sample size. Similar values have been found at bathyal depths in other studies (Snelgrove and Smith 2002). The rarefaction curve in Figure 2.1 represents the entire macro-fauna of all samples combined over the range of *S* and *N* in the individual samples. Nearly all of the observed sample diversities fall below the com-posite rarefaction curve, indicating that relative densities of species have non-random distributions among samples with a tendency toward aggregation. The wide variation in richness among replicate samples taken as close to-gether as navigation permits also suggests a patchy seafloor (Figure 2.1).

The rarefaction curves for combined replicate samples at the nine most well-sampled sites, those with 17 or 18 replicates, are shown in Figure 2.2. The curves have not become asymptotic, indicating that the macrofauna at these sites is not fully characterized; that is, there are presumably more species present locally that have not been collected. Although there are some dif-ferences in levels of diversity among sites, the curves form a coherent group and there is no obvious diversity gradient along the isobath. The sites showed very similar community structure in terms of relative abundance distributions of species, and the most common species were widely shared (Grassle and Maciolek 1992). Rarefaction estimates of diversity exceed Hessler and Sanders' (1967) earlier diversity estimates based on epibenthic sled samples.

How high is deep-sea diversity? Grassle and Maciolek's (1992) data supported the assertions of Hessler and Sanders (1967) and Sanders (1968) that deep-sea diversity is higher than in temperate coastal communities and even higher than the initial estimates, which were based on sampling done with less effective gear. These surprising results and Grassle's (1989) conclusion that "animal diversity in the deep sea may rival that found in tropical rain forests" engendered a vigorous debate on the relative levels of coastal and deep-sea benthic diversity (May 1992, Briggs 1994, Gray 1994, Gage 1996, Carney 1997, Gray et al. 1997). An accurate and meaningful comparison is difficult because it is virtually impossible to control completely for habitat type, sampling methodology, sampling effort, and faunal makeup. Moreover, there are surprisingly few reasonably complete macrofaunal datasets and those that are available represent geographically distant and environmentally distinct ecosystems. Suffice it to say, as all marine ecologists are aware, coastal habitats differ greatly in diversity (Bertness et al. 2001, Levinton 2001); as we will see in Chapter 3, so do deep-sea habitats.

Currently, the most geographically and methodologically controlled comparison possible is between upper bathyal depths sampled in the northern region of the ACSAR Program (Maciolek et al. 1987a) and Georges Bank (Maciolek-Blake et al. 1985) on the adjacent continental shelf of the western North Atlantic. Both studies used quantitative sampling gear and the same sieve mesh size for extracting the macrofauna. Species determinations were supervised by the same taxonomists, and the entire macrofaunal assemblage was recorded. Etter and Mullineaux (2001) compared diversity between Georges Bank (38–167 m) and the upper slope (250–2500 m) by randomly pooling replicate samples at each site to construct species accumulation curves for both sample size and area sampled (Figure 2.3). Diversity is clearly much higher on the slope. Thus, extensive quantitative sampling confirmed Hessler and Sanders' (1967) and Sanders' (1968) contention that diversity increases from coastal to bathyal environments in this part of the western North Atlantic. There is scarcely any overlap in the level of diversity. Species accumulation curves for the slope are also much steeper and show less evidence of becoming asymptotic, indicating, again, that the deeper

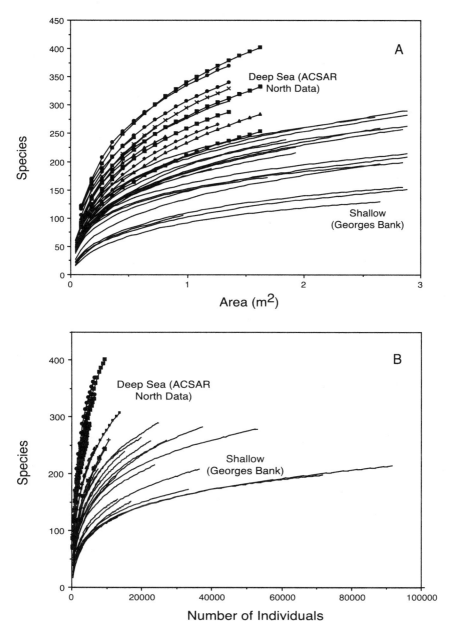

Figure 2.3. Species accumulation curves with area sampled (A) and sample size (B) of benthic sampling stations on Georges Bank and the adjacent continental slope south of New England (United States). From Etter and Mullineaux (2001), with permission of the authors and Sinauer Associates, Inc., based on data from Maciolek-Blake et al. (1985) and Maciolek et al. (1987a).

sites are undersampled and that shelf-slope differences in diversity would be accentuated at a higher sampling intensity. When a subset of slope samples from 1220 to 1350 m is selected to make a regional-scale comparison over similar depth increments, diversity on the slope is much higher (319 spp m^{-2} versus 165 spp m^{-2}, $E(S_{1000})$ = 188 versus $E(S_{1000})$ = 69). The higher rate of species accumulation in the upper bathyal with area sampled (Figure 2.3) also suggests a more patchily distributed fauna, consistent with the aggregational tendency shown in Figure 2.1.

The relative abundance distributions of species differ sharply between shelf and slope environments in the western North Atlantic. Bathyal communities have more rare species and more even relative abundance distributions. The endpoints of commonness and rarity in these distributions are shown in Figure 2.4. Shelf communities display much higher numerical dominance of the most abundant species. They also have fewer species represented by single individuals, although the proportion of these very rare species remains constant between shelf and slope communities. These two basic features of the relative abundance distributions, evenness and the number of rare species, account in significant measure for the steeper rise and lack of asymptote in bathyal rarefaction curves (Figure 2.3). As we show in Chapter 3, evenness and richness vary geographically within the deep-sea realm. It is not possible to generalize these trends (Figures 2.3, 2.4) to other deep-sea basins and their neighboring continental shelf habitats because the data to do so are lacking, but they do provide an exceptionally clear standardized contrast that allows us to ask with confidence: Why is bathyal macrofaunal diversity higher than diversity on the continental shelf in this region?

Much less is known about how meiofaunal diversity differs between the coastal zone and the deep sea. Mokievsky and Azovsky (2002) compared nematode diversity estimated from core samples taken at abyssal and bathyal depths to two independent shallow-water datasets by plotting species richness against sample size (Figure 2.5). Since the samples represent a variety of habitats and are very widely distributed, there is no ecological or geographic control. Nonetheless, the patterns are remarkably consistent. Clearly, diversity is sample-size dependent in both shallow and deep ecosystems. None of the slopes differ and the elevations of the two shallow-water regressions are very similar. At comparable densities, the deep-sea assemblages have about twice the number of species as do shallow-water assemblages. It seems reasonable to conclude that for nematodes, the largest meiofaunal taxon, local deep-sea diversity is considerably higher than local coastal diversity.

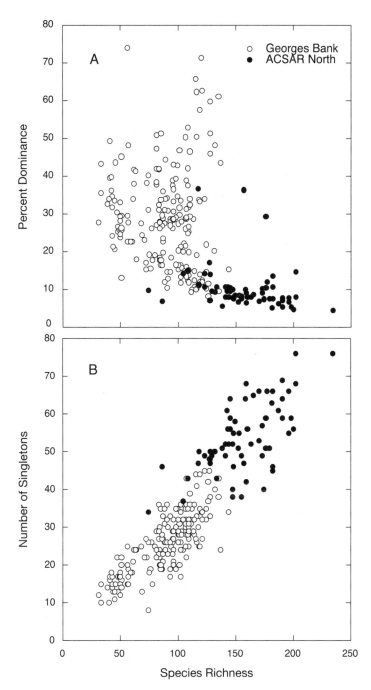

Figure 2.4. The relationships of the percentage dominance of the most abundant species (A) and number of species represented by a single individual (B) to species richness (the number of species) in benthic samples from Georges Bank and the adjacent continental slope south of New England (United States). From Etter and Mullineaux (2001), with permission of the authors and Sinauer Associates, Inc., based on data from Maciolek-Blake et al. (1985) and Maciolek et al. (1987a).

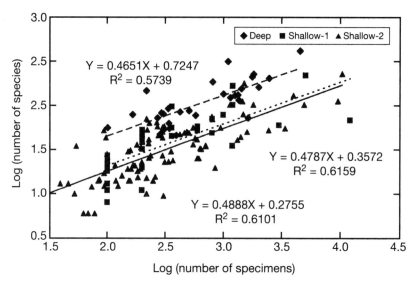

Figure 2.5. The relationship between the number of species and sample size of nematodes from the deep sea and two independent shallow-water datasets. At comparable sample sizes deep-sea diversity is about twice shallow-water diversity. From Mokievsky and Azovsky (2002), with permission of the authors and Inter-Research.

THEORIES OF LOCAL SPECIES DIVERSITY

Theories of local species coexistence have changed dramatically during the past several decades. Early ideas (reviewed by Rex 1983) were quite general and mirrored debates in terrestrial and coastal marine ecology. High species diversity was explained as a characteristic property of the deep-sea eco-system; the principal causes of diversity were inferred independent of scale and geographic variation in community structure. The first major theory was Sanders' (1968) stability-time hypothesis. Sanders proposed that long-term environmental stability in the deep sea allowed diversity to develop through specialized biological interactions—primarily competitive niche differenti-ation. Dayton and Hessler (1972) suggested the alternative possibility that predation by large epibenthic megafaunal species promoted diversity by al-leviating competition among infaunal prey. Grassle and Sanders (1973) re-sponded that the population density and reproductive capacity of most deep-sea species seemed too low to withstand the kind of pervasive preda-tion pressure envisioned by Dayton and Hessler.

When diversity patterns were looked at in a broader geographic con-text, it became apparent that competition and predation are not incompatible mechanisms and that their relative importance depends on other factors, such as productivity, trophic structure, and dispersal (Rex 1976, 1977, Etter and Caswell 1994). But the real significance of Grassle and Sanders' (1973) paper was that it introduced a new idea that was to become the principal paradigm for explaining local species coexistence. Drawing on the non-equilibrial model of Richerson et al. (1970), they suggested that deep-sea diversity is maintained as a spatial mosaic of patches. This concept was later elaborated by Grassle and Morse-Porteous (1987), Grassle (1989), Grassle and Grassle (1994), and Grassle and Maciolek (1992).

The spatial mosaic theory proposed that small-scale disturbances at the seafloor permitted high local diversity by creating successional sequences that are temporally out of phase. Species participating in the sequences were assumed to be adapted to a particular stage of succession. In this sense the theory was a reassertion of Sanders' (1968) emphasis on the importance of niche diversification, now extended to include adaptation to particular ecological circumstances in a temporal succession. The local seascape was predicted to be a nonequilibrial system of small widely separated patches representing different successional stages, with local coexistence maintained by dispersal among patches. The occupation of different patches would reduce species interactions that might depress diversity. The primary events initiating succession were taken to be highly localized temporary increases in food supply superimposed on a food-limited background.

Jumars (1975b, 1976) broadened this concept to include the possibility of stochastic replacement of species following disturbance. The general notion of a temporal mosaic gained credence with the growing awareness that the deep-sea sediment environment is much more textured and dynamic than earlier supposed. Small-scale events that could potentially drive succession include the collection of phytodetritus in depressions (Billett et al. 1983) and burrows (Aller and Aller 1986), plant debris and macroalgae settling on the bottom (Wolff 1979, Suchanek et al. 1985), foraging activities of mega-faunal species (Dayton and Hessler 1972), biogenic microhabitats (Jumars 1976, Levin et al. 1986), fecal casts and trails (Wheatcroft et al. 1989), construction of mounds and burrows (C. R. Smith et al. 1986), and sinking animal carcasses (Stockton and DeLaca 1982), as noted earlier in Chapter 1. It is hard to distinguish the extent to which these perturbations act as a physical disturbance, a biological disturbance, or nutrient enrichment, but all could presumably initiate a localized successional change.

Owing to the physical stability of the deep sea, patches were expected to persist longer than in more energetic coastal habitats. In general, this seems to be a reasonable assumption, but temporal rate processes in the deep sea are still much less well known than spatial patterns. Duration of environmental heterogeneity appears to vary widely. For example, Wheatcroft et al. (1989) observed that persistence of biogenic structures not maintained by animals (tracks and traces in surface sediments made by large mobile epifaunal species) depends on sediment texture, abundance and activity of the benthos, and the intensity of near-bottom currents. Surficial traces in the upper bathyal of Santa Catalina Basin (1300 m) and at the abyssal HEBBLE site (4800 m) lasted days to weeks. Periodic benthic storms at the HEBBLE site completely obliterate traces.

In contrast, at the more quiescent Clipperton-Clarion Fracture Zone of the remote North Pacific (5000 m), tracks made by industrial dredging for manganese nodules (originally 2.5 m wide, 4 cm deep, and bordered by displaced sediment ridges 15 cm high) were still visible 26 years later (Khripounoff et al. 2006). Phytodetritus accumulations on the deep-sea floor persist for weeks to months (Beaulieu 2002); their effects on some elements of the benthos can last appreciably longer, as noted in Chapter 1. Later in this chapter we discuss experimental evidence that successional approaches to background community structure following physical disturbance or nutrient enrichment can take a year or more—much longer than in coastal sediments, where recovery occurs in about a month (C. R. Smith and Brumsickle 1989). Dead fish sinking to the deep-sea floor are stripped to the bone by large highly motile necrophages within hours to days (Dayton and Hessler 1972, Priede et al. 1994). The arrival of a dead whale carcass creates a gradually developing food bonanza that drives local benthic succession for decades (C. R. Smith and Baco 2003). Clearly, a broad range of effects is possible in terms of patch dynamics and community response.

The temporal mosaic theory has been tested in two basic ways: precision-sampling studies to assess patchiness at the expected scales and experiments to examine the temporal dynamics of succession. In the text that follows we present specific examples of each and then summarize the evidence relating to the temporal mosaic theory.

SPATIAL DISPERSION

Jumars (1975a,b, 1976, 1978) tested for evidence of patchiness by measuring species-specific spatial dispersion patterns. He partitioned 0.25 m^2 box corers (see Appendix B) into 25 subcores measuring 10×10 cm and then compared species distributions among subcores and among box-core samples positioned at different distances from one another. The intensity of departures from random dispersion could then be detected with variance-to-mean ratios and the actual scale of patterns identified by spatial autocorrelation along a gradient of intersample distance (Jumars et al. 1977). This work remains among the most technically daunting and analytically elegant research ever conducted in deep-sea ecology.

More is known about spatial dispersion among the macrofauna than for other size categories. Jumars (1975a,b) documented dispersion for 146 species of polychaetes (see Appendix A) among five box-core samples collected at 1230 m in the San Diego Trough of the eastern North Pacific. Variance-to-mean ratios for the species are shown in Figure 2.6. A ratio of 1 conforms to a Poisson series indicating a random dispersion pattern. It is not possible to prove randomness and a number of nonrandom patterns can produce a Poisson series, which means only that the hypothesis of random dispersion cannot be rejected. Dispersion patterns of species within the confidence limits of the random expectation ($S^2/\bar{X} = 1$, or log $S^2/\bar{X} = 0$ as used in Figure 2.6) do not depart significantly from random. Species above the upper confidence limit have aggregated (patchy) dispersion and those below the lower confidence limit have regular (evenly spaced) dispersion. Very few polychaete species depart from random (eight aggregated and two regular), about the same number expected by chance alone given the total number of species examined. Moreover, nearly all of the aggregated cases are attributable to unusually high or low densities in just a single core, making the patterns seem more idiosyncratic than general. The sole exception was a cirratulid worm, *Tharyx luticastellus,* that constructs and inhabits a distinctive "mudball" structure about 1–3 cm in diameter. A similar pattern was found by Gage (1977) for 18 species of bivalves and 105 species of polychaetes from six box-core samples taken kilometers apart at 2875 m in the Rockall Trough of the eastern North Atlantic. Only 11% of these species departs from random expectations, nearly all of which displayed aggregation. A predominance of random patterns and a low degree of aggregation were also found in macrofaunal elements among cores separated by 1–100 km at upper bathyal depths off California (Jumars 1976) and at abyssal depths in the central North Pacific (Hessler and Jumars 1974). At much smaller centimeter scales, among subcores within box cores, Jumars (1975b) found that only 2 of 52 polychaete species deviated from random, again tending toward aggregation.

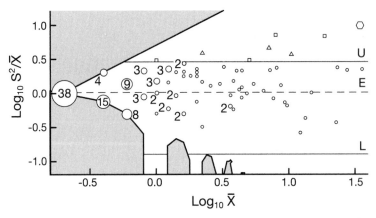

Figure 2.6. The relationship of the log variance-to-mean ratio to log average number of individuals for 146 polychaete species collected in five box-core samples taken at 1230 m in the San Diego Trough of the eastern North Pacific. The dashed line (E) is the Poisson distribution and the solid lines its upper (U) and lower (L) confidence limits. The shaded area is statistically impossible. Numbers indicate more than one species. Values above the upper confidence limit are aggregated distributions, those within the confidence limits do not differ significantly from random, and those below are even distributions. The squares and triangles above U are values from single cores and the hexagon represents the worm *Tharyx luticastellus,* which constructs and inhabits a distinctive mudball structure. From Jumars (1975b), with permission of the author and Springer Science.

Lamont et al. (1995) documented dispersion of bivalves and tanaids (see Appendix A) by using partitioned box cores separated by ten to hundreds of meters at two sites in the eastern North Atlantic, the Tagus Abyssal Plain (8 cores at 5035 m) and Setubal Canyon (11 cores at 3300 m). In the canyon, presumed to be the more heterogeneous environment, 3 of 20 bivalves and 4 of 20 tanaids showed aggregation. In the abyssal plain, 2 of 12 bivalves and 1 of 10 tanaids were aggregated. Species tended to show stronger aggregation at the canyon site. At smaller among-subcore scales within cores there was little evidence of nonrandom dispersion. Cosson et al. (1997) analyzed dispersion of higher taxa in box-core samples taken at upper bathyal, lower bathyal, and abyssal sites in the tropical northeast Atlantic. It is hard to relate their results to studies at the species level, but they did find a trend toward more aggregation at bathyal depths, which tends to support the results of Lamont et al. (1995).

As part of a sampling program specifically designed to measure spatial dispersion at different scales (aptly named Expedition Quagmire), Jumars (1978) used an unmanned submersible to take precisely positioned small cores (20 × 20 cm), each with four subcores measuring 10 × 10 cm within a triangular array at 1220 m in the San Diego Trough. This allowed dispersion to be assessed at scales ranging from 0.1 to about 500 m by using spatial autocorrelation, which takes into account the actual coordinates of organisms to identify the scale of variation in density. Of 13 species of polychaetes studied, only two exhibited significant autocorrelation, one negative at the smallest scale, suggesting spacing of individuals and another showing a more complex pattern with dense patches separated by about 100 m. Neither showed significant variance-to-mean ratios, indicating that spatially explicit measures such as spatial autocorrelation can detect structure that remains obscured by summary statistics.

Jumars (1975a) also developed an extension of dispersion indices for individual species by summing their variance-to-mean ratios to detect an overall average tendency toward nonrandom dispersion. This method accumulates deviations from random expectations by considering species together as replicates. The resulting index is called the total chi-square, which has two components: the pooled chi-square and the heterogeneity chi-square. The former represents variation in the combined abundance of all species and the latter reflects the degree of discordance in abundance of species (relative abundance distributions not proportionately constant among cores). Jumars (1975b) found that polychaetes in the San Diego Trough are on average aggregated among cores because of both variations in total density and discordance among species. There was no evidence of aggregation among subcores within cores. The results, using species as replicates, were similar for the macrofauna of the California continental borderlands (Jumars 1976) and for bivalves and tanaids in the eastern North Atlantic (Lamont et al. 1995).

Even less is known about spatial patterns in the meiofauna (see Appendix A). Thistle (1978) studied dispersion in harpacticoid copepods collected in Expedition Quagmire by examining variance-to-mean ratios at centimeter, meter, and 100-m scales. Individual species showed very little evidence of departure from random patterns at any of these scales. Dispersion patterns at the 100 m scale, shown in Figure 2.7, bear a strong resemblance to polychaetes at among-core scales (Figure 2.6). When species are considered as replicates, there is a significant average trend toward aggregation and both the pooled and heterogeneous chi-squares are significant at all three scales. Bernstein et al. (1978), using different methods of cluster analysis and departure from the assumptions of rarefaction, demonstrated aggregation in agglutinated foraminiferans collected from abyssal depths in the central

North Pacific at centimeter and kilometer scales. Eckman and Thistle (1988), using clusters of smaller (2.56 cm^2) corers, did detect significant aggregation of meiofaunal elements on centimeter scales, but again found little evidence for nonrandom dispersion at larger scales. By comparing the incidence of living and dead foraminiferans, Bernstein and Meador (1979) showed that patch structure must persist between generations at centimeter scales. Rice and Lambshead (1994) found that 25–55% of nematode species were aggregated at centimeter scales at two abyssal sites in the eastern North Atlantic.

What is known about dispersion in the megafauna (see Appendix A) comes from photographic surveys taken by submersibles at upper bathyal depths in the western North Atlantic (Grassle et al. 1975) and eastern North Pacific (C. R. Smith and Hamilton 1983). A grid subdivided into quadrants with built-in perspective was superimposed on successive photographic images of the seabed taken along a transect, so that dispersion could be assessed within and among grids. The results varied considerably for the relatively few species that are abundant enough to analyze. In the western North Atlantic, the brittlestar *Ophiomusium lymani* showed regular spacing at high densities but aggregation at low densities, where it avoids depressions in the sediment. Two deposit-feeding sea urchins, *Phormosoma placenta* and *Echinus affinis,* showed aggregation, the former because it forages in herds and the latter because it forages around clumps of macroalgae deposited on the bottom. The large epibenthic polychaete *Hyalinoecia* and a cerianthid anemone did not depart from random. In the eastern Pacific, the brittlestar *Ophiophthalmus normani* showed a random distribution at scales smaller than 1 m^2, but formed patches 1–4 m across at larger scales. The snail *Bathybembix bairdii* and the sea cucumber *Scotoplanes globosa* formed patches tens of meters in diameter, and the rockfish *Sebastolobus altivelis* was randomly distributed. The aggregated dispersion in echinoids and holothurians found in these studies is very reminiscent of the dense patches in those taxa noted in Chapter 1 on standing stock, although the actual dispersion pattern was not quantified in those studies.

The causes of nonrandom dispersion, where detected, are highly inferential. The most convincing associations are for the megafauna, where abundance patterns and animal activity are more readily observable. For example, aggregation seems reasonably linked to social foraging or mating behavior in *Phormosoma placenta* and *Scotoplanes globosa,* to exploitation of patchy food resources in *Echinus affinis,* and to local population explosions driven by episodic phytodetritus accumulations in *Amperima rosea,* as we discussed in

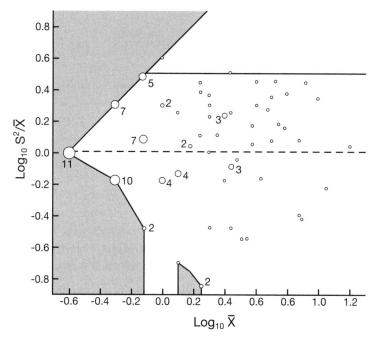

Figure 2.7. The relationship of log variance-to-mean ratio to log average number of individuals for 60 species of harpacticoid copepods collected in 10 cm × 10 cm subcores of a 20 cm × 20 cm Eckman core at 1220 m in the San Diego Trough of the eastern North Pacific. The spatial scale of samples compared is 100 m (median distance among samples). The dashed line is the Poisson distribution and the solid line its upper confidence limit. The lower confidence limit is below the graph. The shaded area is statistically impossible. Numbers indicate more than one species. Values above the upper confidence limit represent aggregated distributions. All other values shown do not depart significantly from random. From Thistle (1978), with permission of the author and the *Journal of Marine Research.*

Chapter 1. Regular dispersion is uncommon, occurring in *Ophiomusium lymani* at densities so high that individuals avoid touching one another. Jumars (1975a,b), using all macrofaunal polychaetes as replicates, found significant negative autocorrelation among subcores, which suggests that intraspecific spacing is possibly a form of territoriality at small scales. A similar pattern was found in the polychaete *Ceratocephale pacifica* (Jumars 1978). In another polychaete, *Polyophthalmus* sp., dense patches were dominated by larger individuals, possibly as a result of earlier settlement events or spatial variation in food resources affecting growth (Jumars 1978).

Patterns of concordance (Hessler and Jumars 1974) or discordance (Jumars 1976) in density among species or higher taxa may indicate biological interactions such as competition, predation, mutualism, or simply specialized adaptation to different environmental circumstances. Jumars and Eckman (1983) reported significant associations between polychaete abundance and sponge fragments in cores, which hints at some kind of direct or indirect interaction; sponges might alter hydrodynamic flow to augment localized POC deposition or act as refuges for worms. A similar phenomenon was reported by Bett and Rice (1992). A negative association between *Tharyx* mudball dwellings and shallow-burrowing paraonid worms may indicate spatial exclusion by these biogenic structures (Jumars 1975b). Thistle (1979a) found significant, though weak, associations between the abundance of meiofaunal harpacticoids and biogenic structures. Harpacticoids also show both positive and negative associations with polychaete functional groups, suggesting that certain worms can generate heterogeneity by either excluding meiofaunal groups or enhancing their abundance (Thistle et al. 1993). Bernstein et al. (1978) found that patches of agglutinated foraminiferans corresponded to both biogenic structures and co-occurring metazoan functional groups. Establishing the causes of population structure experimentally, particularly the diffuse average trends deduced by using species as replicates, would be extraordinarily difficult in this remote environment.

Despite the analytical rigor of studies on spatial dispersion in the deep-sea benthos, the results are inconclusive. There is as yet very little evidence for strong aggregation or patchiness that could explain high local species diversity. However, Jumars and Eckman (1983) caution that, given the low densities of individual species, low sampling intensity, and small sample sizes, the statistical power of tests for departure from random dispersion is quite weak. A large proportion of the species recovered in sampling studies cannot be used to explore dispersion because they occur only as single individuals. Most species are extremely rare. If 0.25 m² box corers are used, the number of samples necessary to detect nonrandom dispersion at local scales for most deep-sea species may be prohibitive (Jumars 1981).

On the other hand, a convincing tendency toward aggregation in some common species has been found at scales ranging from centimeters to hundred of kilometers, suggesting that diversity-generating mechanisms can operate at all scales within this range. There does appear to be a rough positive relationship between the size categories of animals and the scales at which species show a tendency toward aggregation, suggesting that small organisms predictably experience the environment as more coarse grained. Jumars (1976) speculated that important interactions might occur at scales approximating the size of deep-sea organisms. Jumars et al. (1990) conjectured that the deep-sea environment favored a deposit-feeding strategy of macrophagy, one in which relatively small species selectively consume food items one at a time instead of wholesale sediment ingestion. In Chapter 3 we show that macrofaunal species diversity is a positive function of sediment grain-size diversity. This relationship supports the notion that macrofaunal organisms partition sediment food resources or microhabitats and interact with their biotic and abiotic environments at very small spatial scales.

EXPERIMENTAL APPROACHES

The most thorough and well-controlled manipulative experiments on patch dynamics were those conducted by Craig Smith and Paul Snelgrove and their colleagues during the 1980s and 1990s. Performed using deep-sea submersibles under the most challenging circumstances and with a high risk of failure, their experiments were audacious and remarkably successful. We first recount their separate and quite different experimental approaches and then briefly summarize other related evidence on the effects of local disturbance and enrichment events from their review paper (Snelgrove and Smith 2002).

C. R. Smith et al. (1986) and Kukert and Smith (1992) conducted in situ experiments to determine the effects of megafaunal mound construction as a source of disturbance that might cause succession and promote local coexistence through patch dynamics. In Santa Catalina Basin off California deposit-feeding echiurans create fecal mounds that measure about 10 cm high and 30 cm in diameter. They are common features that cover 2% of the seafloor in this region at upper bathyal depths (1240 m). Time-lapse photography showed that mound formation is sporadic, but that during active phases mounds can increase in height 1–2 cm over a period of 5 days (Figure 2.8). Burial rates of glass tracer beads and excess ^{234}Th profiles indicate average fecal deposition rates of 1–2 cm per month, about two orders of magnitude higher than the background sediment accumulation rate.

Since most meiofaunal and macrofaunal organisms inhabit the top few centimeters of sediment (see Chapter 1), burial from mound construction could be a significant localized disturbance. To assess the effects of mound building, Kukert and Smith (1992) used submersibles to emplace artificial mounds of defaunated sediment similar in size to natural mounds and then monitored colonization by macrofauna after 4 days and then at 1, 2–3, 11, and 23 months. The artificial mounds represent an instantaneous and max-

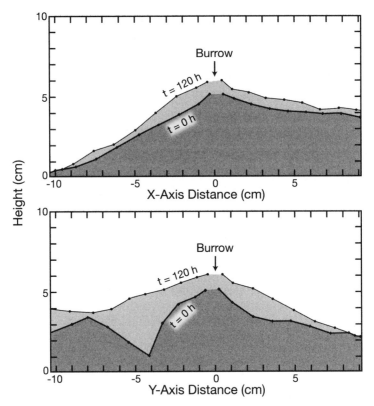

Figure 2.8. Profiles showing the growth of an echiuran mound over a period of 5 days measured from time-lapse stereophotographs taken at 1240 m in the Santa Catalina Basin of the eastern North Pacific. Time difference from $t = 0$ (initial measurement) to $t = 120$ h later is indicated. Mound growth is episodic but can be quite rapid during active phases of mound construction. The burrow indicates the site of fecal deposition by the echiuran. From C. R. Smith et al. (1986), with permission of the authors and Macmillan Publishers Ltd.

imum burial disturbance compared to the more gradual, though episodic, growth of natural mounds. At 4 days, abundance and diversity in artificial mounds were only 30–40% below background levels, indicating that burial disturbance is moderate and recovery quite rapid. Burial impacted all lifestyles and species abundance levels, although rare species seemed to fare most poorly. The recovery community consisted of both burial survivors and lateral colonists, both primarily burrowing deposit-feeding forms. Diversity approached background levels slowly over 1 year and actually exceeded background levels after 23 months (Figure 2.9), by which time the artificial mound

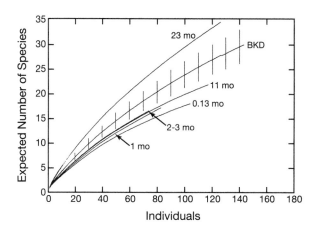

Figure 2.9. Rarefaction curves for the background community (BKD) with 95% confidence limits and artificial mounds sampled at 0.13, 1, 2–3, 11, and 23 months after emplacement at 1240 m in the Santa Catalina Basin of the eastern North Pacific. Diversity in artificial mounds approached background levels gradually over about a year and exceeded background levels at 23 months. From Kukert and Smith (1992), with permission of the authors and Elsevier.

structure had completely disintegrated. Succession continued throughout the 23-month period in terms of habitat types and feeding guilds of constituent species (Figure 2.10). The participants in succession were largely background species rather than exotic opportunists.

Unlike most deep-sea soft-sediment communities, both background and successional assemblages were heavily dominated by a single species, the polychaete *Levinsenia oculata*. Changes in the rest of the community were difficult to detect statistically. Most species that were abundant enough to track individually and that showed a temporal statistical trend in abundance (29 of 148) recovered to background levels within 11 months. Two rare species exhibited significantly enhanced abundance at 23 months, suggesting that there were unusual ecological opportunities during the successional sequence. Rarer background species collectively were better represented at 23 months, accounting for the elevated diversity. Kukert and Smith (1992) concluded, in support of the temporal mosaic theory, that burial disturbance may decrease competition during succession, permitting increased diversity.

In a series of controlled experiments, Snelgrove et al. (1992, 1994, 1996) tested for the effects of nutrient enrichment on patterns of colonization. Earlier efforts to carry out these kinds of experiments (Grassle 1977, Des-

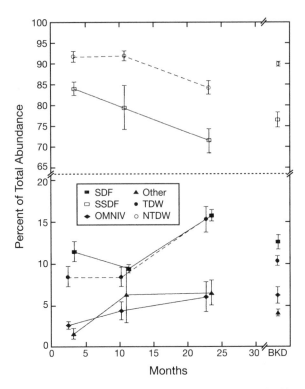

Figure 2.10.   Changes in trophic (solid lines) and domicile groups (dashed lines) during succession within artificial mounds emplaced at 1240 m in the Santa Catalina Basin of the eastern North Pacific. The symbols represent surface-deposit feeders (SDF), subsurface-deposit feeders (SSDF), omnivores (OMNIV), tube dwellers (TDW), non-tube dwellers (NTDW), and others. Background community makeup is shown at the right. Succession appears to continue throughout the duration of the experiment (23 months). From Kukert and Smith (1992), with permission of the authors and Elsevier.

bruyères et al. 1980, Levin and Smith 1984, Grassle and Morse-Porteous 1987) yielded mixed results, in part because the sediment recolonization trays that were used protruded above the seabed, which resulted in a hydrodynamic bias that potentially influenced larval recruitment (C. R. Smith 1985, Snelgrove et al. 1995). Snelgrove et al. (1992) developed improved sediment trays that could be implanted flush with surrounding sediment to minimize flow effects. South of St. Croix at 900 m in the Caribbean Sea, Snelgrove et al. (1992, 1994) used a submersible to deploy and monitor recolonization of trays with unenriched defaunated sediment and sediment laced with diatoms or ground-up *Sargassum*. They also followed recolonization of enriched and unenriched artificial depressions, which were excavated using a small box corer. After 23 days the most striking result was that the enriched trays, especially those with diatoms, attracted significantly higher densities than the unenriched trays or the background fauna (Figure 2.11). By contrast, there were no significant differences among treatment or background levels in depressions (Figure 2.11). The four most abundant colonizers of the enriched trays were opportunistic species, including the notorious pollution-disturbance indicator worm *Capitella,* which were absent or rare in the unenriched treatments and in the background fauna. Most colonists were juvenile individuals. Few background species were found in experimental treatments.

Variation in species diversity among treatments and natural circumstances is shown in Figure 2.12. Background levels, unenriched treatments, and diatom-enriched depressions were all quite similar. Diversity in *Sargassum* and diatom trays and, to a lesser extent, *Sargassum* depressions is conspicuously depressed, indicating a strong negative relationship between diversity and enrichment, which leads to high density at these small scales.

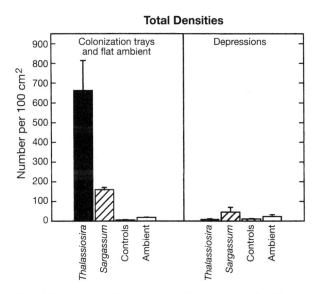

Figure 2.11. Faunal densities after 23–24 days in sediment trays and artificial depressions enriched with diatoms (*Thalassiosira*) or ground-up *Sargassum,* unenriched controls, and ambient (background) conditions. The experiment was conducted at 900 m in the Caribbean Sea south of St. Croix, U.S. Virgin Islands. From Snelgrove et al. (1994), with permission of the authors and the *Journal of Marine Research.*

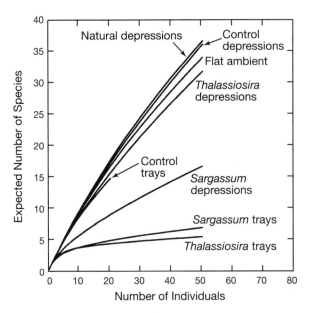

Figure 2.12. Rarefaction curves after 23–24 days for sediment trays and artificial depressions enriched with diatoms (*Thalassiosira*) or ground-up *Sargassum,* unenriched controls, and background conditions (ambient flat sediments and natural depressions). From Snelgrove et al. (1994), with permission of the authors and the *Journal of Marine Research.*

Variation in species composition among background and treatments is shown in Figure 2.13. Trays enriched with diatoms and *Sargassum* form a tight cluster because they are dominated by the same few opportunists. *Sargassum* depressions converge toward them because they also share these species. Natural depressions, unenriched artificial depression, background sediments, and diatom-enriched depressions show considerable variation in community makeup despite their very similar diversities.

Overall, the results show that colonization was quite rapid in the case of enriched trays and that colonists respond differently to nutrient sources and how they are presented (i.e., tray versus depression). Enrichment in trays depresses diversity and favors a small set of opportunists that are rare in the background community. Unlike the Kukert and Smith (1992) experiments, where colonists were adults, colonists of enriched trays were juveniles, suggesting recruitment by larval settlement.

In a much longer deployment (28 months) of the *Sargassum* tray experiment (Snelgrove et al. 1996), density in unenriched controls exceeded both

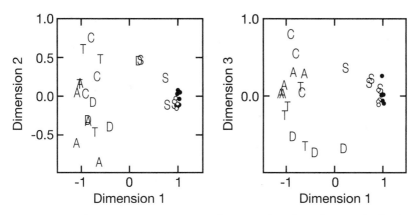

Figure 2.13. Similarity in species composition after 23–24 days (shown by using the first three dimensions of nonmetric multidimensional scaling based on normalized expected species shared) of sediment trays enriched with diatoms (*Thalassiosira*), open circles, or ground-up *Sargassum,* closed circles. Other symbols represent natural depressions (D), unenriched artificial depressions (C), background sediment = ambient (A), depressions enriched with diatoms (T), and depressions enriched with *Sargassum* (S). The diatom- and *Sargassum*-enriched trays form a tight cluster because they are dominated by the same few opportunistic species. The *Sargassum* depressions tend to converge toward the diatom- and *Sargassum*-enriched trays because they share these species. Other treatments and controls show wide variation in species makeup even though they have similar diversities (see Figure 2.11). From Snelgrove et al. (1994), with permission of the authors and the *Journal of Marine Research.*

enriched and background levels. Species composition in enriched trays shifted toward a reduction in opportunists and increased dominance by a background species. Analysis of species composition showed separate distinct clusters of short-term (23 days), long-term (28 months), and background assemblages, but species makeup in the long-term experiment began to approach that of the background community. Diversity in long-term trays remained below background levels. Thus after more than 2 years, enriched treatments continued to experience significant successional changes and still differed in both composition and diversity from background communities, which remained stable throughout the term of the experiment.

Snelgrove and Smith (2002) reviewed 16 macrofaunal and 9 meiofaunal studies in which precision sampling was used to detect the effects of various biogenic structures, phytodetritus, and *Sargassum* on local community structure. Enhanced diversity was observed in only three cases, the macrofauna associated with xenophyophore tests, biogenic mounds, and a whale skeleton. Generally, the associations between community structure and a presumed source of patchiness were weak and involved only a small subset of species. Samples within a site, with and without potential patch-generating phenomena, showed very similar community composition and diversity. Similarly, in 13 experimental colonization studies reviewed that used enriched and unenriched sediments or measured the effects of artificial mounds and depressions, there was only one case of enhanced diversity. Different sources of patchiness elicited different responses, but the number of affected species was small and often included opportunists that were rare or not part of the background community. The authors concluded (p. 313) that, "In summary, biogenic patchiness on the scales studied to date may contribute to, but in themselves are inadequate, to explain the extremely high local levels of species diversity in the deep sea." The results of precision-sampling studies and controlled recolonization experiments are similar to the more qualitative observations on faunal response to enrichment we discussed in Chapter 1. Although the methodologies and objectives differed, measurements of assimilation and population shifts associated with phytodetritus deposition also suggested that responses at the metazoan community level were limited and taxonomically selective.

CONCLUSIONS

In the western North Atlantic, local macrofaunal species diversity on the continental slope exceeds local diversity on the adjacent continental shelf. The generality of this depth-related shift in diversity in other parts of the World Ocean is less well established, but estimates of bathyal diversity from other major deep-sea basins are comparably high, as we show in Chapter 3. Bathyal communities are characterized by higher evenness of relative abundance distributions and more rare species than shelf assemblages. Local nematode diversity also appears to be higher in the deep sea. However, as we demonstrate in Chapters 3 and 4, not all deep-sea habitats support high diversity; moreover, high local diversity may not translate into high regional diversity because of the broad geographic distributions of many deep-sea species.

Local diversity in deep-sea sediments is both high on average and quite variable (Figure 2.1). There are many sources of small-scale habitat heterogeneity at the seafloor that might affect community assembly, including sinking phytodetritus, dead falls, foraging activity, bioturbation, pits, mounds, and minute biogenic structures. The primary theoretical explanation for local species coexistence proposes that periodic disturbances from these sources foster diversity by creating a mosaic of patches among which successional sequences are temporally out of phase (Grassle and Sanders 1973). So far, however, neither precision-sampling studies of spatial dispersion to detect patchiness nor experimental studies of patch dynamics have indicated that the spatial mosaic theory accounts entirely, or even significantly, for high local species diversity.

What then causes high local diversity? Multiple factors operating at different scales of time and space are probably involved (Etter and Mullineaux 2001, Levin et al. 2001). It seems likely that nonequilibrial mechanisms, such as those envisioned by Grassle and Sanders (1973) are important but simply have not been adequately studied. Much more precision sampling and experimental work has to be done at local scales in deep-sea communities to better document patch structure and dynamics. This research is difficult to conduct and expensive, but it is central to understanding community structure and organization in the deep sea. The technology and statistical methods needed to analyze the data are now available. It would be especially informative to do more of this kind of research along depth gradients or among habitats to assess the effects of variation in species makeup, life-history traits, population density, and the nature and spatiotemporal scale of patchiness.

One important variable with many ramifications for diversity is relative body size. Deep-sea species are generally much smaller than their shallow-water counterparts, and the exponential decline in food supply with depth enforces a further decrease in average body size (see Chapter 1). Smaller body size means that individuals experience environmental heterogeneity at smaller scales; that is, their surroundings are perceived as more coarse grained. Most deep-sea species are deposit feeders. Jumars et al. (1990) theorized that decreased body size favored a shift in deposit feeding from microphagy (consuming sediments wholesale) to macrophagy (selective ingestion of individual particles). This could permit finer partitioning of sediment resources, allowing a higher level of species coexistence, which may be one reason why smaller organisms in many environments are typically more diverse than larger ones (May 1988, Blackburn and Gaston 1994, McClain 2004). In Chapter 3, we show that sediment grain-size diversity is, in fact, a significant predictor of macrofaunal species diversity in the deep sea (Etter and Grassle 1992). The slower patch dynamics in deep-sea systems may promote diversity by reducing the rate of competitive exclusion. The limited ability of tiny organisms to respond to enrichment events may also curb population growth and help prevent competitive dominance. Hence it is possible that niche partitioning has a role in maintaining diversity. One promising avenue of research would be to design controlled experiments to reveal the underlying causes of the intriguing relationship between species diversity and sediment grain-size diversity.

Local diversity can also be mediated by processes occurring at much larger scales, such as larval dispersal (Etter and Caswell 1994). The small size of most organisms and low food availability must limit fecundity and production of larvae. This, in turn, constrains recruitment into patches by dispersing larvae, which may help explain why so many species are rare and why relative abundance distributions feature high evenness. In Chapter 5, we show that local diversity can be predicted by the size of the regional species pool, suggesting that regional enrichment through dispersal contributes to maintaining local diversity (Stuart and Rex 1994). Regional species pools themselves are ultimately the products of very large-scale and long-term adaptive radiation.

Although a variety of equilibrial and nonequilibrial mechanisms may structure local communities, we should also recognize the possibility that many species of small macrofaunal deposit feeders could be essentially ecologically equivalent. Diversity may be explained by neutral theory, largely as a consequence of evolutionary dynamics (Hubbell 2001, 2005). In Chapter 6 we show that bathyal environments, which have higher diversity, are more conducive to population divergence and higher species origination rates than abyssal plains. Extinction rates appear to be higher in coastal environments. Consequently, high species diversity on continental margins may partly reflect the accumulation of species from higher rates of origin and lower rates of extinction.

The explanation for local diversity lies in a more complete characterization of small-scale community structure and function and a better understanding of how local processes are integrated with larger-scale ecological and evolutionary phenomena.

3

REGIONAL PATTERNS OF ALPHA
SPECIES DIVERSITY

As we descend deeper and deeper in this region, its inhabitants
become more and more modified, and fewer, indicating our
approach to an abyss where life is either extinguished, or exhibits but a few sparks to mark its lingering presence.

Thomas A. B. Spratt and Edward Forbes (1847)

The Fauna living to a depth of 500 fathoms—3000 feet—100
atmospheres! is WONDERFULLY RICH IN EVERYTHING—Echinoderms, Corals, Ophiurans, Starfishes, Crustacea, Mollusca. . . .

Alexander Agassiz to Fritz Müller (1868)

And so far was I from observing any sign of diminished intensity in this animal life at increased depths, that it seemed,
on the contrary, as if there was just beginning to appear a rich
and in many respects peculiar deep sea Fauna, of which only
a very incomplete notion had previously existed.

George O. Sars (1872)

Even if the precise causes of local species coexistence in the deep sea could
be deciphered, it is unclear whether they could be directly extended to explain patterns at larger scales (Roughgarden et al. 1988, Ricklefs and Schluter
1993, Levin et al. 2001). At regional scales, the considerable variation in diversity observed at small scales becomes resolved into definite geographic
patterns. Most of what is known about deep-sea species diversity comes from
measuring and interpreting these patterns. By regional scale, we mean within
a deep-sea basin, for instance, the West European Basin of the North Atlantic,
bounded by the European continental margin and the Mid-Atlantic Ridge,
or a large topographic feature of a basin such as the Porcupine Seabight,

southwest of Ireland within the West European Basin. These larger areas contain significant environmental gradients on scales of tens to hundreds of kilometers. Regional-scale depth gradients of diversity have been particularly informative in identifying potential causes because they parallel strong changes in biological and physical factors over fairly short distances. Bathymetric gradients in diversity are now recognized as constituting a major class of macroecological phenomena comparable in importance to latitudinal (Roy et al. 1998, Hawkins et al. 2003b, Hillebrand 2004a) and elevational (Stevens 1992, Rahbek 1997, McCain 2007) gradients. In this chapter, we explore bathymetric trends of alpha species diversity, the number of co-existing species. Patterns of alpha diversity at oceanwide scales and beta diversity, the change in species composition along a gradient, are taken up in Chapters 4 and 5.

THE PATTERNS

Macrofauna

The North Atlantic

As with local diversity and spatial dispersion, knowledge of regional bathymetric patterns of biodiversity in the deep sea centers principally on the macrofauna and is strongly biased toward the North Atlantic. Sanders (1968) presented the first geographically and methodologically controlled analysis of diversity-depth relationships for the macrofauna. He showed that normalized diversity of bivalves and polychaetes combined increased from the continental shelf to mid-bathyal depths—a trend now well established for the whole macrofauna in the western North Atlantic (see Chapter 2). Rex (1973, 1976) extended the analysis to abyssal depths for gastropods and found that diversity increased from the shelf to mid-bathyal depths, but then decreased in the abyss. This unimodal diversity pattern across the full depth range of the continental margin and abyssal plain of the western North Atlantic is also found in bivalves, cumaceans, and polychaetes (Rex 1981, 1983). In Figure 3.1, we present an updated version of these patterns with diversity estimates for new samples and diversity data for one additional macrofaunal taxon, the isopods (courtesy of G. D. F. Wilson; see also Svavarsson 1997).

With the exception of some anchor dredge samples for polychaetes (Figure 3.1E), all the diversity values in Figure 3.1 are based on epibenthic sled samples normalized to $n = 50$. Confidence limits for polychaete diversity es-

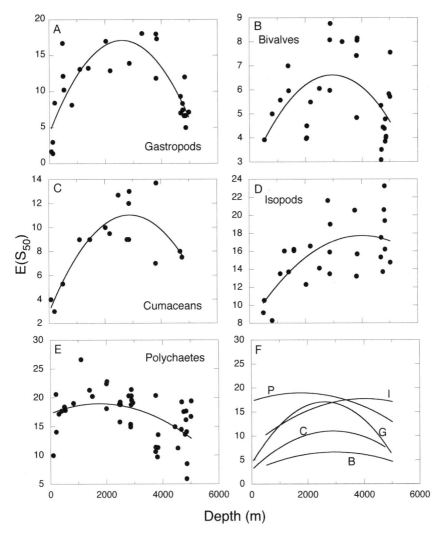

Figure 3.1. Species diversity $E(S_{50})$ for macrofaunal taxa estimated from epibenthic sled samples taken in the western North Atlantic in (A) gastropods, (B) bivalves, (C) cumaceans, (D) isopods, (E) polychaetes, and (F) regressions for all five taxa together. Data are from Rex (1981), with additional estimates, originally from Hartman (1965), Jones and Sanders (1972), Rex (1976, 1977), Allen and Sanders (1996), and courtesy of G. D. F. Wilson (unpublished and in Svavarsson 1997). Regression equations are:

Gastropods: $Y = 4.219 + 0.010X - 1.900 \times 10^{-6}X^2$, $R^2 = 0.641$, $P < 0.0001$
Bivalves: $Y = 2.608 + 0.003X - 4.598 \times 10^{-7}X^2$, $R^2 = 0.232$, $P = 0.0285$
Cumaceans: $Y = 2.931 + 0.006X - 9.699 \times 10^{-7}X^2$, $R^2 = 0.700$, $P = 0.0004$
Isopods: $Y = 7.972 + 0.005X - 6.067 \times 10^{-7}X^2$, $R^2 = 0.426$, $P = 0.0017$
Polychaetes: $Y = 17.248 + 0.002X - 5.629 \times 10^{-7}X^2$, $R^2 = 0.240$, $P = 0.0031$

The quadratic terms (X^2) are significant ($P \le 0.05$) in all cases except for isopods ($P = 0.0684$).

timated from anchor dredge and epibenthic sled samples overlap extensively, so the data from the two sampling devices were combined. Gastropods, bivalves, cumaceans, and polychaetes show clear unimodal diversity-depth trends though the positions of maximum diversity and levels of diversity vary (Figure 3.1F). We consider patterns to be unimodal if the data fit a polynomial better than a linear regression and both the regression and its quadratic term (X^2) are significant. Isopods are a clear exception and show a simple increase in diversity with depth, with the highest diversity recorded at nearly 5000 m (Figure 3.1D).

Etter and Grassle (1992), Etter and Mullineaux (2001), and Levin et al. (2001) presented the diversity-depth pattern for the entire macrofauna recovered from 0.09 m^2 box-core samples taken as part of the ACSAR program (Maciolek et al. 1987a,b, Diaz et al. 1994) at upper bathyal depths south of New England (Figure 3.2). This pattern is also unimodal. The depth of peak diversity is hard to identify precisely because of gaps in sampling, but would appear to be between 1200 and 1500 m, shallower than for any of the macrofaunal taxa in Figure 3.1. Taxa showing a unimodal pattern have maximum diversity in the 2000 to 3000 m range. The shift may be partly due to the predominance of polychaetes, which show peak diversity at the shallowest depth (Figure 3.1F) and must strongly influence patterns in the macrofauna as a whole. It could also reflect differences in the scale of sampling. The much larger sled samples may include more spatial heterogeneity in faunal composition at mid-bathyal depths and consequently represent a larger species pool.

In Figure 3.2 we also include diversity estimates from box-core samples taken in the same general vicinity south of New England at Deep Ocean Stations (DOS) 1 and 2 to assess background community structure as a baseline for recolonization experiments (Grassle and Morse-Porteous 1987). Diversity at DOS 1 falls neatly between the cluster of points at 1250 and 2000 m for the ACSAR program. At the deeper DOS 2 site (3600 m), diversities overlap the cluster at 2000 m and, although there are only three estimates, strongly suggest that lower bathyal diversity remains high and may not continue to decrease with depth at a rate implied by the difference between the 1250 and 2000 m clusters. This is more in keeping with the diversity trends shown in Figure 3.1, where diversity stays high at mid- to lower bathyal depths. Also included in Figure 3.2 are two stations (9 and 10) from off Cape Hatteras, where density is exceptionally high as a consequence of heavy nutrient loading (see Chapter 1). Both stations show very depressed

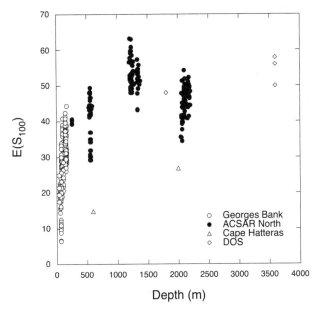

Figure 3.2. Macrofaunal species diversity $E(S_{100})$ estimated from box-core samples taken on the western North Atlantic continental shelf (Georges Bank) and the adjacent bathyal zone to the south. Modified from Etter and Mullineaux (2001) based on original data from Maciolek-Blake et al. (1985) and Maciolek et al. (1987b), with additional data from Grassle and Morse-Porteous (1987) from Deep Ocean Stations (DOS) 1 at 1800 m and 2 at 3600 m south of New England, and from Blake and Grassle (1994) for two sites off Cape Hatteras with heavy nutrient loading and exceptionally high animal abundance, with permission of the authors, Sinauer Associates, and the U.S. Department of the Interior Minerals Management Service.

diversity compared to stations south of New England and other stations off the Carolinas at similar depths (Blake and Grassle 1994).

Another way to envision diversity-depth gradients is to plot depth ranges of species and tabulate diversity as the number of coexisting ranges within depth bins. This is a more conservative method that is insensitive to variation in relative abundance; the main disadvantage is that species may have patchy distributions along their ranges, which means that the actual presence of a species within a depth interval is only inferred. In Figure 3.3, we illustrate this approach for 82 species of prosobranch gastropods in the western North Atlantic whose ranges are well characterized. Diversity increases from upper bathyal depths (≤1000 m) to mid- and lower bathyal depths (1000–4000 m) and then decreases in the abyss. A similar pattern is

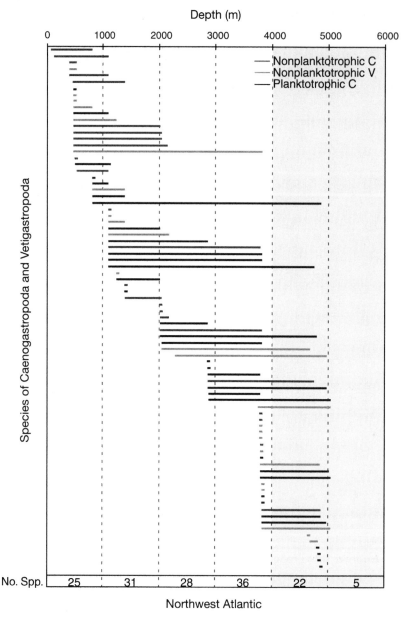

Figure 3.3. Bathymetric ranges for deep-sea gastropods south of New England in the western North Atlantic. Mode of development is indicated. Planktotrophic C (Caenogastropoda) and Nonplanktotrophic V (Vetigastropoda) are assumed to have higher larval dispersal ability than Nonplanktotrophic C. The abyssal zone is shaded. The number of species (bottom) is the number of coexisting ranges in 1000-m depth intervals. A color version of this figure is included in the insert following p. 50. From Rex et al. (2005a), with permission of the authors and the University of Chicago Press. Species names and depth ranges are provided in the online edition of *American Naturalist* (2005), vol. 165, no. 2.

observed for neogastropods in the eastern North Atlantic and protobranch bivalves in both basins (Rex et al. 2005a; and see Figure 4.13).

Diversity in the eastern North Atlantic makes an interesting comparison to the western North Atlantic. Flach and de Bruin (1999) measured macrofaunal diversity in samples collected by corers with diameters of 30 and 50 cm from 200 to 4500 m in the Goban Spur region south of the Porcupine Seabight. Figure 3.4 shows the relationships to depth of animal density, total number of species per sampling station, $E(S_{400})$, and evenness of the relative abundance distribution of species. Density decreases sharply with depth, as expected. The total number of species at each sampling station increases to a peak at 1425 m and then decreases toward the abyss. This pattern is not adjusted for the numbers of replicate samples taken at each, which range from three to five. The number of species in individual samples is similar to what is found in the western North Atlantic (Maciolek et al. 1987a,b). However, unlike in the western North Atlantic, where $E(S_n)$ shows a unimodal depth trend, $E(S_{400})$ in the eastern North Atlantic simply increases monotonically with depth and is highest at abyssal depths. The contradictory patterns between S and $E(S_n)$ can be explained by the fact that $E(S_n)$ is affected by the evenness of the relative abundance distribution. Evenness was expressed as Pielou's J, a permutation of the commonly used Shannon-Wiener diversity index H', which is also influenced by both species richness and evenness (Magurran 2004). The Shannon-Wiener function is calculated as

$$H' = -\sum_{i=1}^{S} p_i \log_2 p_i$$

where $p_i = n_i/N$, n_i is the number of individuals in the ith species, N is the total number of individuals, and s is the number of species. The evenness component J is calculated as

$$J = \frac{H'}{H_{max}} = \frac{H'}{\log_2 s}$$

and ranges from 0 to 1. At $J = 1$, all species in the sample are represented equally. In the Goban Spur benthos J increases steadily with depth from 0.672 at 208 m to 0.926 at 4465 m. In the abyss, there are fewer species than at bathyal depths but they are represented more evenly. The major macro-

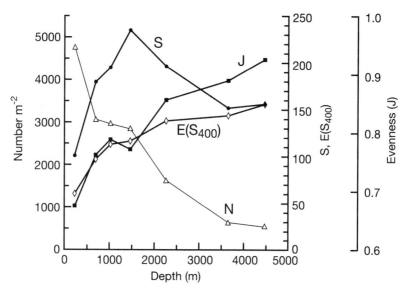

Figure 3.4. Relationships of macrofaunal density, diversity $E(S_{400})$, number of species, and evenness of the relative abundance distribution *J* with depth at the Goban Spur in the eastern North Atlantic. From Flach and de Bruin (1999), with permission of the authors and Elsevier.

faunal taxa—polychaetes, crustaceans, and mollusks—respond in a fashion similar to the macrofauna as a whole in terms of $E(S_n)$, S, and J.

Flach and de Bruin (1999) also calculated $E(S_n)$ at $n = 100$, making their diversity estimates comparable to those from the western North Atlantic ACSAR program (Figure 3.5). Over the more limited depth range sampled in the western North Atlantic, levels of diversity are similar. $E(S_{100})$ continues to increase with depth in the eastern North Atlantic because of the increase in evenness with depth. Evenness has not been calculated over the full depth range for the macrofauna as a whole in the western North Atlantic and data are available for only a few taxa. Polychaetes, the most abundant taxon, show a significant unimodal pattern of evenness with depth ($N = 44$, $F = 4.58$, $P = 0.016$), and gastropods show a similar trend (Rex 1973). In other words, for these two taxa, lower diversity at abyssal depths is associated with a smaller number of species and numerical dominance of one or a few of them. Protobranch bivalves in the western North Atlantic show no significant pattern of evenness with depth ($N = 22$, $F = 0.632$, $P = 0.435$), indicating that depressed abyssal diversity is due to fewer species. Hence, abyssal diversity is somewhat different in the two basins: both basins have fewer

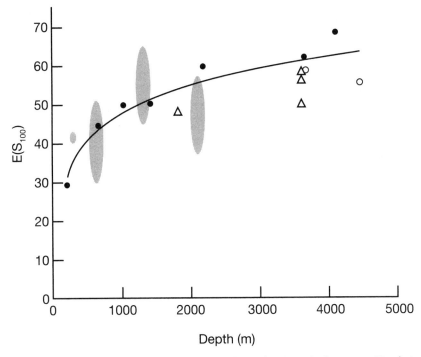

Figure 3.5. Relationship of $E(S_{100})$ to depth at the Goban Spur in the eastern North Atlantic (circles), with data from the western North Atlantic (Shaded areas are dense clusters of points and triangles for DOS stations; see Figure 3.2) superimposed for comparison. From Flach and de Bruin (1999), with permission of the authors and Elsevier.

species in the abyss, but in the eastern North Atlantic abyssal species are more evenly distributed in terms of relative abundance, resulting in a higher $E(S_n)$.

Olabarria (2005) analyzed the diversity-depth pattern for bivalves collected from 500 to 4866 m during extensive British sampling programs in the Porcupine Seabight and Abyssal Plain. The data represent 126 epibenthic sled and trawl samples that yielded 131,334 individuals distributed among 76 species. The depth ranges of the species and an indication of their abundance along the ranges are shown in Figure 3.6. Most species reach maximum abundance at bathyal depths. The apparent faunal discontinuities (general low abundance) at 2000–2500 m and at 3100–3500 m represent a sampling gap and a rocky zone that is hard to sample. Diversity estimated as $E(S_{50})$ is plotted against depth in Figure 3.7. The general trend is for diversity to increase with depth, but there appear to be two peaks at around 1500 and 4000 m separated by a trough of lower diversity. The level of bathyal diversity is similar to values in the western North Atlantic (Figure 3.1B), but abyssal diversity is higher. As with the macrofauna as a whole, the abyssal peak may reflect the evenness of the relative abundance distribution. Most abyssal species are rare (Figure 3.6). We calculated diversity as the number of overlapping depth ranges at 500 m depth intervals using the data in Table 1 in Olabarria (2005). The number of species shows a clear unimodal pattern with a peak at 2500 m and depressed diversity at abyssal depths (Figure 3.8), again suggesting that the elevated abyssal $E(S_n)$ is an evenness effect.

Paterson and Lambshead (1995) found that both the number of polychaete species per box core and $E(S_{51})$ showed a unimodal depth pattern with a peak at about 1500 m in the bathyal (400–2900 m) Rockall Trough north of the Porcupine Seabight. Levels of $E(S_{51})$ were slightly higher, but the overall patterns are otherwise similar to $E(S_{50})$ for polychaetes in the western North Atlantic. Gage et al. (2000) analyzed diversity of the macrofauna (excluding peracarid crustaceans) from 400–1800 m in the same region. Species density (number of species per box core) showed a clear unimodal pattern with a peak at 1000 m. $E(S_{180})$ was more irregular with the highest diversity at 1400 m. Cosson-Sarradin et al. (1998) also discovered a unimodal diversity-depth pattern by using the Shannon-Wiener index for polychaetes in the tropical region (100–4100 m) of the eastern North Atlantic off Cap Blanc, Africa.

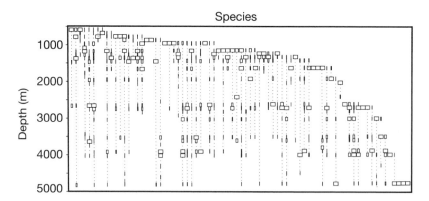

Figure 3.6. Depth ranges of bivalves in the Porcupine Seabight and Abyssal Plain of the eastern North Atlantic. Width of line indicates relative abundance as a percentage of abundance for each species. Species names can be found in Olabarria (2005). From Olabarria (2005), with permission of the author and Elsevier.

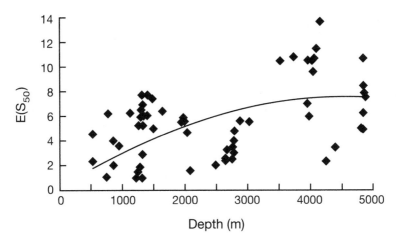

Figure 3.7. Bivalve diversity estimated as $E(S_{50})$ in the Porcupine Seabight and Abyssal Plain of the eastern North Atlantic. From Olabarria (2005), with permission of the author and Elsevier.

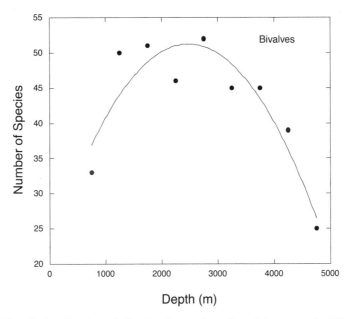

Figure 3.8. Bivalve diversity calculated as the number of coexisting ranges in 500-m depth bins in the Porcupine Seabight and Abyssal Plain of the eastern North Atlantic. Based on data from Table 1 in Olabarria (2005).

*Adjacent Seas: The Norwegian and Mediterranean Seas
and the Gulf of Mexico*

The Norwegian Sea is separated from the North Atlantic by the Greenland-Iceland-Faeroe (GIF) Ridge with a sill depth of 500–850 m. Diversity-depth gradients in this area have been especially well documented in isopods collected by epibenthic sleds. Svavarsson (1997) showed that north of Iceland, diversity increases with depth to around 800 m and then decreases to maximum depths of nearly 4000 m (Figure 3.9). Off the coasts of Greenland and Norway, isopods also show a decrease in diversity from 800 to 3700 m, but upper slope regions have not been sampled (Svavarsson et al. 1990). In the Faeroe-Shetland Channel, essentially the extreme southeastern reach of the Norwegian Sea just north of the GIF Ridge, diversity of polychaetes (Narayanaswami et al. 2005) and the macrofauna as a whole (Bett 2001) both increase to a maximum at 400 m and then decrease in a pattern similar to that of the isopods (Svavarsson 1997). Figure 3.10 shows the dramatic difference in diversity-depth patterns north and south of the eastern GIF Ridge. In the Rockall Trough south of the ridge, diversity increases from 200 m to what appears to be a peak at 1600–1800 m. North of the ridge in the Faeroe-Shetland Channel, diversity peaks at a much shallower depth (~400 m) and then decreases to levels much lower than in the Rockall Trough. A very similar difference was found for isopods (Svavarsson 1997) and amphipods (Weisshappel and Svavarsson 1998) north and south of Iceland in the western sector of the ridge. Wlodarska-Kowalczuk et al. (2004) sampled the macrofauna at a seasonally ice-free site in the extreme northern Norwegian Sea west of Spitsbergen at 79°N. Sampling was carried out from 200 to 3000 m using either 0.1 m^2 van Veen grabs or 0.1 m^2 subsamples of 0.25 m^2 box cores. The upper bathyal part of the diversity trend is hard to discern, primarily because samples taken at less than 400 m were located in the Kongsfjordrenna, a canyon that cuts the shelf. Samples from here showed high variability in standing stock and diversity indices, suggesting a heterogeneous environment. Diversity appears to either peak at around 500 m, as elsewhere in the Norwegian Sea, or simply decrease with depth if we take the mean of aggregate samples at 500 m or less as somehow representative of these depths (Figure 3.11). Standing stock off Spitsbergen is much higher than in the High Arctic, where permanent ice cover inhibits surface primary production and organic carbon flux to the benthos (Paul and Menzies 1974, Kröncke 1994, 1998, Kröncke et al. 2000), but is only

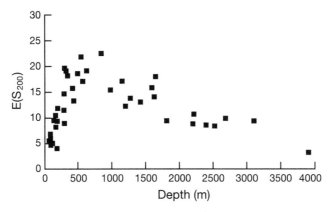

Figure 3.9. Diversity $E(S_{200})$ of isopods from epibenthic sled samples collected in the Nor-wegian Sea north of Iceland. From Svavarsson (1997), with permission of the author and Springer Science.

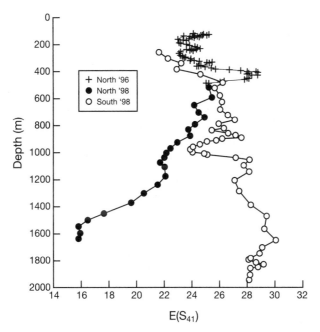

Figure 3.10. Diversity $E(S_{41})$ of macrofauna with depth north and south of the eastern sec-tion of the Greenland-Iceland-Faeroe Ridge (the Faeroe-Shetland Channel north of the ridge and the Rockall Trough south of the ridge). North of the ridge diversity peaks at around 400 m and then declines. South of the ridge, diversity increases from 200 m to around 2000 m. From Bett (2001), with permission of the author and Elsevier.

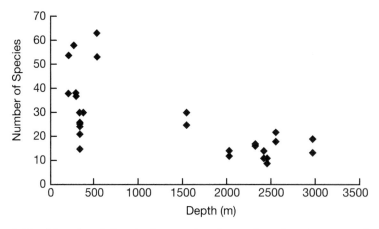

Figure 3.11. Macrofaunal diversity (measured as the number of species per sample from 0.1-m² van Veen grabs and 0.1-m² subsamples of 0.25-m² box cores) with depth, taken off Spitsbergen at 79°N in the Norwegian Sea. Diversity appears to simply decrease with depth. The high variation in samples from depths of less than 400 m were taken in the head of a canyon, the Kongsfjordrenna, and suggests a heterogeneous environment. From Wlodarska-Kowalczuk et al. (2004), with permission of the authors and Elsevier.

somewhat lower than in temperate and tropical regions of the eastern North Atlantic (Gage 1979, Galéron et al. 2000). Diversity exceeded that found in deep basins of the High Arctic (Kröncke 1994, 1998) and was lower than in the eastern North Atlantic (Paterson and Lambshead 1995, Flach and de Bruin 1999).

Tselepides and Eleftheriou (1992) and Tselepides et al. (2000a) sampled the South Aegean in the eastern Mediterranean from 40 to 1570 m using 0.1 m² grabs and box cores. Diversity measured as the number of species simply decreases with depth (Figure 3.12). Over the depth range sampled, diversity is less than half that found in the northern Norwegian Sea (Figure 3.11) and only about 10–20% that in the western North Atlantic (Figure 3.2). Density in the Mediterranean is extremely low (see Chapter 1). From 40 to 1570 m, it drops from an average of 4263 to 191 individuals m^{-2} (Tselepides et al. 2000), which corresponds to the decline in density from mid-bathyal (1500–2000 m) to abyssal depths in the western North Atlantic, a range over which diversity also decreases with depth.

Wei et al. (2009) analyzed macrofaunal species diversity from an extensive survey in the Gulf of Mexico from 200 to 3800 m. $E(S_{100})$ shows a unimodal pattern with depth peaking at about 1400 m (Figure 3.13). Levels of

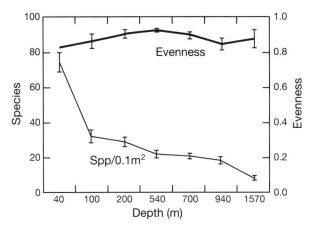

Figure 3.12. Macrofaunal diversity measured as the number of species and evenness against depth in the Aegean Sea of the eastern Mediterranean. Diversity is much lower than in the western North Atlantic (Figure 3.2) and lower than in the Norwegian Sea (Figures 3.10, 3.11). From Tselepides et al. (2000a), with permission of the authors and Elsevier.

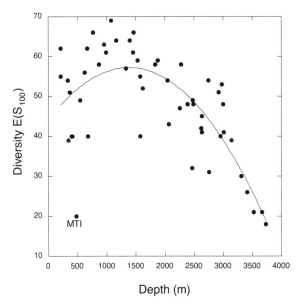

Figure 3.13. Macrofaunal diversity estimated as $E(S_{100})$ with depth in the northern Gulf of Mexico. Diversity peaks at around 1400 m and then drops steeply toward the abyss. Note the depressed diversity at Station MT1 at the head of the Mississippi Canyon. From Wei et al. (2009), with permission of the authors and Inter-Research. The regression equation is: $Y = 44.022 + 0.0192X - 0.000006938X^2$, $R^2 = 0.55$, $P < 0.001$.

diversity are similar to those in the western North Atlantic (Figure 3.2) over the depth range from about 200 to 2000 m. Coastal and upper bathyal diversities in the Gulf are actually higher than in the western North Atlantic, possibly reflecting the well-known latitudinal gradient in species diversity. However, below around 2000 m diversity in the Gulf drops precipitously—at 3800 m (the greatest depth in the Gulf) it is only about one-third its maximum. In the western North Atlantic, diversities of macrofaunal taxa do not appear to decrease as rapidly and remain relatively high at 3800 m (Figure 3.1). Macrofaunal diversities from box-core samples taken at 3600 m are more than twice as high as those in the Gulf (cf. Figures 3.2 and 3.13).

Diversity at Station MT1 at 481 m in the Gulf is much lower than at other stations near the shelf-slope transition (Figure 3.13). MT1 is located at the head of the Mississippi Canyon and receives an unusually high rate of nutrient input from the Mississippi River outflow and high overhead production. Density at MT1 is three to four times higher than at stations at similar depths (21,663 versus 5000–6000 individuals m^{-2}), and it is the only station in the survey with unusually low evenness. The community is heavily dominated by a single amphipod, *Ampelisca mississippiana*.

Oxygen Minimum Zones

Diversity-depth patterns are strongly impacted where oxygen minimum zones (OMZs) intercept the continental slope (Levin and Gage 1998, Levin et al. 2000, Levin 2003). OMZs are found in highly productive regions and impose an environment that combines high organic carbon flux with low oxygen. Near their cores, where oxygen is severely depleted, OMZs favor organisms that can tolerate hypoxia (oxygen concentration < 0.2 ml l^{-1}), mainly meiofaunal elements, several taxa of surface-feeding polychaetes, and the megafaunal spider crab *Encephaloides armstrongi* (Creasey et al. 1997). Other macrofaunal and megafaunal groups are sometimes densely aggregated near the upper and lower margins, where more aerobic conditions permit them to exploit abundant food.

Figure 3.14 shows the depth profile of macrofaunal density in the OMZ of the Oman Margin in the Arabian Sea (Levin et al. 2000). Densities, representing primarily polychaetes, reach very high levels in the core between 400 and 850 m. Sediments here have elevated levels of organic carbon content and surface pigment biomass. Bioturbation is low because of the paucity of large burrowing invertebrates. Both species richness and evenness are depressed in this central zone, where oxygen concentration is

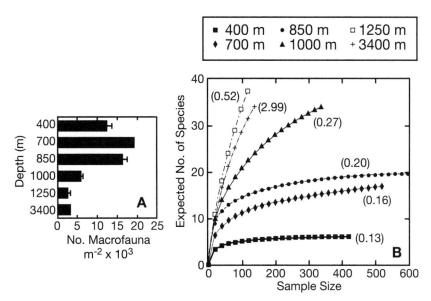

Figure 3.14. (A) Relationship of macrofaunal density to depth in the OMZ of the Oman Margin, Arabian Sea. The high densities at 700 and 850 m are in the core of the OMZ. (B) Rarefaction curves for the different depths sampled across the OMZ. Diversity is depressed at 400–850 m, where densities are high and oxygen concentration is low (0.13–0.20 ml l^{-1}). Diversity increases below the core, where oxygen levels increase (0.27 ml l^{-1} at 850 m and 0.52 ml l^{-1} at 1000 m) and remains high at 3400 m, where conditions are oxic (2.99 ml l^{-1}). From Levin et al. (2000), with permission of the authors and Elsevier.

0.13–0.20 ml l^{-1} (Figure 3.14). Below the core at 1000 m, oxygen concentration increases to 0.27 ml l^{-1} and diversity increases dramatically. It is even higher at 1250 m (0.52 ml l^{-1}) and remains high at 3400 m, where conditions are aerobic (2.99 ml l^{-1}). Levin et al. (2000) suggest that levels of 0.15–0.20 ml l^{-1} may represent a threshold below which most macrofaunal groups are inhibited and diversity is depressed by intolerance to hypoxia. Since OMZs are very widespread in the Indo-Pacific and eastern South Atlantic (Helly and Levin 2004), this physiological constraint on diversity may be quite important.

Meiofauna

Diversity-depth patterns in the meiofauna are less well known, owing largely to the technical and taxonomic difficulties involved in assessing community structure in this group. Foraminiferan diversity typically is determined by subsampling cores and then randomly selecting a constant sample size of individuals from the subcore from which to count the number of species and calculate diversity indices. Most available analyses are based on what is termed the "total" assemblage, meaning live and dead individuals combined (Gibson and Buzas 1973). In this case diversity represents the contemporary biota of a site and those species that have presumably lived there in the recent past. It can be biased by deep-sea currents or downslope transport redistributing dead tests and by selective preservation owing to carbonate dissolution. Earlier studies concentrated heavily on hard-walled (predominantly calcareous) species. Only recently has it been possible to census what is referred to as the "entire" live foraminiferan biota, which includes living hard-walled, organic-walled, and agglutinated species (Cornelius and Gooday 2004). The proportion of soft-walled species can be substantial and varies geographically (Gooday 2002, Gooday et al. 2004). This group may be particularly important below the calcium carbonate compensation depth (Todo et al. 2005).

Here we concentrate on the total assemblage, by far the largest comparative database. Figure 3.15 presents diversity-depth patterns in the Arctic, western North Atlantic, and Gulf of Mexico. South of New England diversity increases across the shelf, decreases initially on the upper slope, and then increases steadily from around 600 m to a maximum at 5000 m in the abyssal plain (Buzas and Gibson 1969, Gibson and Buzas 1973; Figure 3.15A). Farther south along the continental margin (off Virginia and Florida) diversity is higher across the bathyal zone without a dip on the upper slope (Figure 3.15A). Diversities in both regions seem to converge at high levels, but the southern region was not sampled at abyssal depths. The generally high diversity at upper to mid-bathyal depths detected by Gibson and Buzas (1973; Figure 3.15A) was later confirmed by extensive sampling (Cutter et al. 1994). Interestingly, there is also an unusually low value off Cape Hatteras at Site III (Gooday 1999), where nutrient loading is particularly high, which resulted in the highest standing stock of benthos ever recorded at this depth (Schaff et al. 1992). In Figure 3.15B, we show the diversity-depth pattern for the Norway Basin of the Norwegian Sea (Mackensen et al. 1985) and the diversity-depth regression for the central Arctic (Lagoe 1976). Diversities in both regions are lower than in the Atlantic and simply decrease with depth over the bathyal range, where diversity in the western North Atlantic increases. For the living Arctic assemblage, diversity is generally higher in seasonally ice-free areas than under permanent ice cover (Wollenburg and

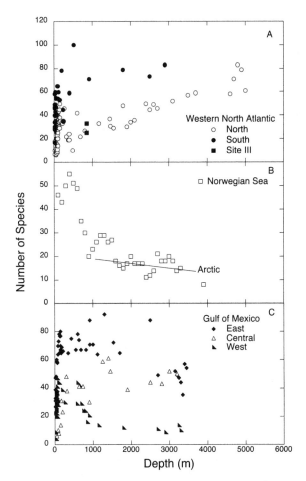

Figure 3.15. Relationships of foraminiferan diversity to depth. (A) Western North Atlantic. The northern region is located south of New England, and the southern region is off Virginia and Florida (data from Gibson and Buzas 1973). In the northern region, diversity dips slightly below the shelf-slope transition and then increases steadily to its highest levels at abyssal depths. In the southern region, diversity is higher at bathyal depths. Site III is an area off Cape Hatteras, where nutrient loading is exceptionally high and diversity is depressed (data from Gooday 1999). (B) Pattern of diversity in the Norwegian Sea (data from Mackensen et al. 1985), and the regression (significant, $P < 0.01$) for diversity against depth in the Arctic Ocean (from Lagoe 1976). Both the Norwegian Sea and the Arctic Ocean show lower diversity than the western North Atlantic and a pattern of decreasing diversity with depth. (C) The eastern, central (including the Mississippi River Delta), and western Gulf of Mexico (data from Gibson and Buzas 1973). The western Gulf is an impoverished region, which shows a decline in diversity below the shelf. In the central Gulf, diversity is depressed at upper bathyal depths in the area of heavy nutrient loading near the Mississippi River Delta, then increases toward lower bathyal depths to levels similar to those found in the eastern Gulf. Surface production is higher in the eastern than in the western Gulf. Diversity shows a unimodal pattern with depth, reaching a peak at mid-slope depths at levels similar to those in the southern region of the western North Atlantic, but then declining toward lower bathyal depths. With permission of the authors and the Geological Society of America.

Mackensen 1998) and is a positive function of organic carbon flux (Wollenburg and Kuhnt 2000).

Diversity trends in the Gulf of Mexico are particularly interesting (Gibson and Buzas 1973). Production and benthic standing stock of macrobenthos in the Gulf are lower than in the western North Atlantic (see Chapter 1), and foraminiferan diversity is generally lower at mid- to lower bathyal depths. There is an east to west decrease in organic carbon flux to the benthos and benthic standing stock (Biggs et al. 2009, Wei et al. 2009). Diversity is highest in the eastern Gulf, where it shows a unimodal pattern with depth. In the western Gulf, where the shelf is narrow and surface production the lowest, diversity is low on the shelf, increases at upper bathyal depths, and then simply decreases to very low values at 3300 m. More recent studies suggest that diversity in the western Gulf was underestimated, but is still lower than in the east (Buzas et al. in press). In the central Gulf, diversity is conspicuously depressed near the Mississippi River Delta, where nutrient loading is very high, and then gradually increases with depth to converge with levels found in the eastern Gulf. Diversity-depth patterns in the much more complex "entire" live foraminiferan assemblage, including the mostly undescribed soft-walled group, have only recently been explored (Gooday et al. 1998, 2004, Cornelius and Gooday 2004, Nozawa et al. 2006, Bernhard et al. 2009) and may change the perceptions based on the "total" biota.

The diversity response of foraminiferans to oxygen depletion is similar to that of the macrofauna (Figure 3.14). In the core of the OMZ (oxygen concentration 0.13 ml l^{-1}) on the Oman Margin of the Arabian Sea, foraminiferan density is high (1.6×10^7 individuals m^{-2}), but diversity [$E(S_{100}) \sim 20$] and evenness are depressed (Gooday et al. 2000). At the center of the Santa Barbara Basin off California in near-anoxic conditions (oxygen concentration 0.05 ml l^{-1}), diversity is even lower [$E(S_{100}) \sim 10$; Bernhard et al. 1997].

Despite the large body of literature on the standing stock of the metazoan meiofauna (see Chapter 1), surprisingly little is known about bathymetric variation in species diversity. Alpha diversity of nematodes, by far the most abundant taxon, is higher in the deep sea than in coastal habitats (Boucher and Lambshead 1995, Mokievsky and Azovsky 2002; also see Chapter 2). Nematodes show no discernible pattern of diversity from 970 to 3294 m in the Norwegian Sea (Jensen 1988) or from 2087 to 4725 m in the Bay of Biscay (Dinet and Vivier 1979). Lambshead et al. (2000) found no depth-related trends among several sites in the North Atlantic, Caribbean

Sea, and Norwegian Sea. Body size decreases with depth (Soetaert and Heip 1995, Vanaverbeke et al. 1997, Soetaert et al. 2002). In an otherwise oligotrophic region of the Norwegian Sea, nematode density and diversity are elevated near the ice edge, where surface productivity and carbon flux are enhanced (Fonseca and Soltwedel 2007). A very similar phenomenon occurs in the abyssal Pacific within a narrow band of higher carbon flux associated with upwelling along the equatorial current (Lambshead et al. 2002). In the Mediterranean Sea, nematodes have a lower density than in the eastern North Atlantic (Soetaert et al. 1991) and are generally smaller in size (Soetaert and Heip 1989). Diversity in the Mediterranean decreases with depth and along the west-to-east decreasing productivity gradient (Danovaro et al. 2008). Even within the highly oligotrophic Aegean Sea, diversity decreases along the north-to-south decreasing productivity gradient (Lampadariou and Tselepides 2006). Diversity is depressed at 7800 m in the Atacama Trench of the eastern South Pacific, where carbon flux from high overhead production is topographically focused at the trench floor, producing extremely high densities (Gambi et al. 2003; see Figure 1.16). Diversity is also low at the HEBBLE site, where food supply is enhanced by erosion and deposition of sediments by strong currents (Lambshead et al. 2001a).

Harpacticoid copepods, the second most abundant taxon, increase in diversity from upper bathyal depths to 3000 m and then decrease to 3940 m in the western North Atlantic (Coull 1972). In the Gulf of Mexico, Baguley et al. (2006) showed that both density and the number of species simply decreased linearly with depth (Figure 3.16). $E(S_{30})$ is weakly unimodal, peaking at about 1200 m with only slightly depressed diversity on the upper slope. Unlike the macrofauna, at Station MT1, for example, there are no hotspots of high density and correspondingly low diversity (Figure 3.13). The number of species is a significant positive function of density ($r = 0.91$), and the highest recorded density corresponds to the highest diversity.

What is the relative level of diversity between the macrofauna and meiofauna? Intensive sampling programs and much more tractable taxonomy have permitted good estimates of the level of species richness in the macrofauna (e.g., see Figures 2.1, 3.1, and 3.2). Richness in a box-core sample taken at mid-bathyal depths is about 100 species in the eastern and western North Atlantic. It is much more difficult to assess this for the meiofauna. However, even a cursory calculation suggests impressive numbers. Foraminiferan diversity in individual samples can approach 100 species for the "total" fauna

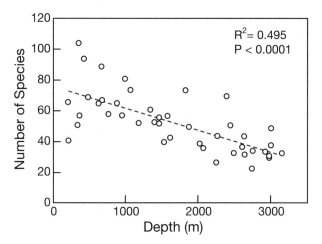

Figure 3.16. Relationship between harpacticoid species richness (number of species per 10 cm²) and depth in the northern Gulf of Mexico. Richness appears to simply decrease with depth. From Baguley et al. (2006), with permission of the authors and Elsevier.

(Figure 3.15). More recent estimates of the living fauna at abyssal depths can exceed 100 species, and one sample from 3400 m in the Arabian Sea yielded over 200 species (Gooday et al. 1998). If to this we add local nematode diversity of roughly 100 species (Dinet and Vivier 1979, Mokievsky and Azovsky 2002), and perhaps 30–60 species of harpacticoid copepods (Coull 1972, Baguley et al. 2006) we reach a rough figure of well over 200 species just for the major groups of meiofauna. Clearly, this comparison must be adjusted to depth and sampling methods, although meiofaunal diversity is typically based on much smaller samples than macrofaunal diversity. Conservatively, it would appear that local meiofaunal diversity is at least comparable to, and more likely higher than, local macrofaunal diversity.

Megafauna

Vinogradova (1958, 1962) was the first to document depth trends in megafaunal species diversity. She pooled depth records of species on a global scale from early twentieth-century Russian and Scandinavian deep-sea expeditions to determine the number of species occurring in successive depth increments. For the fauna as a whole, diversity increased sharply from upper bathyal depths to about 2000 m and then declined toward the abyss and remained low from abyssal to hadal depths (Figure 3.17). She also described

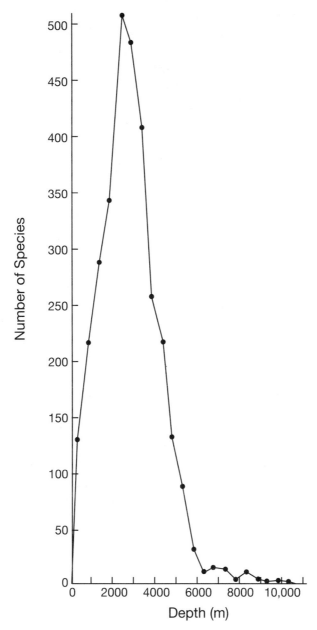

Figure 3.17. Global estimates of the number of megafaunal species versus depth. From Vino-gradova (1962), with permission of Elsevier.

diversity for individual abundant taxa including sponges (and hexactinellids separately), cnidarians, crustaceans (decapods and isopods, the latter usually considered macrofauna), and echinoderms (echinoids; crinoids; elasipod and elpidiid holothurians; forcipulid, spinulosid, and phanerozoan asteroids). All showed clear unimodal diversity-depth trends, most with maximum diversity in the 2000- to 3000-m range, but with elasipod and elpidiid holothurians peaking deeper, at around 4500 m. The data are difficult to interpret precisely in contemporary ecological terms because there is no interregional geographic control and it is unclear how the patterns are affected by sampling intensity and methodology at different depths. All the same, this was a remarkable and very original synthesis, and, as we illustrate below, unimodal diversity-depth patterns do occur in the deep-sea megafauna at regional scales. Vinogradova's work was the first indication in any faunal component that diversity-depth trends might be unimodal.

Haedrich et al. (1975, 1980) conducted an intensive sampling program to document megafaunal community structure on regional scales, collecting 105 trawl samples between 40 and 5000 m in the western North Atlantic. Normalized diversities of demersal fishes and invertebrates are shown in Figure 3.18. Diversity-depth patterns are strongly unimodal, with peak diversities at around 2000 m (~1800 and 2200 m for fishes and invertebrates, respectively), and very similar regressions over the full depth range. Haedrich et al. (1980) showed that the absolute number of species recovered from depth zones defined by a cluster analysis of species composition also showed a unimodal pattern with maximum diversity at around 1500–2000 m. Diversity in both groups is highly variable across the shelf and bathyal zone to around 3000–3500 m, below which it becomes depressed and less variable. The diversity of megafaunal elements is much lower than for the macrofauna of the same region (cf. Figure 3.1), and presumably the meiofauna. In other regions of the Atlantic and Pacific, the diversity of demersal fishes appears to basically decrease with depth, showing some idiosyncratic features but no clear unimodality (Pearcy et al. 1982, Haedrich and Merrett 1988, Powell et al. 2003, Kendall and Haedrich 2006).

British investigators carried out another major megafaunal sampling program in the Porcupine Seabight and Abyssal Plain of the eastern North Atlantic (Rice et al. 1991, Billett and Rice 2001). In this case, patterns of community structure were assessed for individual taxa, but not for the fauna as a whole. Howell et al. (2002) presented diversity-depth patterns for the asteroids based on data from 288 epibenthic sled and trawl samples. Figure

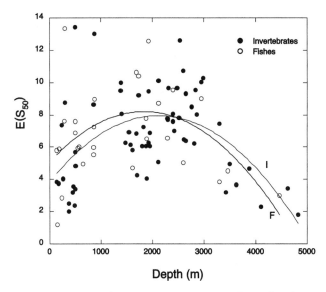

Figure 3.18. The relationship of diversity $E(S_{50})$ to depth of megafaunal invertebrates and fishes in the western North Atlantic. Both show strong unimodal patterns. Modified from Rex (1981), based on original data from Haedrich et al. (1975, 1980), with permission of the *Annual Review of Ecology and Systematics*. The regression equations are:

Fishes: $Y = 5.01 + 0.0035X - 0.000001X^2$, $R^2 = 0.230$, $P < 0.05$
Invertebrates: $Y = 3.87 + 0.0039X - 0.000001X^2$, $R^2 = 0.281$, $P < 0.01$

3.19 shows depth ranges of 47 species collected from 150 to 4950 m. Species often have broad ranges, but tend to be most common in a restricted part of the range, as might be expected if they are adapted to ecological conditions at a particular depth. Figure 3.20 shows diversity normalized to $n = 10$ along the depth gradient for samples with 10 or more individuals. The authors point out that this is an unusually small n value at which $E(S_n)$ is strongly affected by the evenness of the relative abundance distribution. Asteroid diversity peaks at about 1800 m, dips to lower values at around 2600 m, and then increases to about 4800 m. Depressed diversity at 2600 m in samples where $E(S_{10}) \approx 1$ is partly attributable to numerical dominance of one species, *Hymenaster membranaceus*.

The apparent increase in diversity toward the abyss could be related to the increase in evenness that attends widespread rarity at these depths, just as in the macrofauna. In Figure 3.21, we plot the number of coexisting asteroid species in 500-m depth intervals against the midpoints of the inter-

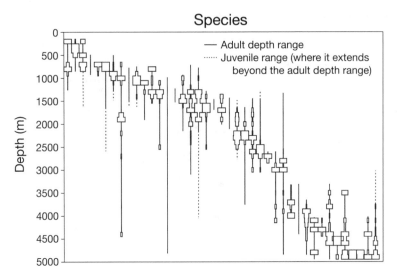

Figure 3.19. Depth ranges of asteroids collected from the Porcupine Seabight and Abyssal Plain of the eastern North Atlantic. Width of line indicates relative abundance as a percentage of abundance for each species. See Howell et al. (2002) for the list of species. From Howell et al. (2002), with permission of the authors and Elsevier.

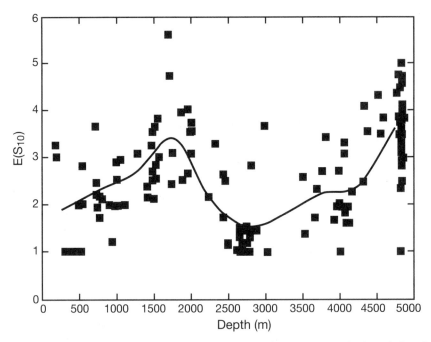

Figure 3.20. Diversity of asteroids estimated as $E(S_{10})$ in the Porcupine Seabight and Abyssal Plain of the eastern North Atlantic. From Howell et al. (2002), with permission of the authors and Elsevier.

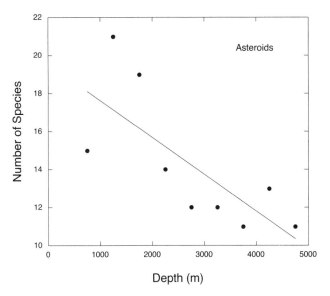

Figure 3.21. Asteroid diversity estimated as the number of coexisting species ranges in 500-m depth bins in the Porcupine Seabight and Abyssal Plain of the eastern North Atlantic. Based on data from Table 1 in Howell et al. (2002). The regression equation is: $Y = 19.539 - 0.0019X$, $R^2 = 0.552$, $P = 0.0218$.

vals by using data from Table 1 in Howell et al. (2002). When species richness is plotted this way, a very narrowly defined but clear maximum still occurs between 1000 and 2000 m, corresponding to the peak in $E(S_{10})$ at 1800 m. However, at lower bathyal and abyssal depths, diversity decreases to values about half those at upper bathyal depths. Overall, even though there is a conspicuous drop in diversity at 500–1000 m and a peak around 1000–2000 m, a negative linear regression fits the data best, suggesting that the simplest interpretation is one of declining diversity from upper bathyal to abyssal depths. Asteroids in the Bay of Biscay to the south, still in the West European Basin, also show a decrease in diversity with depth from 2000 to 4700 m (Sibuet 1977).

Holothurians collected during the British sampling program show a decidedly different pattern (Billett 1991). Using the approach of estimating richness as the number of coexisting ranges in 500 m depth bins showed no statistically significant diversity pattern with depth (Figure 3.22); there is no obvious peak in diversity at intermediate depths and no tendency for diversity to decrease in the abyss. Holothurians in the Bay of Biscay show a clear unimodal diversity-depth pattern with a maximum at about 3100 m (Sibuet 1977). In the eastern North Pacific, they reach maximum diversity at 3500 m and then appear to decline toward the abyss with greater distance from land,

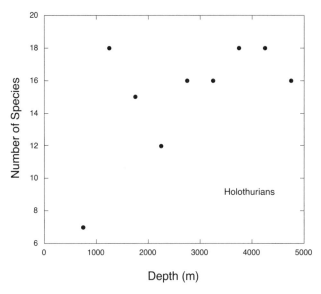

Figure 3.22. Relationship of holothurian diversity, estimated as the number of coexisting ranges, to depth in the eastern North Atlantic. Data from Table IV in Billett (1991). There is no statistically significant relationship between diversity and depth.

though sampling has not taken place below 4000 m (Carney and Carey 1976, 1982). Jones et al. (2007) conducted a photographic survey of megabenthos from 1007 to 1660 m northward along the floor of the Faeroe-Shetland Channel. Although the depth range is fairly narrow, both abundance and species richness decreased significantly with depth, paralleling the pattern observed in the macrofauna north of the GIF Ridge.

Summary

Although there is certainly variation in diversity-depth patterns among taxa, size classes, and regions, the basic patterns can be summarized simply. In the eastern and western North Atlantic, macrofaunal patterns of species richness are primarily unimodal. The Gulf of Mexico shows a similar unimodal pattern with comparable maximum diversity, but with a steeper decline toward the abyss. In the Norwegian Sea, diversity peaks at shallower depths and then decreases with depth to much lower values than those found in the Atlantic. In the Mediterranean, diversity is extremely low and simply drops with depth. Diversity is conspicuously depressed in the core of OMZs and under circumstances of heavy nutrient loading. Richness in forminiferans, the most well-known meiofaunal group, increases with depth in the western North Atlantic and reaches its highest level in the abyss, where macrofaunal diversity is depressed. In the Norwegian Sea and Arctic Ocean diversity is lower than in the Atlantic and decreases monotonically with depth. In the Gulf of Mexico diversity is generally lower than in the Atlantic; diversity-depth patterns in the more productive eastern sector are unimodal, diversity is depressed near the Mississippi Delta, and is lowest overall and decreases with depth in the less productive western region.

It seems likely that when meiofaunal diversity is fully characterized it will exceed macrofaunal diversity. Less can be said about interregional variation in megafaunal diversity. In the western North Atlantic megafaunal fishes and invertebrates show a unimodal pattern and generally have much lower diversity than the macrofauna. Fish diversity in other regions of the Atlantic and Gulf appears to decrease monotonically with depth. It is important to reiterate that we are discussing geographic variation in local (sample) diversity. As we show in later chapters, regional and global estimates of diversity are not simple extrapolations of local diversity.

POTENTIAL CAUSES

Implications of Mid-Domain Effects for Unimodal Patterns

In order to discuss possible ecological causes of diversity gradients, it is first necessary to establish whether the patterns could be a simple consequence of physical limits of the deep-sea ecosystem or if the patterns differ from those generated by random processes within a bounded ecosystem. Pineda (1993) proposed that the vertical ranges of species along bathymetric gradients are strongly influenced by physical boundary constraints imposed by the sea surface and the abyssal seafloor. His theory assumed that species are adapted to an optimal depth where the environment permits maximum abundance and that abundance tapers off symmetrically above and below that depth. Since the optimal depth is seldom known for deep-sea species, it was approximated as the average depth between maximum and minimum depth records. Species with a mean depth of occurrence near each boundary would necessarily have their ranges more compressed than those centered midway between the boundaries. Under these circumstances, the relationship between vertical range and mean depth of occurrence for species living along the depth gradient defines a triangular geometric constraint envelope. Pineda (1993) showed that when vertical ranges of species are plotted as a function of depth for deep-sea isopods, gastropods, sipunculids, polychaetes, and demersal fishes, the shapes of the relationships are triangular. Small ranges occurred at all depths, but were more prevalent toward the sea-surface and abyssal boundaries. Large ranges were concentrated at intermediate depths.

Colwell and Hurtt (1994) developed a similar model and extended it to patterns of diversity that emerge from boundary constraints on ranges. Their model has no underlying assumptions about adaptive properties of species or ecological gradients between boundaries. If a random relationship between breadth and position of ranges is assumed, boundaries produce a triangular constraint envelope as Pineda (1993) showed. Summing overlapping vertical ranges between boundaries produces a unimodal diversity gradient similar to many bathymetric gradients in the deep-sea benthos (e.g., Figure 3.1). The peak in diversity midway between boundaries is called the mid-domain effect. Diversity patterns resulting from random placement of ranges in a bounded domain can serve as a null model to compare with observed diversity gradients. Departure from null expectations might be used to infer environmental causes of diversity other than boundary constraints. Conformity to null expectations means that we cannot discount the possibility that range distributions, and consequently diversity patterns, are environmentally unstructured and simply reflect boundary constraints. Colwell and Hurtt (1994) also showed that predictions of null boundary constraint models are sensitive to the position and effectiveness of boundaries, to the method used to randomly select and cast ranges in the bounded domain, and to sampling bias. Hence, the choice of the appropriate null model is important for meaningful hypothesis testing.

Pineda and Caswell (1998) applied boundary constraint models to assess whether unimodal diversity-depth patterns are merely mid-domain effects or potentially result from ecological causes. Specifically, they tested whether random placement of depth ranges for gastropods and polychaetes between sea-surface and abyssal seafloor boundaries produced diversity-depth trends that differed from observed trends in three ways: elevation and position of peak diversity and the shape (curvature) of the fitted diversity gradients. Both null and observed diversity curves were unimodal, suggesting that boundary constraints might contribute to diversity (Figure 3.23). However, for gas-

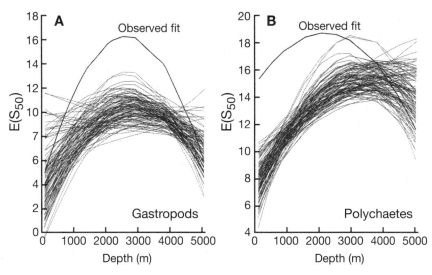

Figure 3.23. A comparison of observed diversity $E(S_{50})$ for gastropods (A) and polychaetes (B) in the western North Atlantic (heavy lines) to simulations (light lines) using random placement of depth ranges between sea-surface and abyssal seafloor boundaries to predict diversity in the context of the mid-domain model. From Pineda and Caswell (1998), with permission of the authors and Elsevier.

tropods, whereas the null and observed curves do not differ significantly in the location of peak diversity, elevation of the peaks and curvature do differ. For polychaetes location and elevation of the peaks differ. Curvature does not, although the shapes of curves are reversed (Figure 3.23). Overall, the observed diversity curves were nonrandom in most of the attributes compared, suggesting that environmental factors other than boundary constraints are responsible.

McClain and Etter (2005) explored six alternative null boundary constraint models to determine how varying their starting assumptions affected predictions and conformity to observed diversity patterns in deep-sea gastropods, bivalves, and polychaetes. Five of the models were proposed by Colwell and Lees (2000), and these use different algorithms to randomly draw species' midpoints and ranges. The sixth model limits possible depth ranges to maximum values actually found in deep-sea species. Based on Pineda and Caswell's (1998) three criteria, predictions of all six models showed poor agreement with observed data. Moreover, null predictions depended heavily upon the position of boundaries, the algorithms used to create ranges, the degree of patchiness at which ranges were occupied, and the proportion of large and small ranges. Predictions tended to match observed trends better when some degree of realism was introduced, either by restricting random range sizes to actual maximum range size or by using algorithms that included actual midpoints or ranges instead of using purely stochastic processes to select range size and placement. Isopods in the deep Gulf of Mexico also show poor conformity to mid-domain expectations (Wilson 2009).

Kendall and Haedrich (2006) compared diversity-depth trends in demersal fishes from three regions of the North Atlantic and the Gulf of Mexico to predictions of boundary constraint models. Null models failed to predict observed patterns; furthermore, in most cases observed patterns were not unimodal. Similarly, diversity patterns of macrofaunal and megafaunal taxa in the Gulf of Mexico are decidedly nonrandom, but they do correlate with a variety of environmental variables (Haedrich et al. 2008).

Evidence for bathymetric gradients of diversity would seem to reject the boundary constraint hypothesis as the primary cause. The model itself has been criticized as being logically inconsistent (Hawkins et al. 2005, Zapata et al. 2005) and difficult to parameterize appropriately and evaluate (McClain et al. 2007, Currie and Kerr 2008). Still, real physiological and geographic boundaries can exert effects on range distributions and consequently patterns of species diversity as Pineda (1993) originally recognized. Of course not all diversity-depth patterns are unimodal. The challenge is to establish the relative influence of boundaries and other environmental factors that shape diversity gradients.

Ecological Causes

There are limits to what can be deduced about causality from comparative data on patterns of diversity along environmental gradients. This is especially true in the deep sea, where the generality of the patterns is uncertain and the potential causes are only measured indirectly or just inferred. There are two broad classes of explanations for diversity gradients, evolutionary and ecological. Speciation, adaptive radiation, and the geographic spread of new taxa ultimately generate the regional pools that contribute species to local communities. Local diversity represents a balance between local extinction rates and rates of immigration from the regional pool (Ricklefs 1987, Karlson et al. 2004, Witman et al. 2004). As ecological rates are generally much faster than evolutionary rates, evolutionary processes become increasingly important in molding geographic variation in diversity at very large inter-basin and global scales. At the scales of bathymetric gradients considered in this chapter, ecological opportunity is probably paramount in structuring communities, although as we show in Chapter 6, continental margins appear to be centers of population differentiation and speciation and this may leave an historical signal in diversity-depth trends. In Chapter 7, we attempt to synthesize ecological and evolutionary causes and identify the spatial and temporal scales at which they operate.

Numerous ecological factors have been advanced to account for diversity-depth patterns, including competition (Sanders 1968), predation (Dayton and Hessler 1972), biological interactions mediated by productivity (Rex 1976), area effects (Osman and Whitlatch 1978), environmental heterogeneity (Etter and Grassle 1992), dispersal (Etter and Caswell 1994), physical disturbance (Gage et al. 1995), oxygen concentration (Levin and Gage 1998), source-sink dynamics (Rex et al. 2005a), and various combinations of these. Undoubtedly, the complete explanation is multivariate, especially for unimodal diversity-depth patterns, which almost necessarily require trade-offs between opposing forces.

Productivity

Recently, energy has emerged as a potential unifying theme in ecology to explain the basic shape of diversity gradients at larger scales and reconcile what appeared to be alternative theories about proximal mechanisms (Rosenzweig 1995, Mittelbach et al. 2001, Hawkins et al. 2003a,b, Evans et al. 2005, Scheiner and Willig 2005, Harrison and Grace 2007). There are two basic versions of the energy theory (Hawkins et al. 2003b, Clarke and Gaston 2006). One is that solar heat energy controls metabolic rate, which affects mutation rate and generation time and hence rates of evolutionary diversification (Rhode 1992, Brown et al. 2004, Allen et al. 2006). Diversity in this case is expected to be a positive function of temperature. There is some evidence for this in the deep sea. Hunt et al. (2005) showed that foraminiferan diversity in sediment cores is positively related to bottom temperature and negatively related to productivity during the Quaternary. Multiple regression showed that temperature had greater predictive power. Moreover, body size in ostracods from cores increased with global cooling during the Cenozoic, suggesting an adaptive response to temperature change

(Hunt and Roy 2006). However, it seems unlikely that temperature directly affects present-day bathymetric patterns of diversity. In the case of unimodal patterns, diversity increases with depth across the upper bathyal zone as temperature decreases. Diversity decreases from mid-bathyal to abyssal depths in a virtually isothermal environment. Stuart and Rex (in preparation) found that bottom temperature was weakly but negatively correlated with deep-sea gastropod diversity on Pan-Atlantic scales. Furthermore, the average size of organisms in the community as a whole decreases with depth and decreasing temperature (Rex et al. 2006; see Chapter 1), and body-size clines within species show conflicting patterns with depth and temperature (McClain et al. 2005). This does not mean that temperature has no effect at regional scales. It may be subordinate to other ecological causes or simply difficult to detect. The elegant analyses of Hunt and colleagues are very intriguing and may help explain the evolutionary development of the deep-sea fauna and consequently diversity at large scales.

The other version of the energy theory involves the conversion of solar energy into food by primary production and how this process influences population dynamics (Rosenzweig and Abramsky 1993, Rosenzweig 1995). Although there is widespread agreement among ecologists that there is an important relationship between species diversity and productivity, both the exact form of the relationship and its underlying mechanisms remain uncertain. Positive, negative, hump-shaped (unimodal), U-shaped (concave upward), and insignificant relationships have been demonstrated, but unimodal patterns appear to predominate for animals at larger geographic scales, particularly across community types (Mittelbach et al. 2001) such as those found along depth gradients at regional scales. We focus here on the unimodal version both because of its obvious relevance to diversity-depth patterns and because positive and negative trends can often be explained by shifting the scale of productivity within the unimodal framework.

The basic unimodal diversity-productivity model is shown in Figure 3.24. We reverse the usual depiction of increasing productivity to correspond to its association with depth. The ascending limb of the curve has been attributed to the accumulation of species with increased population density (Wright 1983), although this positive density-diversity trend can be modified by the distribution of body size among species (Srivastava and Lawton 1998) and adaptive tolerances of species to various climatic conditions (Currie et al. 2004). The basic premise is that low resource supply supports smaller populations, which are more prone to extinction from stochastic events. Depressed diversity at high levels of productivity, the negative relationship along the descending limb (Figure 3.24), may have a variety of causes (Rosenzweig and Abramsky 1993, Rosenzweig 1995). It has most often been attributed to accelerated rates of population growth leading to a higher rate of competitive exclusion (Huston 1979, 1994, Worm et al. 2002). Later we describe other potential causes for circumstances under which the mean and variance of production are positively correlated.

One complication in testing the diversity-productivity theory is that disturbance and productivity interact to regulate diversity because population reduction by disturbance can interrupt competitive exclusion (Huston 1979, Kondoh 2001, Worm et al. 2002). Disturbance can be imposed by consumer

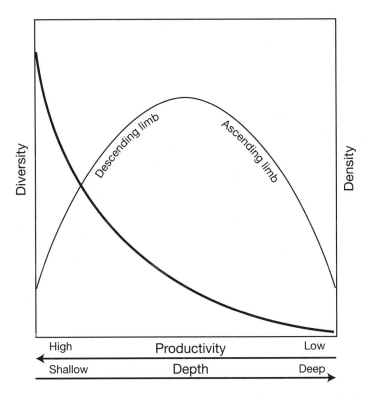

Figure 3.24. The basic unimodal diversity-productivity model as it relates to depth and animal density in the deep-sea benthos. The density (heavy line) of benthos is assumed to reflect organic carbon flux to the seafloor from overhead production. Diversity increases with an increase in population size (the ascending limb of the diversity-productivity curve) and decreases at high levels of nutrient loading (the descending limb).

pressure or by physical environmental perturbations. Its independent relationship to diversity is also thought to be unimodal (Levin and Paine 1974, Connell 1978). Theoretically, a higher incidence of disturbance shifts the peak of the diversity productivity curve toward higher levels of productivity (Kondoh 2001, Worm et al. 2002). In the deep sea, the productivity-diversity relationship can be examined, at least indirectly, by using benthic standing stock as a surrogate for productivity. However, the geography and nature of disturbance can only be vaguely inferred. Rex (1981) attempted to evaluate diversity-depth patterns in the context of Huston's (1979) dynamic equilibrium model, which combines the effects of competitive displacement and disturbance rates. The rate of competitive displacement was assumed to be a function of productivity as reflected in standing stock, and the diversity of predators was assumed to represent consumer pressure. The results were consistent with the dynamic equilibrium model in a very qualitative way, but unfortunately there is still no reliable measure of the incidence of disturbance with depth.

Another difficulty concerns the variance of productivity. The basic diversity-productivity model (Figure 3.24) assumes a steady rate of resource supply. In the deep sea, and perhaps in marine systems in general, there is a positive relationship between the mean and the variance of productivity. High rates of nutrient input in the upper bathyal zone are almost certainly associated with high temporal variability where surface production is seasonal. Chown and Gaston (1999) pointed out that pulsed nutrient loading can limit diversity in marine systems by restricting resource exploitation to only part of the annual cycle of production. If intermediate levels of productivity are more stable, they may allow feeding over a longer period and so support more species. More stable prey resources may also allow predators to diversify by specialization in diet, leading to higher overall diversity (Paine 1966). High rates and variability of production may act in concert to depress diversity, and it is not possible to disentangle their effects. High productivity is accompanied by high temporal variability.

It is also important to recognize that both diversity-productivity and diversity-stability relationships may involve bidirectional interactions: there may be conditions where diversity actually drives productivity and ecosystem stability (Naeem 2002, Worm and Duffy 2003, Gross and Cardinale 2007). The study of deep-sea biodiversity is still largely in an exploratory and descriptive phase and, unfortunately, cannot address these issues, as they require experimental approaches. In the deep sea, the source of primary productivity is extrinsic to the ecosystem, so whereas productivity may determine diversity, benthic diversity is unlikely to affect surface productivity. Secondary productivity in the benthic system is poorly known. Danovaro et al. (2008b) showed that nematode species diversity and trophic diversity were positively correlated with measures of ecosystem functioning (bacterial production and meiofaunal biomass) and efficiency (ratios of biomass to carbon flux). They interpreted this to mean that higher biodiversity augments ecosystem processes and the efficiency of carbon transfer. C. R. Smith et al. (2008) argued that abyssal community structure and function are dominated by POC flux and that climate change affecting carbon flux could strongly influence abyssal ecosystem function.

The diversity-productivity theory has considerable potential to explain the basic features of diversity-depth trends and to account for the large-scale variation in ecological opportunity, which is further mediated at smaller scales by other factors such as sediment grain-size heterogeneity (Etter and Grassle 1992), patch dynamics (Grassle and Sanders 1973), and physical disturbance (Paterson and Lambshead 1995). We stress, however, that the diversity-productivity theory is perhaps the only ecological explanation that can be currently tested with existing information on large scales across the full depth range in the deep sea. There is a large body of information on geographic variation in standing stock (see Chapter 1), but only fragmentary and much more localized information on other potential causes. It is perfectly conceivable that further research to better characterize the deep-sea environment and manipulative experimental approaches will reveal different ecological variables of overarching significance in regulating species diversity gradients at regional scales.

Since the macrofauna provide the largest database with the greatest geographic coverage, we begin by discussing their patterns and how these relate to the productivity theory. Then we take up whether meiofaunal and megafaunal trends are compatible with this. In the eastern and western North Atlantic the basic bathymetric pattern of species richness for the macrofauna as a whole and individual taxa is unimodal. The only apparent exception is for isopods in the western North Atlantic, for which abyssal diversity remains convincingly high. This exception may have an historical explanation. Isopods represent an ancient in situ radiation in the deep sea (Wilson 1998) and may be especially well preadapted to abyssal life (Wilson 1991). The primary difference between the eastern and western basins is that $E(S_n)$ remains high at abyssal depths in the east, only because evenness is high.

Species richness is depressed at upper bathyal and abyssal depths in both basins. A basic feature of macrofaunal diversity patterns for most taxa is that species richness increases from the abyss to bathyal depths. If we can accept standing stock as a proxy for organic carbon flux to the benthos, this decline corresponds to the ascending limb of the unimodal diversity-productivity relationship (Figure 3.24). Rex et al. (2005a) proposed a source-sink (Holt 1985, Pulliam 1988) hypothesis for abyssal biodiversity that embodies the effects of diminishing productivity on population size. As can be seen in Figure 3.3, the abyssal gastropod fauna in the western North Atlantic largely represents deeper range extensions for a subset of bathyal species. The same is true of gastropods in the eastern North Atlantic and protobranch bivalves in both the eastern and western North Atlantic (Rex et al. 2005a).

Abyssal diversity continues to decrease with increasing distance from land. The abyssal segments of species ranges often are very sparsely occupied. In the western North Atlantic, where sufficient quantitative information exists to estimate density at the taxon level, abyssal gastropods and bivalves as a whole average fewer than 1 individual m^{-2} and 5 individuals m^{-2}, respectively. But for both taxa these averages include a dominant species that makes up 60% or more of the assemblage. Most individual gastropod and bivalve species live at densities of about 1–5 individuals 1000 m^{-2} and 1–5 or fewer individuals 100 m^{-2}, respectively. Bathyal densities are orders of magnitude higher. These mollusks are minute organisms (~1–5 mm) with very low vagility, separate sexes, low recruitment rates, low fecundity, continuous reproduction (Rex et al. 1979, Zardus 2002), and life spans on the order of a decade (Gage 1994). It seems very unlikely that this combination of extremely low density and life-history traits would enable abyssal populations of many species to be reproductively self-sustainable.

Fully 73–91% of abyssal molluscan faunas in the eastern and western North Atlantic represent deeper range extensions of bathyal species, which suggests that abyssal endemism is low. In Chapter 4, we extend the analysis of bivalves to the entire North Atlantic. At this much larger scale the degree of abyssal endemism is reduced to a single species because apparent endemics in either the eastern or western basin occur at bathyal depths in the other basin or elsewhere in the Atlantic. The vast majority of abyssal mollusks have long-range larval dispersal ability. Protobranch bivalves have a demersal swimming pericalymma larva (Zardus 2002), and 78–91% of abyssal gastropods in the eastern and western basins, respectively, have either swimming feeding planktotrophic larvae or swimming lecithotrophic larvae (Rex et al.

2005a). Hence most abyssal populations theoretically can be maintained through dispersal from bathyal sources and may essentially represent a mass effect. For some species, particularly those that are rare at abyssal depths, bathyal and abyssal populations may form a source-sink system—abyssal sink populations experience chronic extinction from vulnerability to Allee effects (Rex 1973) balanced by continued immigration from more abundant bathyal source populations.

Ancillary evidence for source-sink dynamics comes from body-size–depth relationships. Average size of gastropods decreases from lower bathyal regions to the abyss in the western North Atlantic, and maximum size within species tends to decrease (McClain et al. 2005). Reduced body size in the abyss may result from unfavorable circumstances for growth. Individuals in abyssal populations may be smaller because they are more immature or because the lower bathyal–abyssal transition favors small taxa. Abyssal and lower bathyal gastropod assemblages exhibit similar ranges of shell architecture, but the abyssal morphospace is more sparsely occupied (McClain et al. 2004). Species with smaller and more energy-efficient shells show an increase in relative abundance in the abyss. These morphological trends suggest that energy constraints become more severe at abyssal depths and that only a few preadapted species can maintain potentially viable population densities (McClain et al. 2004).

Clearly source-sink dynamics do not apply to all abyssal populations, some of which, even megafaunal species, can be abundant and reproduce successfully in the abyss, as seen in Chapter 1. Moreover, even sparse abyssal populations may not function exclusively as sinks. Long-term cycles of carbon flux (K. L. Smith and Kaufmann 1999) or sporadic occurrences of phytodetrital accumulation on the seafloor (Billett et al. 2001) may allow population growth and reproduction and, consequently, reciprocal dispersal between source and sink (Gonzalez and Holt 2002). It is also possible that populations existing at low density at some abyssal localities are more common at others, so that persistence is supported as a large-scale metapopulation. If this is the case, then rare abyssal populations may be independent of bathyal sources.

The source-sink theory of abyssal diversity can be tested in other taxa by determining whether abyssal faunas consist largely of attenuated ranges of bathyal species. Using a nested analysis of species composition, Moreno et al. (2008) showed that the abyssal polychaete assemblage in the eastern South Pacific is a significant subset of the bathyal fauna at species, genus, and family levels. Olabarria (2005), using independent and more extensive collections of bivalves than those available to Rex et al. (2005a), also proposed that abyssal diversity of the eastern North Atlantic represented a source-sink effect. That the abyssal assemblage is mostly composed of deeper range extensions can be clearly observed in Figure 3.6. Another approach would be to examine the reproductive status of bathyal and abyssal populations of the same species. Abyssal populations of rare species should show less evidence of reproductive maturity and mating than bathyal populations of the same species. The most conclusive test would be a comparison of genetic population structure. Abyssal populations should have nested sets of the bathyal gene pool and presumably no unique haplotypes, since they are assumed to be nonreproductive and would therefore have no way to perpetuate new mutant genes.

The descending limb of the diversity-productivity curve corresponds to the upper bathyal zone (Figure 3.24). In the western North Atlantic there is enhanced seasonal productivity at the shelf-slope break caused by an oceanographic frontal system (Ryan et al. 1999) and high carbon flux to the bottom because of weak currents (Martin and Sayles 2004). Rex (1976, 1983) proposed that high productivity would drive higher rates of population growth, which could accelerate competitive exclusion, as predicted in Huston's (1979) dynamic equilibrium model. There is no direct evidence for this, but evenness is also typically lower at upper bathyal depths, indicating the kind of distortion of the relative abundance distribution that might take place if some species increase in abundance at the expense of others.

As we discussed earlier, the temporal variability of productivity may also be implicated. Depressed species richness and evenness under circumstances of unusually high organic carbon flux to the seafloor and elevated benthic standing stock appears to be a general phenomenon in deep-sea communities. In addition to upper bathyal zones, it also occurs where topography concentrates sinking material in submarine canyons (Gage et al. 1995, Vetter and Dayton 1998) and in trenches (Jumars and Hessler 1976, Danovaro et al. 2002, Gambi et al. 2003). Lateral advection of organically rich sediments (Schaff et al. 1992, Blake and Hilbig 1994, Gooday 1996) and stimulation of bacterial growth by cycles of sediment erosion and deposition in strong current regimes (Thistle et al. 1985, Gage et al. 1995, Aller 1997) have the same effect. In OMZs below the depth of their cores, where dense communities develop in more oxic conditions, diversity and evenness remain low (Levin and Gage 1998, Levin et al. 2000, Levin 2003). Enrichment caused by fluvial input (Wei et al. 2009) and high productivity at marginal ice zones (Schewe and Soltwedel 2003, Fonseca and Soltwedel 2007) also augment abundance and lower diversity. Whale carcasses, which introduce a huge localized supply of food, increase density severalfold and suppress richness and evenness in the immediately surrounding soft-sediment community (C. R. Smith and Baco 2003). Even occasionally observed unexplained spikes of standing stock along depth gradients are associated with markedly reduced richness and evenness (e.g., Narayanaswamy et al. 2005). Levin and Gage (1998) found a significant negative relationship between sediment organic content and the diversity $E(S_{100})$ of the total macrofauna and polychaetes in the Indo-Pacific. Since the data were all bathyal (154–3400 m), they suggested that the relationship represented the descending limb of the diversity-productivity curve.

How do macrofaunal diversity-depth patterns in adjacent seas compare to those in the North Atlantic? To better understand the potential role of productivity, we have plotted density-depth curves for the North Atlantic, Gulf of Mexico, and the Norwegian and Mediterranean seas in regions as near as possible to where diversity measurements were taken (Figure 3.25). Standing stock is the end product of organic carbon flux, which is ultimately manifested as species diversity. The eastern and western North Atlantic density-depth patterns are similar and relatively high. All three adjacent seas have lower density in the following order of decreasing density: Gulf of Mexico > Norwegian Sea > Mediterranean Sea.

In Figure 3.26, we illustrate the relative positions of diversity-depth patterns for the North Atlantic (east and west combined), Gulf of Mexico, Norwegian Sea, and Mediterranean Sea. The curves are meant to represent species richness. Since there was no universally common measure of diversity among the studies, we used relative proportional estimates of $E(S_n)$ and S to construct the curves. Their positions with respect to one another are reasonably accurate. The eastern and western North Atlantic have similar unimodal diversity-depth patterns across the full depth range from the shelf-slope transition to the abyss. The Gulf of Mexico shows a unimodal pattern of diversity with peak diversity at a depth similar to those found in the eastern and western North Atlantic and a similar maximum diversity (1000–2000 m for quantitative samples; cf. Figures 3.2, 3.13). However, diversity then drops sharply to the deepest seafloor (3800 m), a depth at which diversity in the North Atlantic remains relatively higher (Figures 3.1, 3.2). Interestingly, Wei et al. (2009) note that average organism size is smaller in the Gulf of Mexico

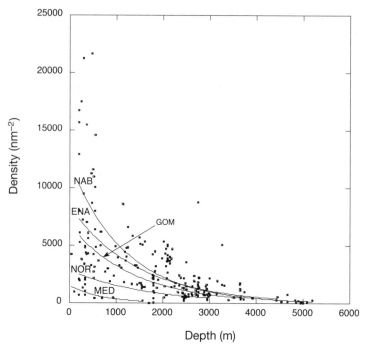

Figure 3.25. Density-depth relationships for macrofauna in the western North Atlantic (data from Johnson et al. (2007), eastern North Atlantic (Flach and de Bruin 1999), Gulf of Mexico (Wei et al. 2009), Norwegian Sea (Wlodarska-Kowalczuk et al. 2004), and Mediterranean Sea (Tselepides et al. 2000a). The regression equations are:

Western North Atlantic: $Y = 12154e^{-0.00084079X}$, $R = 0.81$
Eastern North Atlantic: $Y = 8390.1e^{-0.00068845X}$, $R = 0.99$
Gulf of Mexico: $Y = 6788.3e^{-0.00073898X}$, $R = 0.62$
Norwegian Sea: $Y = 2840.48e^{-0.00062732X}$, $R = 0.60$
Mediterranean Sea: $Y = 1493.9e^{-0.0014606X}$, $R = 0.69$

All regressions are significant at $P < 0.01$, except the Mediterranean shown only to indicate the general trend.

than in the Atlantic. To some extent this may permit higher density and hence diversity, allowing maximum diversities between the Atlantic and Gulf of Mexico to be similar. But with increased depth, the severe drop in food supply in the Gulf may exceed the adaptive limitations of size reduction and more rapidly drive many species to critical population densities below which they are not reproductively viable. In the Norwegian Sea, diversity peaks even shallower at 500–1000 m and then decreases steeply (Figures 3.10, 3.11). In the Mediterranean, diversity simply decreases with depth (Figure 3.12).

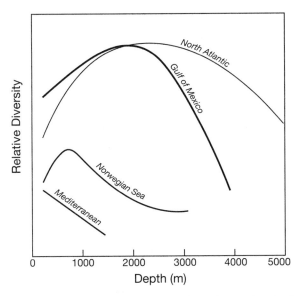

Figure 3.26. Relative macrofaunal diversity-depth relationships in the North Atlantic (eastern and western combined), Gulf of Mexico, Norwegian Sea, and Mediterranean Sea. The composite was created from information in Figures 3.1, 3.2, 3.4, 3.5, and 3.11–3.13.

We suggest that these differences in macrofaunal diversity can be explained by shifting the scale of productivity as reflected in standing stock. The eastern and western North Atlantic are more productive regions than the adjacent seas. Organic carbon flux to the seabed and benthic standing stock decrease exponentially with depth. Diversity shows a unimodal pattern similar to the classic diversity–productivity relationship (Figure 3.24). As regional productivity decreases, the elevation of the density–depth relationship also decreases. As the range of productivity shifts toward the left, peak diversity occurs at shallower depths (as in the Norwegian Sea) and is finally eclipsed so that the diversity–depth relationship represents only the ascending limb (Mediterranean Sea). If this scenario is correct, then a conspicuous depression of deep-sea diversity at upper slope depths is expected only when density approaches levels of 5000–10,000 individuals m^{-2}, some reduction is seen in the neighborhood of 2500 m^{-2}, and there is no effect below this level. The exact differences are not resolvable. The important point is that with progressively lower levels of productivity, the diversity–depth relationship shifts from a fully realizable unimodal pattern toward just the ascending limb. In terms of density, and presumably productivity, the monotonically decreasing pattern of diversity with depth in the Mediterranean is tantamount to the mid-bathyal to abyssal range of the diversity pattern in the North Atlantic, as can be readily seen by comparing Figure 3.25 to Figures 3.1 and 3.12.

This scaling helps explain an apparent exception to unimodal patterns for abyssal polychaete diversity and abundance. Glover et al. (2002) showed that polychaete abundance increased across the abyssal floor of the Pacific along a gradient of increased density toward a narrow zone of localized upwelling along the equatorial current, which caused higher productivity and enhanced organic carbon flux to the seafloor. A similar positive relationship between abundance and diversity existed among abyssal samples from both the Atlantic and the Pacific. The authors point out that this diversity-productivity relationship is not unimodal as they anticipated. The reason is that the maximum recorded density was about 350 individuals m^{-2}. Since polychaetes typically make up the largest proportion of macrofaunal abundance, total macrofaunal abundance was perhaps in the 500–1000 individuals m^{-2} range. Looking at Figure 3.25 we can see that their maximum density at enhanced abyssal sites is comparable to mid-bathyal density in the North Atlantic. In other words, the range of densities observed corresponds to only the ascending limb of the unimodal diversity-productivity curve. Densities at the most diverse abyssal stations are much lower than levels that correspond to depressed diversity at upper bathyal depths.

The meiofaunal and megafaunal patterns are less well documented but generally fit the suggested diversity-productivity trends proposed for the macrofauna. Fishes represent a higher trophic level and consequently obtain less energy from organic carbon flux than the macrofauna. In the western North Atlantic their diversity-depth pattern is unimodal with peak diversity at a shallower depth than macrofaunal elements (e.g., Figure 3.18). The overall level of diversity is much lower. Elsewhere in the Atlantic Ocean and the Gulf of Mexico, their diversity simply declines with depth. We suggest that their higher trophic position essentially shifts them into a lower range of food supply, so that they occupy either the ascending limb of the unimodal productivity-diversity curve or a truncated version of the curve where regional productivity is sufficiently high, such as in the western North Atlantic. Megafaunal invertebrates show a unimodal pattern in the western North Atlantic, again, as expected, with a shallower peak and lower overall diversity than the macrofauna (Figures 3.1, 3.18). An exception to the trend is found in holothurians, a group that is highly evolved in the deep sea, with members that are clearly specialized to quickly exploit sinking phytodetritus and reproduce early, even at abyssal depths (Billett et al. 2001, Wigham et al. 2003a,b).

The foraminiferan diversity-depth patterns also make sense in the context of the unimodal diversity-productivity model. As we discussed in Chapter 1, diversity patterns are hard to relate to density because the latter is seldom measured and can represent different components of the foraminiferan biota. Diversity is clearly depressed at sites that experience unusually high nutrient loading, such as the continental shelves, the Mississippi Delta, and Site III off Cape Hatteras (Figure 3.15), and in OMZs, where foraminiferans are able to withstand low oxygen concentrations and still benefit from high organic carbon flux. In the High Arctic, where productivity is certainly lowest among the regions, diversity is lowest and decreases with depth, suggesting that this represents the ascending limb of the diversity-productivity curve. In the Norwegian Sea diversity also decreases with depth and is somewhat higher than in the High Arctic, probably owing to the absence of permanent ice cover, which permits higher productivity. Diversity in the Gulf of Mexico is highest and shows a unimodal pattern in the eastern regions. In the central Gulf diversity is depressed near the Mississippi Delta. At abyssal depths diversity converges closely with levels in the east. Diversity is depressed and decreases with depth across the bathyal region in the west, where surface productivity is lower.

Foraminiferan diversity may remain high at abyssal depths in the western North Atlantic because food supply (primarily bacteria and small organic particulates) and, consequently, population densities remain high enough that diversity is not depressed by inverse density dependence. Foraminiferans are less vulnerable to Allee effects because they have asexual reproduction. Another factor may be ecological release. The very reduced abundance of macrofauna and megafauna with increased depth may make a relatively higher proportion of nutrients available for foraminiferans. Since many metazoans consume foraminifera, there may be a decrease in predation pressure as well. Depressed diversity in the western Gulf and a unimodal pattern in the east with lower abyssal diversity indicate that food supply to the deepest Gulf is not sufficient to support a diverse community below around 3000 m, as is found in the abyss of the western North Atlantic. The unimodal foraminiferan pattern in the eastern Gulf mimics the macrofaunal pattern in the western North Atlantic, possibly because production in the Gulf is shifted to lower values over the full depth range. In other words, when the abyssal environment is more severely impoverished, even meiofaunal elements experience a reduction in diversity.

Jorissen et al. (1995) proposed the TROX (*trophic conditions and oxygen concentration*) model of foraminiferan microhabitat availability in sediments that may be directly relevant to geographic variation in alpha diversity. Foraminiferans live not only at the well-oxygenated and relatively nutrient-rich sediment-water interface, but also penetrate the nutrient-poor dysoxic and anoxic layers (Figure 1.9). According to the TROX model, microhabitat space (depth in the sediment) is a balance between food supply and oxygen concentration. Mesotrophic and oligotrophic environments are predicted to be food limited; microhabitat occupancy is a positive function of organic carbon flux to the seabed. Under extreme eutrophic circumstances, nutrient loading can deplete sediment oxygen content, reducing infaunal microhabitat space. A similar set of constraints may govern microhabitat space in nematodes (Soetaert et al. 2002).

Gooday (2003) suggested that the TROX model may account for regional differences in foraminiferan diversity, implicitly by equating microhabitat space to niche space. He pointed out that both extremely oligotrophic regions, such as the Eastern Mediterranean and Central Arctic, and extremely eutrophic sites, such as the OMZs of the Arabian Sea and the Santa Barbara Basin, have depressed diversity as predicted. Well-oxygenated continental margins and abyssal plains support higher diversity. The TROX model is also consistent with most of the foraminiferan diversity gradients shown in Figures 3.15. Two areas with heavy nutrient loading, Site III off Cape Hatteras (Figure 3.15A) and the continental shelf off the Mississippi Delta (Figure 3.15C), have conspicuously depressed diversity. The Arctic Ocean, which is highly oligotrophic, also has very low diversity. In the Gulf of Mexico, a mesotrophic basin, diversity appears to be largely a function of food supply (based on depth and location in the absence of density data). The decrease in diversity with depth in the Norwegian Sea, where metazoan standing stock is lower (Figure 3.25), may also be related to reduced food supply. All of these diversity trends are reasonably consistent with predictions of the TROX model, at least in relative terms.

What the model as adapted by Gooday (2003) does not readily explain is the general increase in diversity with depth in the western North Atlantic (Figure 3.15A). Apart from Site III off Cape Hatteras there do not appear to be areas of extreme nutrient loading and presumably periodic oxygen depletion. The TROX model predicts that diversity should decrease with depth as oligotrophic abyssal conditions are approached. However, this may be a matter of scale. As for the metazoan meiofauna (Figure 1.12), data on foraminiferan density (Figure 1.18) do not indicate a decrease with depth nearly as dramatic as for the megafauna and macrofauna. Perhaps foraminiferans do not experience the abyss of the western North Atlantic as a highly oligotrophic environment comparable to the Central Arctic. In other words, the place on the mesotrophic-oligotrophic axis where microhabitat space (diversity) is depressed has simply not been reached. Productivity and benthic standing stock are higher in the western North Atlantic than in the Gulf of Mexico (Figure 3.25), possibly indicating that abyssal populations in the latter do experience extreme oligotrophic conditions. Another possibility is that other factors may modulate the relationship between food supply and microhabitat space. Abyssal foraminiferan diversity may remain high because of a release from predation pressure by larger organisms or because of source-sink dynamics in which survival in abyssal sinks is enhanced by asexual reproduction, dormancy, and high dispersal ability.

We stress that evidence for the role of productivity in shaping regional patterns of species diversity in the deep sea is indirect. It is based on the assumption that organic carbon flux to the benthos is productivity in this system and that productivity, in turn, is reflected in standing stock. Even to the extent that this is accurate and meaningful, we still have few studies in which standing stock and diversity are measured simultaneously across the full depth range between the shelf-slope transition and the abyss. There are also many aspects of the relationship between productivity and diversity that remain unexplained theoretically. Although the connection between productivity and standing stock is reasonable, what is less clear is how standing stock is partitioned among species. Why is food supply simply not monopolized by a single species? Furthermore, as we point out later in our discussion of disturbance and environmental heterogeneity, other potentially important causes of diversity vary with depth and are not easy to dissociate from productivity. Our more extensive treatment of productivity devolves largely from there being much more information available about patterns of diversity and standing stock on regional scales.

The large-scale bathymetric gradient of productivity in the form of organic carbon flux to the benthos may govern the general form of the relationship between local diversity and depth. But whereas it may determine the range of ecological opportunity within which local diversity can vary, it cannot account for the considerable variation in local diversity at particular depths (Figure 3.2), which must be a function of variation in environmental conditions at scales similar to sample size. Some of this variation might be due to patchiness in food supply or disturbance (see Chapter 2), but there is little evidence that patch structure affects species richness, and the incidence and intensity of patchiness with depth are unknown.

Sediment Particle-Size Diversity

One factor that does correlate with local diversity across a broad depth range is sediment particle-size diversity (Etter and Grassle 1992). Most deep-sea species are deposit feeders. Coastal deposit feeders selectively consume different particle-size distributions as a mechanism of competitive resource partitioning (Fenchel and Kofoed 1976, Self and Jumars 1988). Etter and Grassle (1992) were able to collect benthic fauna to measure species diversity and sediment samples simultaneously from subcores of 558 box cores collected during the ACSAR program. Species diversity was estimated as $E(S_{100})$. Sediment particle size diversity was assessed by applying the Shannon-Wiener index to the percentage of standard sediment grain-size classes. Species diversity was generally significantly correlated with sediment particle-size diversity over the full range of particle sizes. To maximize feeding efficiency, deposit feeders selectively consume the smallest particles and species diversity was particularly highly correlated with particle-size diversity within the silt fraction of sediments.

Figure 3.27 shows the relationships of species diversity to depth and silt diversity for the northern part of the ACSAR program. Both relationships are highly significant ($r = 0.653$ and 0.665, respectively, $P < 0.001$). To determine which variable best predicts diversity, the alternative variable was held constant by partial regression. When one controls for silt diversity, the relationship between species diversity and depth collapses. However, when depth is statistically removed, the relationship between species diversity and silt diversity remains highly significant. The same basic result obtains for the mid- and southern regions of the ACSAR program. Although the exact underlying causality of these statistical relationships is unknown, they strongly suggest that small-scale resource variability in terms of food, habitat complexity, or both affects local species diversity. Unlike the nonequilibrial patch-dynamic model discussed in Chapter 2, the correlations between diversity and sediment grain size suggest that species coexistence depends to some extent on resource partitioning in a more equilibrial system.

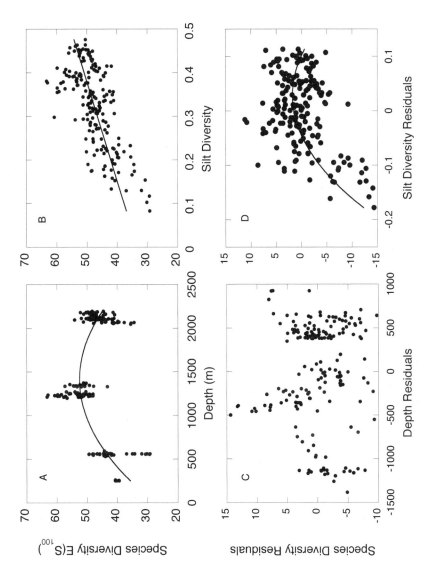

Figure 3.27. Relationships of species diversity $E(S_{100})$ to depth (A) and sediment particle-size diversity for the silt fraction (B) for box-core samples taken in the northern region of the ACSAR program. Residuals analyses with the effect of silt diversity held constant (C) and depth held constant (D). The relationship in C is not significant. Regressions in A, B, and D are significant ($P < 0.001$). From Etter and Grassle (1992), with permission of the authors and Macmillan Publishers Ltd.

Physical Disturbance

Physical disturbance generated by near-bottom currents is another factor that influences diversity (Gage 1997, Levin et al. 2001), in this case on much larger geographic scales of perhaps 1–100 km. Intermittent currents strong enough to resuspend deep-sea sediments occur globally, primarily along continental margins associated with western boundary currents, internal tides, and storm-driven eddies, but also sometimes far out at sea in abyssal plains (Gardner and Sullivan 1981, Hollister et al. 1984, Richardson et al. 1993, Cacchione and Pratson 2004). Cycles of erosion and deposition can potentially affect benthic communities in complex ways. Scouring can regrade sediments, homogenize patch structure, increase emigration rates of small light species, and increase extinction rates by physical damage to organisms and burial. However, stirring sediments stimulates bacterial activity, and depositional events can introduce new organic matter and augment immigration rates, which is why chronically disturbed deep-sea sediments support much higher standing stock, often dominated by juveniles (see Chapter 1).

Gage (1997) compared diversity at five sites representing a gradient of hydrodynamic disturbance based on current measurements and photographic observations of bed forms. The most disturbed was the HEBBLE site at 4820 m in the western North Atlantic, described in Chapter 1. This is a region of episodic benthic storms, where currents can reach 73 cm s^{-1}. The Scottish Marine Biological Association permanent station at 2900 m in the Rockall Trough of the eastern North Atlantic is another area of high kinetic energy, where currents reach 20 cm s^{-1} and rippled bed forms indicate periodic high flow, but probably not as severe as at the HEBBLE site. The third site was the Setubal Canyon at 3400 m on the continental margin off Portugal, where rippled bed forms indicate currents similar to those in the Rockall Trough, which appear to be caused by vigorous tidal currents. The fourth and fifth are very tranquil abyssal regions, the Tagus Abyssal Plain (5035 m), about 200 km west of the Portuguese continental margin, and the very remote central North Pacific (5500–5800 m).

Rarefaction curves for polychaetes collected at the five sites are shown in Figure 3.28. The Tagus and central North Pacific sites have the highest normalized diversity, followed by the two sites of intermediate disturbance, the Rockall Trough and Setubal Canyon. This is the exact opposite of the more usual situation where bathyal diversity exceeds abyssal diversity. HEBBLE, the most disturbed site, has the lowest diversity. Strong hydrodynamic disturbance

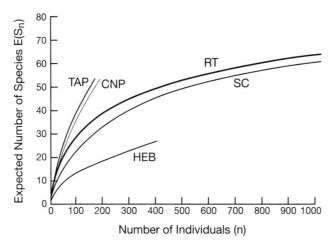

Figure 3.28. Rarefaction curves for polychaetes collected from box corers in the Tagus Abyssal Plain (TAP), Central North Pacific (CNP), Rockall Trough (RT), Setubal Canyon (SC), and HEBBLE (HEB). Diversity is highest at the most tranquil sites (TAP and CNP), decreases with an increased severity and incidence of hydrodynamic disturbance (RT and SC), and is lowest at the highly disturbed HEBBLE site. From Gage (1997), with permission of Cambridge University Press.

is associated with both lower richness and evenness (Figure 3.28). The comparisons also highlight the difficulty of separating the effects of disturbance and productivity. Polychaete density is 5–9 times higher in the disturbed sites than in the Tagus Abyssal Plain, and 18–36 times higher than in the central North Pacific. Density of the macrofauna as a whole at HEBBLE is an order of magnitude higher than in normal communities at comparable depths (Figure 1.14). What remains unclear is whether the degree to which diversity is depressed is due more to physical disturbance or to unusually high productivity in the form of food supply.

Paterson and Lambshead (1995) provide the most convincing correlative evidence for the effects of disturbance on community structure. Polychaetes collected with box corers from 400–2900 m on the Hebridean Slope and the floor of the Rockall Trough display a unimodal diversity–depth pattern. However, abundance showed no significant relationship with depth, suggesting that nutrient input was not the cause. As a measure of evenness, they used the V-statistic derived from the Ewens-Caswell neutral model (Caswell 1976). More negative V indicates lower evenness. Figure 3.29 shows the relationship of the V-statistic to the frequency of currents exceeding

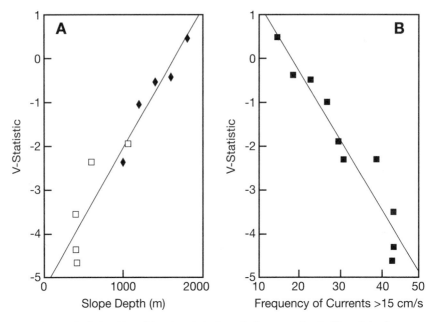

Figure 3.29. Relationship of the *V*-statistic for polychaetes to depth (A) and the frequency of currents that exceed 15 cm s⁻¹ (capable of resuspending sediments) (B) on the Hebridean Slope of the Rockall Trough, eastern North Atlantic. More negative values of the *V*-statistic indicate lower evenness. From Paterson and Lambshead (1995), with permission of the authors and Elsevier.

15 cm s^{-1}, those strong enough to mobilize fine sediments. On the Hebridean Slope (400–1800 m) the *V*-statistic increases with depth and is a negative function of the frequency of >15 cm s^{-1} currents. Disturbance events from strong currents are highest on the upper slope and decrease with depth to the more stable slope base. Paterson and Lambshead (1995) attributed depressed diversity at upper slope depths to hydrodynamic disturbance. They also showed that upper slope polychaete assemblages are dominated by species belonging to families that are considered to be opportunistic and are often found associated with disturbed habitats.

CONCLUSIONS

Bathymetric patterns of alpha diversity can provide important clues about the causes of community structure because they parallel steep environmental gradients over short distances. At first glance, there seems to be a bewildering array of trends among taxa, functional groups, and geographic regions. The patterns do not reflect null boundary constraints, which suggests that they are actively shaped by ecological and evolutionary causes. Much of the large-scale variation in diversity with depth and among ocean basins appears to be related to food supply and its effects on population density and population growth.

Diversity-depth patterns are best known for the macrofauna. Their most fundamental feature is that diversity is high in the narrow bathyal zone and decreases in the abyss. In basins with relatively moderate to high surface production and off-shore productivity gradients, for example, the eastern and western North Atlantic, the rate of food supply to the benthos as reflected in animal abundance decreases exponentially with depth. Diversity-depth patterns across the gradient are unimodal, resembling the well-known macroecological relationship between diversity and productivity at large scales. Many abyssal populations are attenuated range extensions of bathyal species with larval dispersal ability. At abyssal depths, diversity is probably depressed by vulnerability to Allee effects—population densities become reduced to a level that is not reproductively viable. For many species, bathyal and abyssal populations may function as a source-sink system in which chronic local extinction from inverse density dependence in the abyss is balanced by immigration from bathyal populations.

Depressed diversity at upper bathyal depths is more difficult to explain. Based on population densities, the rate of nutrient input must be orders of magnitude higher and is probably more variable because of close proximity to seasonal plankton production nearer the coast. Higher population growth rates may accelerate competitive exclusion or populations may be destabilized by seasonality. Although the proximal mechanism is not known, all circumstances of high nutrient loading in the deep sea, irrespective of depth, result in high population density and depressed diversity compared to neighboring habitats. Examples include concentration of sinking material by topographic features, lateral advection of organically rich sediments, stimulation of bacterial growth by sediment erosion and deposition, close associations with OMZs, fluvial input, and productive marginal ice zones. Even small-scale enrichment studies, discussed in Chapter 2, show similar effects.

Among-basin differences in macrofaunal diversity-depth trends can be explained by shifting the scale of productivity as reflected in standing stock. For example, density-depth curves show the following progressive decrease in elevation: North Atlantic > Gulf of Mexico > Norwegian Sea > Mediterranean

Sea. Correspondingly, overall diversity declines in the same sequence, and the mode of the diversity-depth curve shifts toward the descending (more productive upper bathyal to mid-bathyal) limb. In the impoverished Mediterranean the mode is lost and diversity simply decreases with depth, just as it does in the ascending (less productive abyssal to mid-bathyal) limb of the fully realized unimodal curve. In other words, the overall density-depth relationship in the Mediterranean is comparable to the much deeper mid-bathyal–abyssal segment in the North Atlantic; predictably, diversity simply decreases across both gradients. The Mediterranean fauna below upper bathyal depths is very sparse. Source-sink dynamics may be displaced upward toward bathyal depths.

Diversity-depth patterns in the meiofauna and megafauna are less well known, but seem consistent with our interpretation of macrofaunal patterns. Megafaunal species are much larger organisms and many are carnivores, which makes them more sensitive to the decrease in food supply with depth. Their diversity is generally much lower than for the macrofauna, and either the diversity-depth patterns are unimodal or diversity simply decreases with depth. Meiofaunal diversity is still not well characterized, but is potentially very high in terms of alpha diversity. Foraminifera are particularly interesting. Their diversity-depth trends may span the full range of the diversity-productivity curve. In the North Atlantic, diversity increases with depth to its highest level in the abyss. The reasons foraminiferans apparently do not suffer a decline in diversity at abyssal depths may include decreased mortality from metazoans and a less-diminished food supply through ecological release. Moreover, they are less vulnerable to Allee effects than macrofauna because they can rely on asexual reproduction, remain dormant, and have high rates of dispersal. With decreasing overhead production, foraminiferan diversity does decline with depth, possibly representing the ascending limb of the unimodal pattern. It is difficult to link their diversity patterns to productivity, except in a very general way, because there are so few data on density-depth trends.

While food supply to the benthos might account for the basic shapes of diversity-depth trends, it is clearly not the only ecological factor involved. Diversity-depth regressions leave much of the variance in diversity unexplained. Also, the relationship is indirect since datasets on density, a presumed surrogate of productivity, and diversity are seldom concurrent. In most cases we are left using depth as the independent variable representing relative food supply. At the very least, we must assume that diversity-depth patterns are strongly modulated by other ecological variables. OMZs play a major role in interrupting diversity–depth trends. There is also compelling evidence that large-scale trends in alpha diversity are influenced by sediment characteristics and physical disturbances, but less is known of their effects across the full depth range and at among-basin scales.

There are clear exceptions to what appear to be the basic diversity–depth trends. Macrofaunal isopods and megafaunal holothurians do not show depressed abyssal diversity, at least in some regions. Holothurians are particularly well adapted to exploit abyssal phytodetritus accumulations. Isopods underwent a major ancient radiation in the deep sea and may be especially well adapted to life in the abyss. In Chapters 4 and 6 we discuss this further and show how evolutionary rates of diversification may influence diversity–depth patterns.

4

OCEANWIDE VARIATION IN ALPHA SPECIES DIVERSITY AND LONG-TERM CHANGE

There is absolutely nothing to restrict the geographical range of animals in the deep sea. . . . We got quite tired on the *Challenger* of dredging up the same monotonous animals wherever we went.

Henry N. Moseley (1880)

The greater cumacean diversity found at the low latitude deep-sea stations suggests that the bathyal and abyssal populations there are inherently more diverse than those at higher latitudes. The reasons for this possible pattern are not intuitively obvious since the physical regime is monotonously constant everywhere in the World Ocean below the permanent thermocline.

Norman S. Jones and Howard L. Sanders (1972)

Soon after Hessler and Sanders' (1967) discovery that deep-sea species diversity was much higher than expected, evidence began to emerge that diversity varied not only with depth, as Sanders (1968) first showed, but also geographically on among-basin scales (Jones and Sanders 1972, Jumars 1975b, Sibuet 1979). We reviewed variation in diversity-depth patterns in the North Atlantic and adjacent seas in Chapter 3. At larger oceanwide and global scales diversity trends are much less well documented and their causes remain highly conjectural. It is likely that the evolutionary processes of speciation, adaptive radiation, and geographic spread of higher taxa play a more important role at very large scales in the deep sea, as they do in other environments. In this chapter, we describe Pan-Atlantic variation in alpha diversity and discuss what can be inferred from paleoceanography about the historical development of the deep-sea fauna.

There is now general agreement among deep-sea ecologists that there are significant regional differences in diversity. Recently, the central questions have become: Are there regular latitudinal gradients of diversity in the deep-sea benthos like those found in other ecosystems; and, if so, why? Latitudinal gradients of species diversity, the decrease of species richness from the tropics toward the poles, are perhaps the most dominant large-scale biogeographic patterns on Earth. Recognized since the great eighteenth- and nineteenth-century European voyages of discovery (Forster 1778, von Humboldt 1808) and the object of intense investigation since then, latitudinal gradients are ubiquitous features of terrestrial environments (Hillebrand 2004a). They are also found in coastal (Roy et al. 1994, 1998) and pelagic (Angel 1997, Macpherson 2002) marine environments, although they appear to be somewhat irregular in shape, due to strong regional oceanographic influences, and can be asymmetrical between hemispheres (Hillebrand 2004b, Rex et al. 2005b). Numerous possible ecological and evolutionary causes have been proposed for latitudinal gradients (summarized by Rohde 1992). Recent attention has focused on energy and history as unifying themes (Ricklefs and Schluter 1993, Hawkins et al. 2003b, Jablonski et al. 2006). Latitudinal gradients of diversity were not expected in the deep-sea benthos because it was long assumed that the thick overlying water column would buffer against the climatic forcing that seems to be so influential in shaping the biodiversity of surface environments. Yet there is growing evidence that a number of deep-sea taxa show latitudinal trends (Stuart et al. 2003). Understanding these patterns, if they are real, could contribute significantly to understanding the general phenomenon of latitudinal gradients because the deep sea is such a distinctive environment.

GEOGRAPHIC PATTERNS

Macrofauna

The Woods Hole Benthic Sampling Program (Sanders 1977) and the NORBI expedition (Dahl et al. 1976) provided data to test for the existence of latitudinal gradients in the North and South Atlantic and the Norwegian Sea in three taxa: gastropods, bivalves, and isopods (Poore and Wilson 1993, Rex et al. 1993). The data represent 97 epibenthic sled samples collected from 500 and 4000 m between 77°N and 37°S (Figure 4.1). Geographic coverage of sampling, although it includes both eastern and western corri-

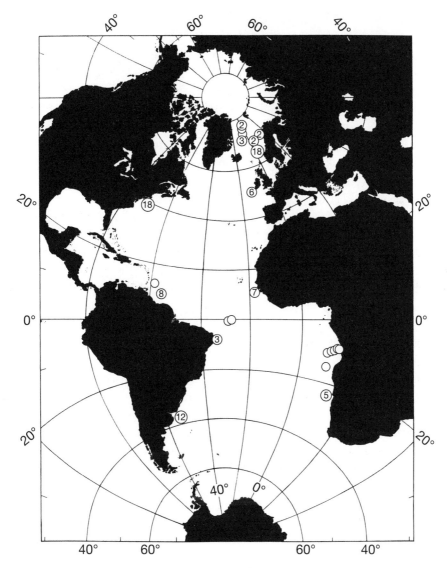

Figure 4.1. Locations of bathyal (500–4000 m) epibenthic sled samples taken in the Atlantic Ocean and the Norwegian Sea used to examine latitudinal patterns of diversity in isopods, gastropods, and bivalves (see Figures 4.2 and 4.3). Circles represent individual samples and numbers in circles represent the number of samples in a region. The basins sampled are indicated in Figure 4.2. From Rex et al. (1993), with permission of the authors and Macmillan Publishers Ltd.

dors of the Atlantic, is nevertheless still very weak compared to typical ter-restrial, coastal, and open-ocean pelagic studies. The diversity patterns, based on $E(S_n)$, are shown in Figure 4.2. All three taxa show statistically significant poleward decreases in diversity in the North Atlantic Ocean and the Nor-wegian Sea. The South Atlantic is characterized by strong regional variation: particularly low diversity in the Cape Basin associated with an oxygen min-imum zone and unusually high diversity in the Argentine Basin. Nonethe-less, despite the shorter span of latitude and noisy data, gastropods and bivalves show weak, but significant, poleward decreases in diversity. Isopods show no significant trend in the South Atlantic, and, indeed, exhibit their highest diversity at temperate latitudes in the Argentine Basin.

Much of the variation in diversity within basins in Figure 4.2 represents the strong depth-related changes we analyzed and discussed in Chapter 3. To determine whether there is an independent latitudinal signal in diversity, it is necessary to statistically remove the effect of depth by partial regression. Otherwise the apparent latitudinal trend could simply be a spurious conse-quence of uneven depth coverage among regions. For example, an extreme case would be if tropical samples were all bathyal, where diversity is usually higher, and high-latitude samples were abyssal, where diversity is usually lower. A simple plot of diversity against latitude would be negative for the uninter-esting reason that more-poleward samples were deeper. That is not in fact generally the case, but depth must still be held constant to test for an inde-pendent latitudinal effect. In Figure 4.3, we do this by plotting the residuals of diversity and latitude with depth removed. The diversity gradients remain negative and highly significant with the exception of isopods in the South Atlantic, which were not significant anyway (Figure 4.2). We return to the case of isopods later in this chapter because their taxonomic makeup and bio-geography reveal a potential historical explanation for their contrary pattern.

Not surprisingly, given the sparsity of samples and the earlier assump-tion that latitudinal gradients are unlikely to occur in deep sea, the validity

---

Figure 4.2.   The relationships between diversity, estimated as $E(S_n)$, for isopods, gastropods, and bivalves, and latitude in the Atlantic Ocean and the Norwegian Sea. Sampling localities are shown in Figure 4.1. The major basins sampled are indicated below. Diversities of all three taxa are negative and significant ($P < 0.0001$) functions of latitude in the North Atlantic and the Norwegian Sea. In the South Atlantic, gastropods and bivalves are negative and signifi-cant ($P = 0.025$ and $P = 0.003$, respectively), but isopods are not significant ($P = 0.118$). Re-gression equations and complete statistics are given in Table 1 in Rex et al. (1993). From Rex et al. (1993), with permission of the authors and Macmillan Publishers Ltd.

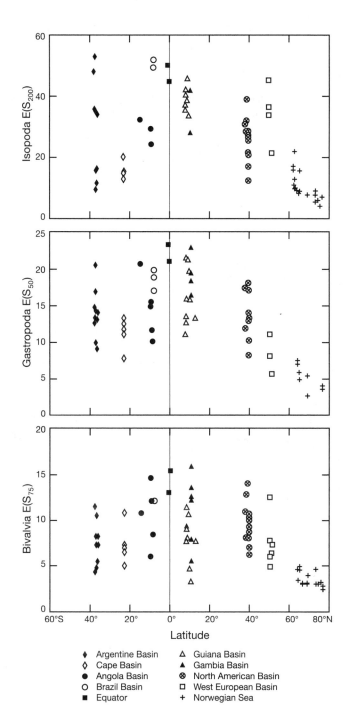

Isopoda E(S_{200})

Gastropoda E(S_{50})

Bivalvia E(S_{75})

60°S   40°   20°   0°   20°   40°   60°   80°N

Latitude

♦ Argentine Basin   △ Guiana Basin
◇ Cape Basin   ▲ Gambia Basin
● Angola Basin   ⊗ North American Basin
○ Brazil Basin   □ West European Basin
■ Equator   + Norwegian Sea

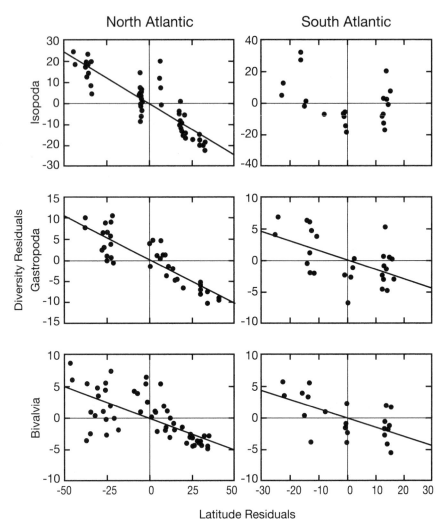

Figure 4.3. Relationships of diversity to latitude with the effects of depth removed by partial regression for isopods, gastropods, and bivalves. All of the relationships with fitted regression lines are significant ($P = 0.0001$ for all taxa in the North Atlantic Ocean and the Norwegian Sea, and $P = 0.003$ to 0.004 for mollusks in the South Atlantic), except for isopods in the South Atlantic. The analysis shows that variation in sampling with depth has no significant effect on the diversity-latitude patterns shown in Figure 4.2. Regression equations and complete statistics are given in Table 1 in Rex et al. (1993). From Rex et al. (1993), with permission of the authors and Macmillan Publishers Ltd.

of these initial results was questioned vigorously. Gage and May (1993) pointed out that $E(S_n)$, the measure of diversity used, is influenced by both the number of species and the evenness of the relative abundance distribution. Might the decrease in $E(S_n)$ at higher latitudes reflect only a decrease in evenness, rather than the decline in richness implicit in established latitudinal gradients? Rex et al. (1993) demonstrated that Pielou's evenness measure, *J*, was uncorrelated with latitude for bivalves and more weakly correlated than $E(S_n)$ for gastropods and isopods in the Northern Hemisphere, where the gradients are clearest. Rex et al. (2000) reanalyzed data for the Northern Hemisphere using the Shannon-Wiener index, *H'*, which is highly correlated with $E(S_n)$ in these samples. *H'* and the number of species in the samples decrease significantly with latitude. Again, *J* was insignificant for bivalves and weakly significant for gastropods and isopods. When the effects of evenness were removed by partial regression, species richness remained highly correlated with latitude. Conversely, when richness was held constant, evenness was insignificant for bivalves and weakly significant for gastropods and isopods. In other words, the latitudinal gradients do appear to represent mainly species richness, and there is an independent statistically subordinate decrease in evenness with latitude for two of the taxa. Rex et al. (2000) also showed that in the Turridae, the largest family of deep-sea gastropods, richness estimated as the number of coexisting species ranges in 3° latitudinal bands showed a strong latitudinal gradient in the eastern North Atlantic. The gradient is significant within depth increments of 1000 m throughout the bathyal zone, suggesting that it is not confounded by uneven depth coverage. Latitudinal gradients detected in the deep sea of the North Atlantic Ocean and the Norwegian Sea do reflect variation in species richness, just as they do in other environments.

Gray (1994) asserted that the significant patterns in the Northern Hemisphere demonstrated by Rex et al. (1993) could be attributed only to low values in the Norwegian Sea and that there was no obvious pattern within the North Atlantic. The trends are, in fact, strongly anchored by low values in the Norwegian Sea (Figure 4.2). This is problematic because low diversity in the Norwegian Sea has often been attributed to special historical circumstances. It was impacted by Quaternary glaciation and experienced catastrophic submarine landslides within the last 6000–50,000 years (Bugge et al. 1988). Its deep-sea fauna appears to be relatively young and shows a strong affinity to the Atlantic fauna (Dahl 1979, Piepenburg 2005), suggesting that it may have been only recently recolonized. It is also a smaller basin that is partially isolated from the Atlantic by the Greenland-Iceland-Faeroe Ridge. Thus, the Norwegian Sea might not represent a simple continuum along an environmental gradient with the North Atlantic and its fauna may have had a different evolutionary development. However, when the Norwegian Sea data are removed from the analysis, the latitudinal gradients in the North Atlantic remain significant for gastropods and isopods, and if two outliers in the Guiana Basin are removed, they are significant for bivalves as well (Rex et al. 1997). The trends are not due solely to depressed diversity in the Norwegian Sea.

The South Atlantic is much less well sampled. All three taxa show pronounced regional variation and only gastropods and bivalves exhibit significant latitudinal trends, albeit weaker than in the North Atlantic (Figure 4.2). There is a long-standing debate about whether latitudinal gradients of diversity exist in the South Atlantic, particularly when the Southern Ocean off Antarctica is included (Clarke 1992, Clarke and Crame 1997, Gray 2001). The seas around Antarctica present an extremely hostile environment and have been explored only recently, with great difficulty. Brey et al. (1994) compared the diversities of isopods, bivalves, and gastropods from the deep Weddell Sea to values shown in Figure 4.2 from the Tropical Atlantic. Diversities were similar, which led Brey et al. (1994) to conclude that there was no latitudinal gradient, but the comparison is methodologically and analytically flawed. The Weddell Sea study combined data from Agassiz trawl samples collected over an enormous area, 8° of latitude and 500–2000 m in depth, to calculate diversity. This regional-scale diversity estimate was then compared to local diversities from individual epibenthic sled samples that each cover around 1000 m² of seafloor. Furthermore, the Agassiz trawl uses much coarser mesh size (10–20 mm) than the epibenthic sled (1 mm). Not surprisingly, the Weddell Sea samples were dominated by megafaunal elements, whereas the sled samples yielded primarily macrofauna. Thus Brey et al. (1994) were comparing diversity estimates based on gear designed to sample different components of the benthos, as well as between areas of seafloor that differed in size by orders of magnitude. Recent intensive sampling in the Scotia Sea just north of the Weddell Sea, using epibenthic sleds modified to better collect small organisms living at the sediment–water interface (see Appendix B), suggests that isopod diversity is high in the Southern Ocean, but no direct statistical comparisons to the South Atlantic were made (Brandt et al. 2007). The more recent ANDEEP (*An*tarctic benthic *deep*-sea biodiversity) intensive sampling program conducted by Angelika Brandt and

Brigitte Ebbe and their colleagues (Brandt and Hilbig 2004, Brandt and Ebbe 2007) is enhancing our knowledge of this region.

Gage et al. (2004) analyzed large-scale patterns in cumaceans (a group of peracarid crustaceans) using 122 epibenthic sled samples, including those taken at the stations shown in Figure 4.1 and additional material collected by French and British expeditions. The data provided considerably better coverage of the eastern North Atlantic. Again, there were far fewer samples collected over a more limited latitudinal range in the South Atlantic. The relationship of $E(S_{50})$ to latitude in both hemispheres is shown in Figure 4.4. A polynomial regression provided the best fit and is significant (including the quadratic term; C. T. Stuart, personal communication), suggesting that diversity is high in the tropics and decreases poleward, although the variability is obviously quite high. The effect of depth was not removed from this oceanwide analysis, but depth coverage seemed fairly even across the full range of latitude considered. The South Atlantic showed no significant latitudinal gradient. In the north, diversity in the Norwegian Sea was severely depressed (see Figure 4.4). The eastern North Atlantic data showed a significant negative gradient with or without data from the Norwegian Sea, but

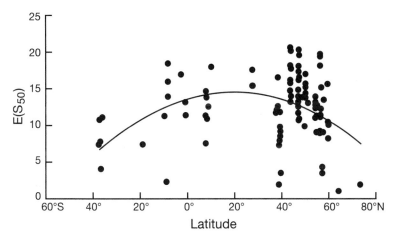

Figure 4.4. The relationship of diversity estimated as $E(S_n)$ in cumaceans to latitude in the North Atlantic Ocean, the Norwegian Sea, and the South Atlantic Ocean. The overall relationship provides a significant fit ($P < 0.01$) to a polynomial regression with peak diversity at tropical latitudes and very depressed diversity in the Norwegian Sea (>60° N). However, the variance in diversity accounted for by latitude is low ($R^2 = 0.155$), and the only region showing a significant pattern is the eastern corridor of the North Atlantic (see text). From Gage et al. (2004), with permission of Inter-Research.

including these data improved the relationship considerably. However, the western North Atlantic, where there was a smaller latitudinal range available, showed no significant pattern. The Berger-Parker evenness measure also showed a strong decrease with latitude in the eastern North Atlantic, indicating that dominance increased poleward as richness declined.

The cumacean results are similar to those for gastropods, bivalves, and isopods in showing less evidence of a latitudinal gradient in the South than in the North Atlantic, although it is important to remember that the South Atlantic is still much less well sampled. In the North Atlantic there appear to be significant regional effects in cumaceans with only the eastern corridor showing a convincing latitudinal decline in diversity and with dramatically depressed diversity in the Norwegian Sea. Moreover, the cumacean data exhibit a lot more variation, so the latitudinal signal, where it does occur, is much weaker than for mollusks and isopods.

Meiofauna

Lambshead et al. (2000) searched for a latitudinal signal in deep-sea nematodes collected from seven sites in the North Atlantic Ocean, the Caribbean Sea, and the Norwegian Sea. The sites varied in depth from 545 to 8380 m and included typical soft-sediment habitats, turbidites, the chronically disturbed HEBBLE site, and the Puerto Rico Trench. The Norwegian Sea showed conspicuously depressed diversity and was eliminated from the main analysis. In the North Atlantic and Caribbean, most of the variance in the number of species (82%) was accounted for by sample size. A small but significant part of the residual variance (8%) correlated with latitude. Interestingly, the relationship of diversity to latitude was positive; that is, diversity increased poleward. Rex et al. (2001) showed that when the effect of depth was removed, nematode diversity was no longer significantly related to latitude (see also Lambshead et al. 2001b). The interesting result is that in the North Atlantic, unlike for the macrofauna, there is no evidence for a simple linear latitudinal trend in deep-sea nematodes. Mokievsky and Azovsky (2002), using a different dataset, showed that regional deep-sea nematode diversity might peak at temperate latitudes. Coastal nematodes also show no evidence of a latitudinal gradient (Boucher and Lambshead 1995, Mokievsky and Azovsky 2002).

Lambshead et al. (2002) reported that abyssal nematode diversity increased along a productivity gradient in the central Pacific. Their analysis included 17 box-core samples taken at four sites (0°, 2°, 5°, and 9° N) along the 140° W meridian and four samples from a site at 23° N, 2300 km to the northwest off the Hawaiian Islands. The series of tropical stations (0°–9° N) represents a decline in productivity extending away from upwelling in the equatorial current system (C. R. Smith et al. 1997). The site at 23° N is located in the oligotrophic North Pacific subtropical gyre, where productivity is low (Karl et al. 1996). The authors predicted that nematode diversity should be positively associated with organic flux to the benthos. This seems quite reasonable because the abyss is clearly a food-limited environment, and (as we pointed out in Chapter 3) a positive relationship between diversity and productivity appears to be a nearly ubiquitous feature of natural communities at relatively low levels of productivity. In the deep sea, the increase in macrofaunal diversity from the abyss to the bathyal zone along a gradient of increasing standing stock has been attributed to an increase in food availability.

The abyssal central Pacific is an interesting venue for examining a relationship between diversity and productivity that is not compromised by large-scale topographic relief. Lambshead et al. (2002) used both the number of species and rarefaction diversity as dependent variables, but preferred the former because it is not potentially confounded by variation in the evenness of relative abundance distributions (cf. Gage and May 1993, Lambshead et al. 2000). Latitude was used for the independent variable as a surrogate for productivity. Both measures of diversity were found to be negatively correlated with latitude, implying a positive association between diversity and productivity. The effect, while significant, was slight (about 20%), with diversities at the oligotrophic site overlapping with those at enriched equatorial sites. Given the relatively short latitudinal span and special localized circumstances of equatorial upwelling, the results do not shed light on latitudinal gradients per se. They are, however, very interesting in that they suggest that nematode diversity does respond to changes in food availability, although this may be difficult to detect. Little is known about how nematode diversity varies with depth (see Chapter 3) or at large spatial scales in the Atlantic, but the same systematic variation with productivity observed in the macro- and megafauna has not been found. Patterns of nematode community structure in the deep sea may be fundamentally different from other groups. Local nematode diversity may be relatively high throughout the deep sea except in isolated basins, unusually stressed habitats, and highly oligotrophic conditions.

Culver and Buzas (2000) showed that deep-sea foraminiferans displayed latitudinal gradients of species richness in both the North and South Atlantic

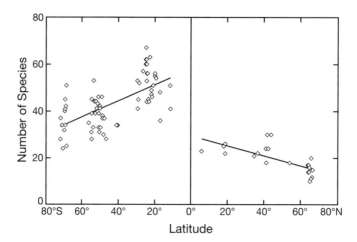

Figure 4.5.   The relationship between species richness and latitude for modern foraminifer-ans collected by quantitative coring devices in the North Atlantic Ocean, the Norwegian Sea, the South Atlantic Ocean, and the Weddell Sea. All samples are lower bathyal from depths restricted to 1995–3736 m. Relationships of diversity to latitude are negative and significant ($P < 0.001$) in both hemispheres. From Culver and Buzas (2000), with permission of the au-thors and Elsevier.

(Figure 4.5). The data extend from the Norwegian Sea to the Weddell Sea. To remove the potential effect of depth, which is important for foraminifer-ans (see Chapter 3), samples were restricted to the 2000 to 4000 m depth range. Their estimates of polar diversity off Antarctica were recently con-firmed by Cornelius and Gooday (2004). Since sample size varied somewhat, Culver and Buzas (2000) also used Fisher's α diversity measure, which is in-dependent of N if species abundance distributions fit a log series, which they do in deep-sea foraminiferans (Hayek et al. 2007). Fisher's α also showed a latitudinal gradient in both the North and South Atlantic. Diversity in the two hemispheres is asymmetrical, being consistently higher in the south. Buzas et al. (2007) recently also found a poleward decrease in deep-sea foraminifera diversity over a 20° latitudinal span in the South Pacific around New Zealand.

There are strong latitudinal shifts in the taxonomic makeup of foramin-iferans that reflect feeding mode and are associated with productivity and car-bon flux to the benthos. Sun et al. (2006) distinguished two foraminiferan assemblages in the North Atlantic. The *Epistominella exigua–Alabaminella wed-dellensis* group consists of species known to rapidly colonize and consume sinking phytodetritus and to undergo rapid population growth. Their relative

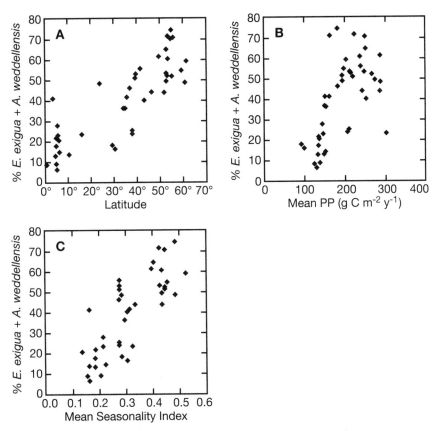

Figure 4.6. The relationships between the percentage of the *Epistominella exigua–Alabaminella weddellensis* foraminiferan assemblage and latitude (A), mean surface productivity (B), and seasonality of surface productivity (C) in the North Atlantic. All of the relationships are significant (*P* < 0.01). *E. exigua* and *A. weddellensis* are species that opportunistically exploit sinking phytodetritus. The increased proportion of this group with increasing latitude is assumed to reflect a poleward increase in high seasonal phytodetritus deposition on the seafloor. Maps of average productivity and seasonality of productivity in the North Atlantic are given in Figure 1.1. From Sun et al. (2006), with permission of the authors and Elsevier.

abundance in the foraminiferan biota increases with latitude and is positively correlated with the mean and seasonality of surface productivity (Figure 4.6; see Figure 1.1 for maps of surface productivity and seasonality from Sun et al. 2006). The other assemblage, the *Globocassidulina subglobosa–Epistominella umbonifera* group, named for its most abundant representatives, consists of nonopportunistic species. It dominates the foraminiferan biota in the tropics and subtropics, and its relative abundance shows negative relationships to latitude, mean primary production, and seasonality—exactly the opposite of the first group. Corliss et al. (2009) have shown that diversity in the deep-sea foraminiferan assemblage decreases with latitude in the North Atlantic and that diversity is a negative function of seasonality in surface production.

HISTORICAL PATTERNS

In Chapters 1 and 2, we discussed short-term changes in the deep-sea fauna at seasonal, interannual, and decadal scales. These were essentially changes in community composition, population dynamics, and reproductive patterns in response to variation in local ecological circumstances linked to episodic changes in surface production or climatic cycles such as El Niño and the North Atlantic Oscillation. Oceanographers long assumed that the deep-sea ecosystem had been stable on geological time scales, and this was a key feature of Sanders' (1968) stability-time hypothesis of deep-sea biodiversity. However, recent advances in paleoceanography show clearly that the deep sea is not immune to effects of climate change and that it has experienced catastrophic environmental perturbations throughout its history. Huge areas of seafloor have been obliterated intermittently by massive submarine landslides (Rothwell et al. 1998), ash from volcanic eruptions (Hess and Kuhnt 1996), and asteroid impacts (Gersonde et al. 1997). Global deep-sea anoxia at the Permo-Triassic boundary may have persisted for 20 million years (Knoll et al. 1996, Isozaki 1997). The Cretaceous-Tertiary mass extinction event resulted in a drastic decrease in surface production and export flux, the effects of which lasted 3 million years (D'Hondt et al. 1998). Rapid warming and oxygen depletion at the Paleocene-Eocene transition resulted in the extinction of 30–50% of benthic foraminiferan species (Kennett and Stott 1991). This more recent event has been studied intensively. Its ultimate causes are still being debated, but appear to involve complex interactions among temperature change, release of methane from the seafloor, reorganization of the deep thermohaline circulation, and shoaling of the calcium carbonate compensation depth (Thomas et al. 2000, Zachos et al. 2005). Kaiho (1994) and Horne (1999) reviewed extinction episodes in the deep sea during the last 100 million years and their potential causes.

A great deal has been learned about the effects of Pliocene and Pleistocene glaciation on the deep-sea environment and the diversity of fossil benthic ostracod and foraminiferan assemblages preserved in seabed cores. Within 10^4- to 10^5-year cycles of the advance and retreat of polar ice sheets, there were significant changes in deep-water temperature and thermohaline circulation that occurred on decadal, centennial, and millennial time scales (Adkins et al. 1998, Oppo et al. 1998, Raymo et al. 1998). These events are linked to surface climate change and presumably involve changes in pelagic-benthic coupling. They were associated with marked changes in the diversity of foraminiferans and ostracods and bathymetric range shifts for individual species (Cannariato et al. 1999, Cronin et al. 1999, Rodriguez-Lazaro and Cronin 1999). Global sea level fluctuated by as much as 120 m during the Pleistocene (Bintanja et al. 2005). Low sea level stands would have exposed much of the shelf and extended the euphotic zone downward to current upper bathyal depths. It was an era of very active canyon formation and submarine landslides, which were undoubtedly major sources of disturbance to the bathyal community.

Recent reconstructions of the deep-sea environment indicate that it experienced significant changes through time, often on surprisingly short scales, and may not even have been continuously occupied by a deep-sea fauna through the Phanerozoic. Far from being an isolated ecosystem, the deep sea appears to be an integral part of Earth's biosphere that is intimately connected to surface climate systems.

For the macrofauna, most evidence of an evolutionary-historical imprint is inferential and based on the geographic distribution of higher taxa and the relationship of local to regional diversity. Very little fossil evidence exists. However, for two meiofaunal groups, foraminiferans and ostracods, a detailed fossil record is available from seabed cores taken by the Deep-Sea Drilling project. The cores contain a continuous sedimentary record that can be precisely dated, and concurrent stable-isotope ratios from the carbonate shells of microfossils provide environmental proxy variables to reconstruct past climates.

Macrofauna

Deep-sea isopods are particularly well studied and have been very informative in regard to morphological adaptation and diversification in the deep sea. As shown in Figure 4.1, isopod diversity appears to peak at temperate latitudes in the western South Atlantic. Wilson (1998, 1999) pointed out that deep-sea isopods consist of two fundamentally different taxonomic groups. Most species belong to the Asellota, a suborder with many endemic families that colonized the deep sea early, possibly during the late Paleozoic or early

Mesozoic; survived mass extinction events; and radiated in situ. Unlike many other taxa, their diversity increases from bathyal to abyssal depths in some regions (see Chapter 3). Members of the large asellot superfamily Janiroidea have highly evolved genitalia that permit mating over a longer part of their lifespan (several instars) than in other isopods where mating is confined to the prebrooding molt. This may be a preadaptation that permits reproductive success at the low population densities found in deep-sea species (Wilson 1991). Flabelliferans, the second group, represent a more recent invasion of the deep sea centered in the Southern Hemisphere. They have not yet evolved endemic families and their diversity decreases with depth. Diversities of both are plotted against latitude in Figure 4.7. The flabelliferans show a south to north decrease in diversity across the entire Atlantic Ocean. Their invasion and spread northward may be partly responsible for elevated isopod diversity in the South Atlantic and a decreasing latitudinal gradient in the North Atlantic.

The relationship between regional and local diversity also depends partly on evolutionary history. There is growing evidence that local diversity in terrestrial (Ricklefs 2004) and marine (Karlson et al. 2004, Witman et al. 2004) environments is not a function of strict local ecological determinism. Rather, local communities appear to be loosely structured systems in which diversity represents a balance between local extinction through biological interactions and density-independent events and continued colonization from the regional species pool. Since the regional pool is ultimately generated by regional- and global-scale adaptive radiation, the connection between regional and local diversity has an historical component. If regional enrichment is important, local diversity should develop to reflect relative regional pool size, resulting in a positive relationship over a broad range of regional diversities (Ricklefs 1987). If the relationship becomes asymptotic, the implication is that local ecological opportunity limits diversity more than regional enrichment.

Stuart and Rex (1994) examined the relationship between local and regional diversity in deep-sea gastropods of the bathyal (500–4000 m) North Atlantic and Norwegian Sea. As we have already discussed, gastropods as a taxonomic class show a latitudinal gradient of local diversity (Figure 4.1). Stuart and Rex (1994) repeated the analysis using only prosobranch gastropods, the largest order, because their dispersal ability can be ascertained from larval shell morphology. Prosobranchs also show a latitudinal gradient (Figures 4.8A). Regional diversity was estimated by combining species lists

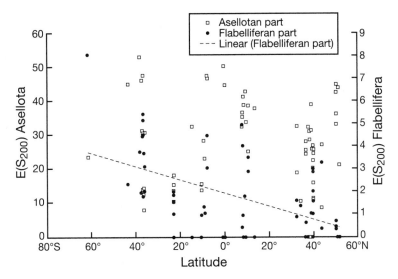

Figure 4.7. The relationship of Asellota isopod diversity (left) and Flabellifera isopod diversity (right), estimated as $E(S_n)$, to latitude in the North and South Atlantic. All samples were collected at bathyal depths (500–4000 m) by epibenthic sleds. Flabelliferan diversity (represented by the dashed line) decreases from south to north across the entire Atlantic Ocean. The Asellota represent an ancient in situ radiation, and the Flabellifera a relatively recent invasion centered in the Southern Hemisphere. From Wilson (1998), with permission of the author and Elsevier.

for individual samples and then normalizing the regional pools to a common sample size. To avoid problems of statistical redundancy and to make local and regional diversities independent, half of the local samples were randomly selected to make up the regional pool and the remaining half was used for local diversity. This procedure was repeated to minimize the effect of sampling error on the relationship. One potential difficulty not addressed is that bathyal faunas show strong depth-related turnover in species makeup (see Chapter 5). Hence, the degree to which all species in the regional pool are adaptively suited to actually participate in all local communities is uncertain.

Regional diversity decreases with latitude and is positively and significantly related to local diversity (Figure 4.8B). Dispersal potential was assessed by determining the percentage of species with planktotrophic (swimming-feeding) development. Nonplanktotrophic (lecithotrophic) species were assumed to have less dispersal ability. The proportion of planktotrophic species does not show a latitudinal gradient, but is regionally specific. Local diversity was significantly predicted by a multiple regression that used regional

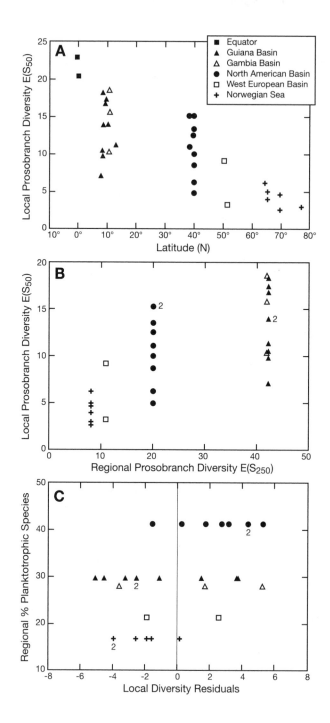

pool size, percentage of planktotrophic species, and depth as independent variables. Regional diversity explains nearly half (48%) of the variation in local diversity, followed by the percentage of planktotrophic species (11%) and depth (9%). All of the independent variables were significant and positively related to local diversity.

When the effects of regional diversity and depth are removed, the residuals of local diversity show an interesting relationship to the percentage of species with planktotrophic development (Figure 4.8C). Where diversity is underpredicted by regional diversity (i.e., local diversity is higher than otherwise expected on the basis of the regional pool, hence positive residuals), there is a higher incidence of planktotrophic development, as seen in the North American Basin. Conversely, where diversity is overpredicted, there is a low incidence of planktotrophy, as observed in the Norwegian Sea. This explains the positive significant role of percentage of planktotrophic species in the multiple regression. In other words, local diversity is augmented where dispersal potential among sites is high and depressed where dispersal potential is low. It is important to note that percentage of planktotrophic species is a crude measure of the dispersal potential of an assemblage, but it is the only one available. It is now known that some archaeogastropods with lecithotrophic development may have an extended demersal swimming phase (Rex et al. 2005a). If so, long-range dispersal is even more prevalent in gastropods. The analysis provides comparative support for regional enrichment, including the essential connection by dispersal between local communities and the regional pool. This is indirect evidence that latitudinal gradients in deep-sea gastropods have, at least in part, an evolutionary-historical explanation.

Figure 4.8.   The relationship of prosobranch gastropod diversity, estimated as $E(S_n)$, to latitude (A) and regional species pool size (B) in the North Atlantic; and the regional proportion of species with planktotrophic (swimming-feeding) larvae to local diversity with the effects of depth and regional diversity statistically removed (C). Local prosobranch diversity is a significant negative function ($P < 0.001$) of latitude and a significant positive function ($P < 0.001$) of regional diversity. Local diversity can be predicted by a multiple regression with regional diversity, the regional proportion of species with planktotrophic development, and depth as significant independent variables. Panel C shows that dispersal potential of the regional species pool, measured as the percentage of species with planktotrophic development, is high where local diversity is underpredicted by regional diversity and low where it is overpredicted, suggesting that dispersal modulates local diversity by linking regional to local diversity. From Stuart and Rex (1994), with permission of the authors and Columbia University Press.

Meiofauna

Hunt et al. (2005) related Quaternary foraminiferan species richness from 10 deep seabed cores taken in the Atlantic and Pacific to proxy variables for temperature and productivity. The benthic foraminiferan accumulation rate in the cores was used as a proxy for productivity and bottom-water temperature curves derived from magnesium-calcium paleothermometry as a proxy for temperature. An example of data from one core with especially good relationships between diversity and environmental variables is shown in Figure 4.9. Species richness is a positive function of temperature and a negative function of productivity. However, among all 10 cores the results were mixed. For each core, a multiple regression was used to predict species richness, with temperature and productivity as independent variables. Both variables showed positive and negative relationships to diversity in the multiple regressions and varied widely in their predictive strength. A meta-analysis of regressions from the cores showed that, overall, temperature had a significant positive association with species richness, but that productivity was insignificant. Temperature was not thought to influence the evolutionary rate because divergence time in foraminiferans appears to exceed the 130,000-year time span analyzed, but it may affect species richness by causing shifts in the bathymetric or geographic distributions of species.

Hunt and Roy (2006) found that the body size of ostracods in seabed cores increased significantly during the last 40-million-year cooling phase of the Cenozoic. Similar changes occurred in multiple lineages, and the amount of change corresponded to size-temperature relationships in geographically distributed modern populations. This suggests that evolution at the level of adaptation within species is related to temperature change in the deep sea, although it remains unclear how much of this phenotypic change is genetically based and how much may be an environmentally induced plastic response.

Cronin and Raymo (1997) documented fluctuations of benthic ostracod diversity through 11 glacial-interglacial cycles of the late Pliocene in a core taken at 3400 m in the North Atlantic (Figure 4.10). Diversity was measured as the Shannon-Wiener information function. Coincident changes in ice volume and temperature were assessed by oxygen isotope ratios in foraminiferan tests and magnesium-calcium ratios in ostracod shells. The oscillation of ostracod diversity matches the orbitally driven 41,000-year cycle of glaciation, being high during interglacial phases and low during ice formation. Diversity correlated positively with temperature and negatively

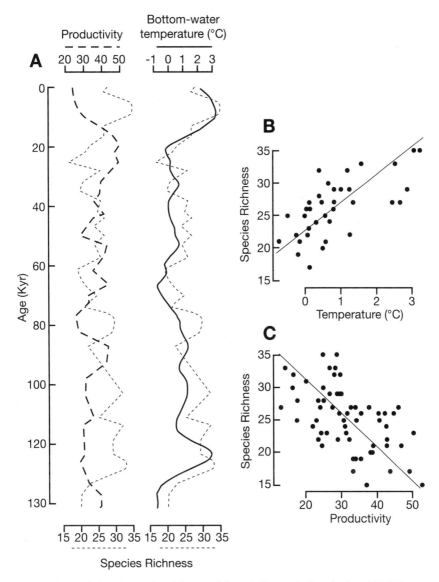

Figure 4.9. Variation in species richness of foraminiferans during the last 130,000 years of the Quaternary in a South Atlantic seabed core plotted along with productivity and bottom-water temperature (A) and the overall relationships of species richness to temperature (B) and productivity (C). Foraminiferan accumulation rate (an estimate of the number of benthic foraminifera deposited cm^{-2} 1000 year^{-1}) is used as a proxy for productivity. In this core, diversity is positively correlated with temperature and negatively with productivity. However, a meta-analysis of 10 cores found a significant association of diversity with temperature, but not productivity. From Hunt et al. (2005), with permission of the authors and Blackwell Publishing.

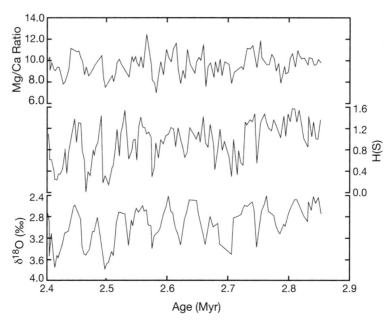

Figure 4.10. Benthic ostracod diversity measured as the Shannon-Wiener information function *H*(*S*) in a North Atlantic seabed core during a 450,000-year period of the late Pliocene, plotted with oxygen isotope ratios, which reflect ice volume, and Mg/Ca ratios, which reflect bottom-water temperature. Oscillations of diversity match the 41,000-year obliquity cycle of Pliocene glaciation. Higher Mg/Ca ratios indicate warmer bottom-water temperature, and lighter oxygen isotope ratios indicate lower ice volume. Diversity is high during warmer interglacial phases with low ice volume and low during colder glacial phases with high ice volume. From Cronin and Raymo (1997), with permission of the authors and Macmillan Publishers Ltd.

with ice volume. The relationships are remarkably consistent through the nearly half-million-year period. The fluctuation in diversity did not represent origination and extinction of species. Species lost during glacial advances had refugia elsewhere in the North Atlantic and recolonized during interglacial phases. Cronin and Raymo (1997) proposed that these cyclical changes in diversity were not caused by temperature, but by variation in food availability that attended large-scale shifts in overhead productivity associated with glacial-interglacial cycles. Productivity and export flux were higher during warm interstadial periods and lower during cold phases (Thomas et al. 1995, Rasmussen et al. 2002), either because of relative ice cover or disruption of overturning circulation (Schmittner 2005).

Deep-sea ostracods also show marked changes in diversity at centennial-millennial time scales in the North Atlantic during the last 20,000-year interglacial period (Yasuhara et al. 2008). Diversity in the normal background community is reduced by as much as half during abrupt surface cooling events that force a reorganization of the deep circulation. Phases of depressed diversity coincide with a temporary intrusion of opportunistic species, suggesting that changes in productivity might be implicated, but the exact proximal causes of diversity shifts are unclear. Yasuhara and Cronin (2008) provide a remarkably detailed record linking changes in deep-sea ostracod diversity to climate change during the last 3 million years.

Thomas and Gooday (1996) were actually able to demonstrate the development of latitudinal gradients in foraminiferans during the Cenozoic by analyzing diversity in cores from the Weddell Sea and equatorial Pacific (Figure 4.11). The Eocene (40 million years ago) marked the onset of global cooling that led to the growth of massive polar ice sheets (Eldrett et al. 2007). Initially, polar and tropical diversities overlapped extensively, but then they

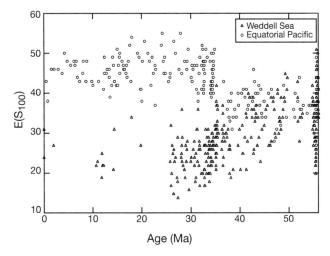

Figure 4.11. Deep-sea foraminiferan species diversity $E(S_{100})$ from seabed cores taken in the Weddell Sea and equatorial Pacific during the post-Paleocene Cenozoic. Beginning with the late Eocene initiation of global cooling, polar and tropical diversities diverged to establish modern latitudinal diversity gradients. Depressed polar diversity is associated with increased dominance of species that opportunistically exploit phytodetritus, suggesting that it was caused by the onset of seasonal productivity at high latitudes. From Stuart et al. (2003), modified from Thomas and Gooday (1996), with permission of Elsevier.

diverged as diversity at high latitudes declined. Thomas and Gooday (1996) attribute this decrease in diversity to the more seasonal productivity that attended cooling and the formation of sea ice. The post-Eocene decrease in high-latitude diversity coincides with an increase in *Epistominella exigua* and *Alabaminella weddellensis,* the same two opportunistic phytodetritus-exploiting species that still dominate the temperate and polar North Atlantic, where their abundance correlates with the mean and seasonality of productivity (Figure 4.6).

POTENTIAL CAUSES

Interbasin variation in deep-sea species diversity is now well established, but whether there is a consistent regular latitudinal signal in this variation remains open to question. In the macrofauna, four major taxa show statistically significant poleward decreases of diversity in the North Atlantic. These patterns are intriguing, but given the irregular geographic distribution of sites and low sampling intensity (e.g., Figure 4.1), they might just represent a fortuitous combination of regional effects. Regional effects are particularly strong in the South Atlantic and Southern Ocean, where latitudinal diversity gradients are weaker and where one group, the isopods, reaches peak diversity at temperate latitudes. Whatever drives interbasin variation in diversity, different macrofaunal taxa appear to respond quite differently. Not only are large-scale patterns taxonomically restricted, but they are known exclusively for bathyal depths. It would be extremely interesting to search for patterns at abyssal depths, but there are currently too few abyssal samples. Among meiofaunal elements, the most convincing latitudinal gradients have been demonstrated for foraminiferans. Nematodes appear to show no evidence of latitudinal gradients, but are much less well sampled. Unfortunately, we know of no large-scale analysis for the megafauna, which would make for an interesting comparison because there are specific theoretical predictions for the relationship of latitudinal gradients and body size (Hillebrand and Azovsky 2001) that have not been tested in deep-sea organisms.

Any explanation of large-scale patterns in this remote environment based on so few data is tentative. One obvious question is: Can the diversity-energy relationships discussed in Chapter 3 for regional-scale bathymetric gradients of diversity be extended to larger scales? Not in a direct way because most larger-scale macrofaunal data come from epibenthic sled samples, so diversity cannot be related to abundance, which is the most reliable measure of local food supply. The meiofaunal diversity data are based on taking subsamples of a constant number of individuals from cores. This procedure standardizes diversity comparisons to sample size, but tells us nothing about differences in abundance among samples. For the macrofauna, it is interesting that the decrease in species richness with latitude in the North Atlantic is associated with a decrease in evenness. Rex et al. (1993, 2005b) suggested that lower evenness could be an ecological response to pulsed nutrient loading from sinking phytodetritus, since surface productivity is higher and more seasonal at higher latitudes (see Figure 1.1). Diversity might be depressed at high latitudes for the same reason that it is depressed at upper bathyal depths and under the other circumstances of pulsed nutrient loading discussed in Chapter 3. The same explanation has been invoked for poleward declines in foraminiferan diversity (Culver and Buzas 2000, Corliss et al. 2009). High-latitude communities are dominated by species known to exploit phytodetritus (Sun et al. 2006), and the same opportunistic species are implicated in the decrease in polar diversity during Cenozoic global cooling and the emergence of seasonal environments (Thomas and Gooday 1996).

There is also some evidence for the role of evolutionary-historical processes in the development of interbasin diversity patterns. For gastropods, the positive relationship between local and regional diversity implies that regional-scale adaptive radiations, which create the regional species pool, ultimately influence local diversity. In isopods, the recent colonization and radiation of flabelliferans in the Southern Hemisphere may have shaped oceanwide patterns of diversity. Foraminiferan diversity was heavily impacted by global anoxic extinction events. Both ostracod and foraminiferan assemblages underwent cyclical changes in diversity during glaciation, although these apparently involved range shifts more than origination and extinction of species.

DEEP-SEA DIVERSITY AT VERY LARGE SCALES

What is the contribution of the deep-sea benthos to total global biodiversity? The answer involves scaling up diversity from local to regional to global levels as a function of the geographic ranges of species. Given the paucity and sparse distribution of deep-sea biological samples (Rex et al. 2006), this is hard to do in a critical way. Our attention was first drawn to this problem by several anecdotal, but nonetheless striking, examples in deep-sea gastropods. Two common species, *Benthonella tenella* and *Xyloskenea naticiformis,* are found throughout much of the Atlantic (Bouchet and Warén 1993, Rex et al. 2002) and in the Pacific off Japan (Hasegawa 2005). The same appears to be true of many species that have high or intermediate abundance somewhere in their range, although the global proportion of widely distributed species is unknown. An extraordinary example is *Palazzia planorbis,* an unusually small (~1 mm) snail, even by deep-sea standards, which occupies a geographic range that includes both the eastern and the western corridors of the North and South Atlantic and a depth range from 241 to 5216 m and is very rare everywhere it has been sampled (Rex 2002). This combination of features places it in the most uncommon form of rarity known in nature (Rabinowitz 1981, Gaston 1994). A core group of 30–40 widespread species make up 90% of the polychaete assemblages at eight abyssal sites in the central Pacific separated by 200–3000 km (Glover et al. 2002). Paterson et al. (1994) found a similar result for abyssal polychaetes in the eastern North Atlantic. The majority of named polychaetes in the deep Southern Ocean either have distributions outside Antarctic waters or are cosmopolitan (Schüller and Ebbe 2007). Some 6–50% of the protobranch bivalve fauna is shared among 10 major basins of the eastern and western North and South Atlantic (Allen and Sanders 1996). Asteroids (Sibuet 1979) and tunicates (Monniot and Monniot 1978) also appear to have broadly distributed elements. These extremely broad distributions suggest that diversity might not build with increasing spatial scale as rapidly as it does in coastal systems, where geographic ranges appear to be more restricted.

Grassle and Maciolek (1992) made the first quantitative attempt to scale up diversity by calculating species accumulation curves for the macrobenthos as a function of the number of individuals collected along the 2100 m isobath off Delaware and New Jersey as part of the ACSAR program. Nine stations with replicate box-core sampling were spaced along a 176 km transect, in total covering about 21 m^2 of seafloor. The most common species were widely shared among the sites. Most species were extremely rare. The rate of species accumulation along the isobath extrapolated to the entire deep-sea floor below 1000 m produced a global diversity estimate of 10^8 species, or perhaps 10^7 since density and diversity were recognized to be lower in the abyss. Even though Grassle and Maciolek (1992) carefully qualified their estimate, it was widely criticized on methodological grounds (e.g., May 1992, Briggs 1994, Bouchet et al. 2002). Perhaps the most parsimonious explanation of their species accumulation curve, considering the ubiquity of common species, is that the apparent addition of rare species with distance along the transect simply represented encountering more rare species that were actually widespread as well, but harder to sample. The actual geographic rate of species accumulation is probably somewhere between this scenario of little accumulation and Grassle and Maciolek's projection of about 100 species per 100 km, especially since in nature common species often have broader distributions than rare species (Brown 1995). In any event, Grassle and Maciolek (1992) asked an important question: How many species are there in the deep sea? It cannot be answered definitively, but new data on the large-scale distribution of species have begun to provide a perspective for considering global levels of deep-sea diversity.

Carney (1997) made an interesting comparison of commonness and rarity of species between deep-sea and coastal faunas that sheds light on how diversity scales up with increased area in the two environments. The same 2100 m dataset of Grassle and Maciolek (1992) discussed above was used for the deep sea and a baseline study on the continental shelf of Texas (Flint and Rabalais 1981, Carney 1995) for the coast. Both studies used roughly comparable quantitative sampling methods and found similar overall diversities (851 and 799 species, respectively). The deep-sea samples had on average about twice the species richness of the coastal samples, which corresponds well with the difference between Georges Bank and the upper bathyal zone south of New England we described in Chapter 2. Carney proposed two definitions of rarity. For a sample, it was the number of singletons, species represented by a single individual. For the entire set of samples, it was the frequency of occurrence: species found in 10% or fewer of the samples were considered rare and those found in 50% or more of the samples were considered common.

Figure 4.12 shows the ordered cumulative frequency of occurrence of species in the species pools plotted against the cumulative proportion of singletons contributed by those species. The dashed lines represented rare and common levels of occurrence among samples. In the deep sea, 73.9% of species were rare and contributed about 25% of the singletons. Common species made up 8.81% of the pool and contributed another quarter of the singletons. The singletons, therefore, are widespread among species having different abundance in the pool. By comparison, 92.1% of coastal species are rare in terms of occurrence among samples and these contribute half of the singletons. Only 0.62% of species are common and they provide only about

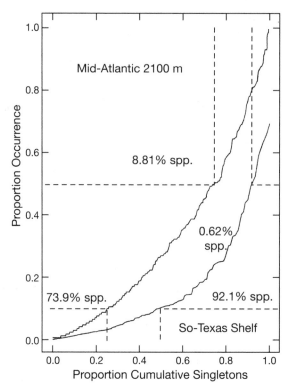

Figure 4.12.   A comparison of commonness and rarity of macrofaunal species in communities on the continental slope off New Jersey and Delaware (left curve) and the continental shelf in the Gulf of Mexico off Texas (right curve). The graph plots the ordered cumulative proportion of species among all samples in each environment against the proportion of singletons (single occurrences within a sample) that they contribute. Species occurring in 10% or fewer of all the samples are considered rare and those occurring in 50% or more of the samples are considered common (dashed lines). In the deep sea, rare and common species contribute about equal proportions (25%) of singletons. On the shelf, half of the singletons are contributed by rare species and a few by common species. Relatively speaking, the implication is that in the deep sea, species rare at one site are common elsewhere in the region. Local diversity is high, but regional diversity is composed of widespread species. By contrast, local shelf diversity is lower and regional diversity represents an accumulation of more heterogeneous local communities. From Carney (1997), with permission of the author and Springer Science.

8% of the singletons. Carney did not explicitly relate these patterns of commonness and rarity to geographical distributions of species. But the general implication is that in the deep sea there is a greater tendency for species that are locally rare to be common at other sites in the region. In other words, the deep sea may have relatively higher local diversity but lower regional diversity. On the shelf, more species are consistently rare and may have more limited distributions, so that diversity builds more rapidly with increasing area. Only 5 of 799 (0.6%) shelf species occurred in half the samples compared to 75 of 851 (8.8%) in the deep sea.

McClain et al. (2008) extended the analysis of diversity and bathymetric ranges in protobranch bivalves (Rex et al. 2005a) to compare the eastern and western North Atlantic (Figure 4.13). Both basins show a high rate of species turnover with depth. A surprising result was that 43% of the species are shared between the basins and that 88% of these have overlapping depth ranges. Even the few shared species with disjunct ranges are separated by less than 1000 m. The proportion of shared species increases from 44% at upper bathyal depths to 60% in the abyss, confirming for the North Atlantic the general finding of Allen and Sanders (1996) that abyssal species tend to be more cosmopolitan. Abyssal endemism in the North Atlantic appears to be low. Two species that would be classified as abyssal endemics in the eastern basin alone are present at bathyal depths in the western basin. Two other species that would be considered abyssal endemics in the western North Atlantic are found at bathyal depths in the western South Atlantic. Fully 88% of abyssal species are range extensions of bathyal species, providing additional evidence for the source–sink hypothesis of abyssal biodiversity we discussed in Chapter 3. These results suggest, at least for this one important macrofaunal taxon in the North Atlantic, that many species are widely distributed even among major basins, particularly at abyssal depths. Rates of species accumulation along geographic axes in one basin or within one depth zone cannot be used to accurately project global diversity because much of the fauna is redundant on large scales.

A similar pattern is found in deep-sea nematodes. Mokievsky and Azovsky (2002) examined changes in marine nematode diversity with depth at two spatial scales: local studies of fewer than 15 samples taken over a few square kilometers and regional studies of 15 to more than 50 samples representing a coastal feature such as a bay or a whole sea (1 to >1000 km^2). Diversity residuals with the effect of sampling area removed are plotted against depth in Figure 4.14. At local scales (Figure 4.14A), deep-sea (>900 m) diversity exceeds shallower (<900 m) diversity, as was also illustrated in Figure 2.5. However, at larger regional scales (Figure 4.14B), deep-sea diversity is less than shallower diversity. As with Carney's (1997) analysis of the macrofauna, the inference is that in coastal environments species have more restricted distributions, and regional diversity increases rapidly by accumulating

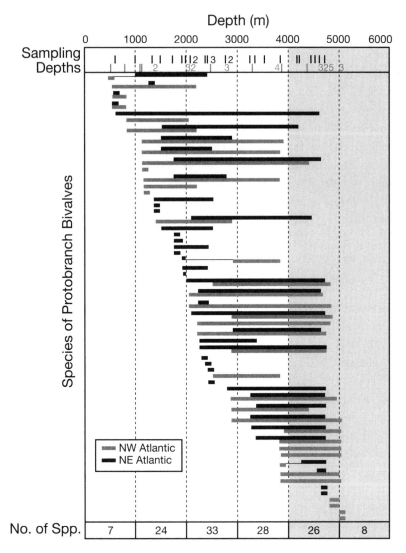

Figure 4.13. A comparison of the depth ranges of protobranch bivalves between the eastern and western North Atlantic based on data from Allen and Sanders (1996). Where ranges do not overlap, they are connected by a thin line. Shown at the bottom is the number of co-existing ranges within 1000-m depth intervals. At the top are depth locations of the samples on which the bathymetric ranges of species are based (ticks are single samples and numbers are multiple samples). Names of individual species and their depth ranges are provided in McClain et al. (2009). The deep-sea protobranch faunas of the eastern and western basins of the North Atlantic overlap extensively. From McClain et al. (2009), with permission of the authors and the University of Chicago Press.

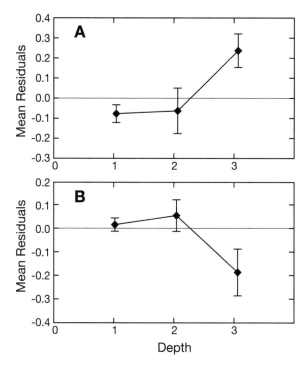

Figure 4.14. The residuals of nematode diversity (with the effect of sampling area removed) plotted against depth (depths 1, 2, and 3 represent <100 m, 100–900 m, and >900 m, respectively) at local scales (A) and regional scales (B). At local scales, diversity at more than 900 m exceeds shallower diversity (see also Figure 2.5), but at regional scales, diversity at more than 900 m is lower. This indicates that whereas local diversity in deep-sea samples is higher than in shallower samples, regional diversity in the deep sea is lower. From Mokievsky and Azovsky (2002), with permission of the authors and Inter-Research.

distinctive local communities. In the deep sea, local diversity is composed more of species that are broadly distributed. Lambshead and Boucher (2003) came to the same conclusion by analyzing local and regional diversity of nematodes collected in the deep equatorial Pacific.

As we have already seen, some foraminiferans, particularly phytodetrital assemblages, are very widely distributed in both space and time. Gooday et al. (1998) compared the species composition of the 20 most abundant species at each of three sites in the Atlantic (Porcupine, Madeira, and Cape Verde Abyssal Plains) and one in the Arabian Sea. Of the 78 abundant species only 17 were endemic to one site, 22% were shared among all four sites, 26% among three sites, and 31% between two sites. More than 65% of the calcareous species collected in the deep Weddell Sea off Antarctica are shared with the Porcupine Abyssal Plain of the eastern North Atlantic (Cornelius and Gooday 2004). Clearly, while local diversity of deep-sea foraminiferans is often quite high, this does not translate into high global diversity since many species are cosmopolitan.

We caution that all of these estimates of faunal redundancy at large scales are based on morphospecies, that is, species designations that rely on similar phenotypic appearance. As we show in Chapter 6, molecular genetic studies have now shown that some deep-sea morphospecies may be complexes of cryptic (sibling) species. To the extent that morphospecies are real species and that the taxa discussed above are representative of the deep-sea fauna, it seems likely that regional diversity is lower in the deep sea than in coastal environments and that estimates of total deep-sea diversity as high as 10^7 or 10^8 will be revised downward.

If deep-sea species are more widely distributed than their shallow-water counterparts, it must mean that dispersal mechanisms are more pervasive in the deep-sea fauna, that there are fewer barriers to dispersal, and that the selective regime is more uniform at large scales. All of these conditions may obtain for some taxa. Chapter 6 includes more information about selective gradients, gene flow, and geographic barriers. Here we briefly summarize dispersal potential.

Early theories about reproduction and larval dispersal in deep-sea benthic invertebrates predicted continuous gamete production, because the environment seemed so constant (Orton 1920), and direct development, because there appeared to be insufficient food at depth for planktotrophic (swimming-feeding) development, and ontogenetic vertical migration of larvae through the deep-water column to reach phytoplankton resources seemed implausible (Thorson 1946, 1950). Bouchet (1976a) first pointed out that many deep-sea gastropods have, in fact, distinctive larval shell morphologies that clearly indicate planktotrophic development. Rex and Warén (1982) showed that the incidence of species with planktotrophic development actually increases with depth in prosobranch gastropods of the western North Atlantic. The same trend occurs in the eastern North Atlantic (Potter and Rex 1992). Planktotrophic larvae of deep-sea snails have been recovered from surface plankton samples (Bouchet 1976b, Bouchet and Warén 1994), and isotopic studies of larval and adult shells show that planktotrophic larvae develop in warm surface waters (Bouchet and Fontes 1981, Killingley and Rex 1985).

Rokop (1974, 1977, 1979) carried out the first deep-sea sampling program specifically designed to assess reproductive patterns in bathyal species. Most showed continuous year-round reproduction, but a minority had annual cycles of gamete production. Rex et al. (1979) analyzed gamete development and size-frequency distributions of individuals in the abyssal snail *Benthonella tenella* (see Appendix A). Populations had an improbable combination of traits, including planktotrophic development, continuous reproduction, and variable and infrequent recruitment, that seemed not to fit neatly into any model of life-history tactics. Somewhat at a loss to explain this, the authors concluded, "It may be no more meaningful to speak of a typical deep-sea reproductive strategy than it would be to typify a single shallow water and terrestrial strategy" (p. 187). Indeed, this turned out to be the case. The 1980s and 1990s saw very intensive investigation of life-history tactics in deep-sea species, most of it conducted by Paul Tyler, John Gage, Craig Young, and their colleagues. This body of work, among the most important and impressive advances in deep-sea ecology, is summarized in Gage and Tyler (1991), Young and Eckelbarger (1994), and a recent comprehensive review by Young (2003).

Deep-sea species do show a wide range of reproductive and developmental modes including internal and external fertilization; seasonal, asynchronous, and continuous reproduction; brooding; direct development in egg capsules; pelagic lecithotrophic development; and planktotrophic development. However, the dominant pattern is continuous reproduction and pelagic lecithotrophic larvae. This is the most important generality for the present discussion about maintaining broad geographic ranges. Pelagic lecithotrophic larvae are provisioned with yolk to undergo development and do not feed. They probably disperse in near-bottom currents. At extremely cold deep-sea temperatures and low metabolic rates they may be able to disperse for around 6 months to 1 year or even longer, especially if they can absorb dissolved nutrients as some larvae do (Jaeckle and Manahan 1989, Hoegh-Guldberg et al. 1991, Shilling and Manahan 1994). Even at very sluggish abyssal current velocities, this translates into potential dispersal distances on the order of 100 km. One important deep-sea taxon, peracarid crustaceans, is an exception because they brood their young. Perhaps early hatched stages are dispersed by near-bottom currents, although, to our knowledge, there is no evidence for this in the deep sea. Some species are epifaunal and swim as adults (Thistle and Wilson 1987). We return to peracarids in Chapter 6 to see what their molecular population structure tells us about dispersal and divergence.

Deep-sea harpacticoid copepods show an unexplained phenomenon known as "emergence," in which adults leave the sediments for excursions up into the benthic boundary layer, where they are exposed to bottom currents (Thistle et al. 2007). Foraminiferans can disperse, much like invertebrate larvae, as a protoculus, the minute first-formed chamber of their test. The protoculus can also remain dormant in sediments, which could increase colonization potential at settlement (Alve and Goldstein 2003). Much remains to be learned about dispersal in the deep sea, but it is becoming clearer that at some stage in their life cycle the vast majority of species have the dispersal ability needed to establish and maintain a broad geographic range.

CONCLUSIONS

On oceanwide scales, there are interbasin differences in species diversity, which indicates that the deep-sea environment is not uniform throughout, as was long assumed. Recently, attention to large-scale patterns has focused on the question of whether latitudinal gradients of species diversity exist in the deep sea. The answer is potentially important for an understanding of latitudinal gradients in general because the deep sea is a unique ecosystem, physically and biologically, where only a subset of factors thought to cause latitudinal gradients can operate. Three macrofaunal taxa (gastropods, bivalves, and isopods) are known to show poleward declines in diversity in the North Atlantic Ocean and the Norwegian Sea. In the South Atlantic, where there has been less sampling, latitudinal patterns are weaker in mollusks, and isopods show peak diversity at temperate latitudes. Cumaceans show an overall pattern of elevated tropical diversity and poleward declines across the entire Atlantic. However, there is no significant independent pattern in the South Atlantic; in the North Atlantic only the eastern corridor, where sampling has been more extensive, shows a significant latitudinal trend. Gastropods, isopods, and cumaceans in the North Atlantic show not only a decrease in species richness at higher latitudes, but also a decrease in the evenness of species relative abundance distributions.

Among the meiofauna, foraminiferans display latitudinal gradients in both the North and South Atlantic and an increased incidence of opportunistic phytodetritus-consuming guilds at high latitudes. The proportion of opportunistic species is positively correlated with both the mean and seasonality of surface production. It remains unclear as to whether nematode diversity shows latitudinal gradients, although on regional scales they do respond to productivity, having depressed diversity under circumstances of extremely low and high trophic input (see also Chapter 3).

Existing data suggest that depressed species richness and increased dominance toward the poles may be related to pulsed nutrient loading from high and seasonal production at higher latitudes, the same process that is associated with depressed diversity in the upper bathyal zone at regional scales (see Chapter 3). However, we stress that knowledge of interbasin- and global-scale patterns of diversity is still very fragmentary, confined as it is to only several macrofaunal taxa and foraminiferans in the Atlantic. It would be especially informative to document patterns in the polychaetes, the largest macrofaunal taxon, and in the megafauna. This could be at least partly accomplished by using archived data and collections. It should also be a consideration when planning the geographic destinations of future sampling programs. Virtually nothing is known about large-scale patterns in the great abyssal plains that constitute most of the deep-sea floor or for any physiographic feature of the Indo-Pacific.

It has also become clear that, contrary to an important assumption in early theories of species diversity, the deep sea has not been a constant environment over geological time scales. The deep-sea fauna has experienced major mass extinction events, and diversities of foraminiferans and ostracods recovered from deep seabed cores vary on orbital, millennial, centennial, and decadal time scales. These changes in community structure are linked to surface climate change, indicating that the deep sea is not an isolated ecosystem immune from outside influence.

At oceanwide scales, evolutionary-historical processes may become more influential in shaping diversity patterns. Isopod diversity decreases from the South Atlantic to the North Atlantic. This may be the result of a recent invasion of taxa in the Southern Hemisphere superimposed on an older oceanwide in situ radiation. In gastropods, a positive relationship between local and regional diversity that is significantly modulated by dispersal potential implies that regional processes, including evolutionary radiation to produce the regional species pool, influence local diversity. Diversity fluctuations and range shifts in foraminiferans and ostracods correspond to cycles of glaciation and their effects on surface production and the deep thermohaline circulation. The origin of latitudinal gradients in deep-sea foraminiferans can be traced to the onset of global cooling in the Cenozoic and is characterized by an increase in opportunistic phytodetritus-consuming species at high latitudes.

Clearly, there is much more to be learned about large-scale patterns of diversity and what the taxonomic makeup of assemblages and the natural history of species can reveal about environmental controls. Current evidence is consistent with the growing consensus among terrestrial and coastal marine ecologists that a confluence of productivity and evolutionary-historical processes are primary determinants of large-scale biogeographic trends. Phylogeographic studies based on molecular genetics, discussed in Chapter 6, have just begun and should reveal how evolution has unfolded in the deep sea and how this process affects large-scale variation in diversity.

Documenting large-scale patterns of diversity is essential to determining the contribution of the deep-sea fauna to global biodiversity. The total diversity of the benthos cannot be estimated by extrapolating local and regional patterns of species accumulation. Distributional patterns of deep-sea species appear to differ fundamentally from those in coastal marine systems. Local diversity in the deep sea can be impressively high, but many of the constituent species appear to have broader ranges, particularly in the abyss, than do coastal species. In deep-sea taxa for which the systematics and biogeography are reasonably well known, much of the fauna appears to be redundant over quite large interbasin scales. Broad geographic ranges may reflect high dispersal ability, fewer effective barriers, and more muted environmental gradients than in coastal systems.

5

BETA DIVERSITY ALONG DEPTH GRADIENTS

We have not yet acquired sufficient knowledge of the factors regulating vertical distribution to be able to divide the different parts of the Atlantic into vertical zones, and a division of this kind will, I fancy, always be more or less a matter of personal opinion.... In any case there is no clearly defined boundary between archibenthal and abyssal areas.

John Murray and Johan Hjort (1912)

On very long timescales, the question arises as to why populations have not adapted to extreme conditions via natural selection and expanded their range. This is an especially relevant question in the deep sea where the distance between 200 and 4000 m bottoms can be relatively short, contain no obvious physical barriers to dispersion and more food is available up slope. In other words, why aren't all species present everywhere, especially across the seemingly homogeneous soft mud bottom between the shelf break and the abyssal floor?

Robert S. Carney (2005)

As we have seen in earlier chapters, there are dramatic changes in standing stock and alpha species diversity with depth. From the earliest exploration of the deep sea, it was observed that species composition also changes with depth (Wyville Thomson 1873, 1878, Agassiz 1888, Murray and Hjort 1912). Initially, interest in bathymetric patterns of species turnover centered on placing biogeographic boundaries at certain depths by identifying what appeared to be homogeneous species assemblages separated by abrupt transitions. Sequences of slow and rapid faunal change were termed "zonation." Recognizable zones were often designated by their dominant megafaunal elements (Le Danois 1948), in much the way that early terrestrial ecologists characterized biomes by a few abundant charismatic species. Elaborate schemes were developed to define and further subdivide zones and to determine whether they were regional phenomena or showed horizontal continuity along isobaths on oceanwide scales (Menzies et al. 1973).

Since the 1960s, much more intensive sampling along depth gradients and a greatly improved ability to collect the whole fauna, including the macro- and meiofauna, have revealed that discrete zones as originally envisioned were largely illusory. Rather, the deep-sea fauna changes continuously with depth, showing some variation in the rate of species replacement (Sanders and Hessler 1969, Haedrich et al. 1980, Carney et al. 1983, Haedrich and Merrett 1990, Carney 2005). The coastal-shelf fauna gives way to a distinctive endemic deep-sea fauna across the shelf-slope break at around 200 m throughout most of the oceans, but deeper toward the poles. Within the deep sea there is a rapid but variable rate of species turnover in the bathyal region (200–4000 m), followed by a more uniformly distributed fauna in the abyss. The most rapid change occurs almost universally at upper to mid-bathyal depths (Carney 2005). Few species are shared between the abyss and the upper bathyal. As we will see, the pattern of continuous depth-related change can be interrupted by oxygen minimum zones, strong bottom currents, and associated changes in food supply and the sedimentary regime, as well as by major topographic features. But generally, on uncomplicated passive margins, species composition changes continuously with depth with only modest variations in the rate of change. Some exceptions, primarily in the megafauna, are noted as we consider the case studies below.

The term "zonation," still in widespread use in the deep-sea literature, is unfortunate because it implies a nonrepeating sequence of slow and abrupt change. In deep-sea soft-sediment habitats, species turnover is nothing like the zonation seen in the more familiar rocky intertidal, where there exist sharply delineated monospecific bands of barnacles, mussels, and algae. The deep-sea patterns are much messier, involving many species with widely overlapping ranges. We prefer to call these changes beta diversity—the rate of species replacement along an environmental gradient (Whittaker 1960).

Beta diversity in the deep sea has been attributed to many biotic and abiotic factors, some of which we review below. However, just as in other ecosystems, changes in species assembly have proven much more difficult to explain than alpha-diversity gradients. What is seldom recognized in deep-sea studies is that beta diversity is closely connected to standing stock and alpha diversity. These are not independent aspects of community structure. Although the exact environmental causes of variation in beta diversity in the deep sea can seldom be determined with confidence, the rate of species replacement itself provides clues to understanding the underlying ecological causes of biodiversity. Just as important, as we show in Chapter 6, beta diversity mirrors the selective gradients that drive population differentiation and speciation, the evolutionary processes that ultimately generate deep-sea biodiversity.

THE PATTERNS

We have already seen some indication of depth-related beta diversity in the compilation of species depth range data in Figures 3.3, 3.6, 3.19, and 4.13. Beta diversity is more typically represented by multivariate statistical methods that resolve the variation in distributional data into a few abstract dimensions. The most common of these is cluster analysis (e.g., Haedrich et al. 1980), which is based on a matrix of faunal similarity among sampling stations. The patterns of similarity in the matrix are summarized visually in a branching diagram that is constructed by sequentially extracting samples with the highest similarity and then recalculating the matrix based on similarity relationships of remaining samples to the extracted groups. The resulting cluster diagram shows the overall hierarchical pattern of similarity among samples. The groups of samples formed (i.e., the clusters) consist of samples more similar to one another than to members of other clusters.

Interpreting cluster analyses requires considerable caution. The cluster diagram reduces a highly complex multivariate pattern of faunal similarity to just two dimensions (three actually, since the clusters rotate on their stems like a mobile sculpture). Hence, there is a great deal of distortion of similarity relationships among samples. Both the coefficient for calculating the similarity matrix and the algorithm used to generate the cluster diagram can have profound effects on the grouping of samples. Basing the coefficient of similarity on the simple presence or absence of species in the samples or incorporating relative abundance data can also have a very significant impact on the outcome, especially when abundance distributions are uneven.

A variety of ordination methods have also been used, including factor analysis (e.g., Rex 1977), reciprocal averaging (e.g., Hecker 1990b), correspondence analysis (e.g., Weisshappel and Svavarsson 1998), and nonmetric multidimensional scaling (e.g., Howell et al. 2002). These techniques are all based on some measure of calculated similarity among samples and suffer from many of the same interpretational problems as cluster analysis. They do, however, have the advantages of being able to portray patterns of similarity in several dimensions, making it possible to detect more subtle variations in beta diversity, and of identifying the species or ecological factors most responsible for forming the resulting groups.

The wide range of cluster and ordination methods used in deep-sea studies makes it very difficult to compare taxa, size categories, and regions in a critical objective way. In our view, many analyses are overinterpreted, especially those that attempt to define "zones" and "boundaries." Multivariate methods are specifically designed to identify groupings, and they do so even when patterns are fairly gradual as long as there is some variation in the rate of species turnover. Often not enough attention is paid to the level of similarity within and among groups. In cluster analysis, very low levels of association within clusters cause cascading of samples into the next cluster, presenting the appearance of two groups and an intervening boundary when the fauna is continuously and rapidly changing. Moreover, cluster and ordination methods are extremely sensitive to the sampling design. Spatial gaps in sampling along depth gradients will create what appear to be discontinuities in the analysis when change is actually gradual simply by interrupting the sequence of species replacement.

Differences in sampling intensity are also important. For example, more samples at one depth will inevitably yield more species, and the resulting differences in alpha diversity constrain calculated faunal similarity between well-collected and less well-collected areas. Suppose, for example, that thorough sampling at one depth provided 100 species and that less sampling at another locality yielded only 50 species. Even if these 50 species made up a perfect subset of the 100 species, maximum calculated percentage similarity can only be 50%, even though the two faunas may actually have identical species lists when adequately characterized by the same sampling intensity. Transitions in beta diversity can also inflate alpha diversity by generating mass effects: diversity is elevated by sharing many species through dispersal among adjacent faunas. The deep-sea literature is replete with such cases, mainly because of the logistical difficulties of equable or random sampling and the use of gear with different sampling efficiencies.

Even given all of these problems of sampling, analysis, and interpretation, there are some excellent and very informative case studies of beta diversity in the deep-sea benthos. In the pages that follow we review examples for the mega-, macro-, and meiofauna, again relying on well-sampled regions of the Atlantic and adjacent seas and on studies that minimize the sampling and methodological pitfalls.

Megafauna

Rowe and Menzies (1969) documented an exceptionally clear case that we use here to illustrate the basic nature of beta diversity in the deep sea and its relationship to standing stock and alpha diversity (Figure 5.1). They used bottom photography to assess actual population densities of megafaunal invertebrates between the shelf-slope break and 5500 m off North Carolina. Individual species do occupy restricted depth ranges; however, widths of ranges and levels of abundance within the ranges vary considerably. Although there is some indication of a more rapid change in species makeup at around 1000–2000 m, there are no detectable "zones" in the classical sense. There is also no evidence of related species with similar lifestyles ecologically displacing one another along the depth gradient—a pattern often associated with competitive resource partitioning in terrestrial and coastal systems. For example, four anemones (Species 5–8) in the upper bathyal, four sea urchins (13–16), and four brittle stars (18–21) in the mid-bathyal have widely overlapping depth ranges. Abundance decreases with depth in the megafauna (Figure 5.1), a global trend that we described in Chapter 1. Alpha diversity, calculated as the number of overlapping ranges within 1000-m depth bins (Figure 5.2) increases to a peak at around 1000–2000 m and then decreases sharply toward the abyss. Only two species, an ophiuroid (21) and a sea pen (11) hang on at low density in the abyss. Much of the abyssal seafloor supports no epibenthic megafaunal species that are visible in the photographic survey. This simple plot of species abundance shows much that is often obscured in more complex multivariate analyses.

Hecker (1990b) conducted what remains the most insightful survey and analysis of beta diversity in the deep-sea megafauna. She carried out four photographic transects that documented epifaunal invertebrate and demersal fish abundances at slope depths in the western North Atlantic. The transect sites differed in topography and surficial geology (Figure 5.3). The USB region south of Georges Bank is steep and the upper slope is littered with glacial erratics and exposed rock outcrops. The other three sites are located to the west along the slope south of New England. One (WS) is steep and has similar rocky geological features but located at greater depths; the two others (WT and ES) have gentler slopes and contain primarily soft-sediment

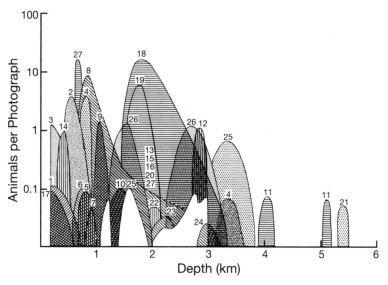

Figure 5.1. Population density distributions of megafaunal invertebrate species off North Carolina (United States) assessed by photographic survey. Numbers indicate individual species, the names of which are provided in Rowe and Menzies (1969). The density of the megafaunal community as a whole decreases with depth. Species have broadly overlapping depth ranges and replace one another with depth in a gradual progression. From Rowe and Menzies (1969), with permission of the authors (G. T. Rowe) and Elsevier.

habitats. At all four sites, slopes are steeper in the middle sections and more gradual in the upper and lower sections. The survey made it possible to examine both vertical (~300–2000 m) and horizontal (~400 km) changes in community composition and to relate these changes to substratum type, slope declivity, currents, and food availability.

Variation in faunal makeup is shown in Figure 5.4. The cluster diagram represents the overall pattern of faunal similarity within and among sites. The identities of samples, in terms of depth and location, are indicated to the left. In this case, samples are the combined fauna of 100 m depth bins along the transects. Most of the variation in species composition is related to depth. Contiguous clusters overlap in species composition, indicating a continuous change from the upper to the middle to the lower slope. With increasing depth, the clusters become tighter: samples within (and among) clusters show higher similarity, meaning that the fauna becomes more uniform. However, there is also interesting horizontal variation among sites. The upper middle slope of USB (3) and elements of its upper slope (1a,c) and

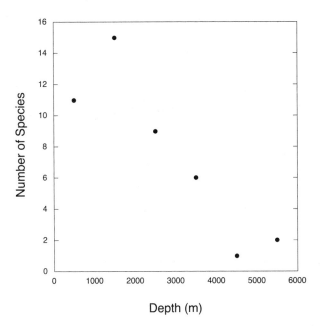

Figure 5.2.   Species richness measured as the number of coexisting depth ranges in 1000 m depth bins estimated from the density distributions in Figure 5.1. Diversity peaks at 1000–2000 m and then decreases rapidly toward the abyss.

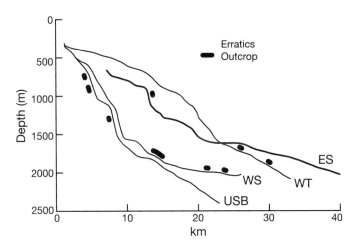

Figure 5.3.   Topographic profiles for four photographic survey transects sampled by Hecker (1990b) on the continental slope south of Georges Bank (USB) and New England (ES, WS, WT east to west, respectively). The plots show variation in slope declivity and the presence of rock outcrops and glacial erratics. Beta diversity of megafauna observed along the four transects is shown in Figures 5.4 and 5.5. From Hecker (1990b), with permission of the author and Elsevier.

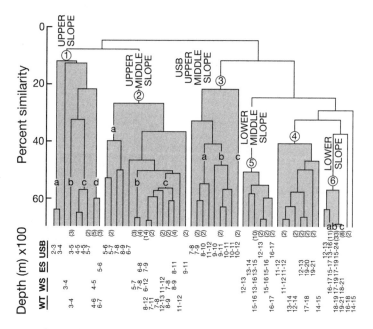

Figure 5.4. Cluster diagram showing the pattern of beta diversity of megafauna (invertebrates and fishes) along the four photographic survey transects shown in Figure 5.3. The diagram is based on the percent similarity coefficient and the unweighted pair-group clustering algorithm. On the left, under the transect headings, are the samples (100 m depth intervals) that are included in the clusters. In general, species replacement with depth is gradual and species similarity among samples increases with depth. However, parts of the steep rocky UBS transect (1a,c; 2a), at upper and mid-slope depths, cluster out separately because of a shift in dominance of some species. From Hecker (1990b), with permission the author and Elsevier.

upper middle slope (2a) cluster out separately because of a shift in dominant species between the steep rocky area and the other three transects, none of which show conspicuous faunal differences (i.e., their participation in the clusters in Figure 5.4 is mixed throughout the slope).

The patterns can be viewed in three dimensions by reciprocal averaging (Figure 5.5). The first axis, which accounts for the most variance, represents depth. The continuous change along the depth gradient from the upper to the lower slope in western sites is reflected in how the samples are arrayed sequentially along the axis. Axis 2 includes depth and locality. Reflections on Axis 2 clearly show the displacement of the USB middle upper slope from the western transects. Axis 3 separates gradually sloping (high values) from steep (low values) terrain. More pronounced changes in composition with

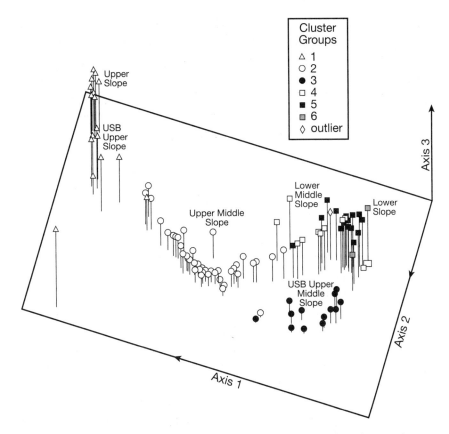

Figure 5.5. Ordination by reciprocal averaging of samples from the four photographic survey transects of megafauna shown in Figure 5.3. The cluster groups (upper right) refer to Figure 5.4. Most of the variation in species turnover is depth related, represented by the reflections of samples on Axis 1. Horizontal variation in species composition is shown by the separation of USB upper middle slope samples on Axis 2. Axis 3 differentiates flatter areas from steeper areas (see Figure 5.3). Depth-related change is gradual, but more rapid where slope declivity changes. From Hecker (1990b), with permission of the author and Elsevier.

depth coincide with changes in slope declivity and shifts in animal abundance. In general, turnover rates were higher in steep areas and lower in more gradually sloping areas, as can be seen by the convex relationship of western samples to depth. Samples show a greater rate of change (are more dissimilar) at intermediate depths, where slopes are generally steepest (Figure 5.5) and group more tightly at upper and lower slope depths because they are more similar. This suggests that the environmental gradients driving species turnover are a function of the rate of change in depth. The pattern of rapid species replacement of megafauna at slope depths shown for this region by Hecker (1990b) accords well with submersible observations (Grassle et al. 1975) and trawl studies (Haedrich et al. 1975, 1980) in the same area. Haedrich et al. (1980) extended the analysis of beta diversity to lower bathyal and abyssal depths (2500–5000 m), where they found a more impoverished (see Figure 3.18) and redundant fauna, again suggesting greater faunal uniformity at greater depths.

Hecker (1990b) suggested that the changes in species assembly associated with slope topography might ultimately be related to food supply. The higher faunal densities observed on the upper and lower slopes were partly attributable to an increase in suspension feeders. Near-bottom current velocities are higher where the grade is less, and this may augment the POC flux necessary to support suspension feeders. In addition, sediments are coarser on the upper and lower slopes.

Hecker (1994) later surveyed seven parallel depth transects from 157 to 1924 m spaced along a 50 km stretch of continental slope off Cape Hatteras. As we discussed in Chapter 1, this region is a depocenter that supports unusually high faunal densities. It is primarily a soft-sediment habitat, but the topography can be complex with exposed clay outcrops and evidence of active sediment slumping. The fauna differs from south of New England in that four normally rare species (three fishes and an anemone) become dominant at middle slope depths. A novel feature is the megafaunal foraminiferan *Bathysiphon filiformis* (see Appendix A), which is quite common here but not observed in the New England transects [except in certain canyons (B. Hecker, personal communication)]. Otherwise, species composition of the slope fauna was similar between the two regions, which are separated by roughly 550 km. Hecker suggested that the unusual abundance of several species off Cape Hatteras was caused by high POC flux. The anemone and *Bathysiphon* are deposit or suspension feeders, and the three fishes may benefit from more abundant infaunal prey populations.

The basic pattern of faunal change off Cape Hatteras is shown in Figure 5.6. The figure does not include upper slope areas (<400 m), which had low-density patchy faunas and were not sampled in some of the transects. Again, most variation in species composition is associated with depth, represented by Axis 1. There is a continuous fairly rapid turnover from the upper middle to the lower slope. In the upper middle slope there is also some variation with location (Axis 2), especially noticeable by the displacement of

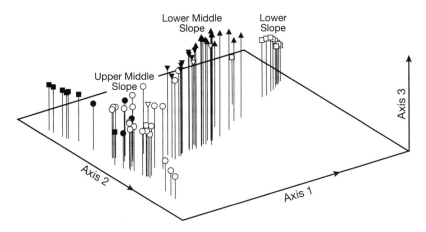

Figure 5.6.   Ordination by reciprocal averaging of megafauna along seven parallel photo-graphic survey transects at continental slope depths off Cape Hatteras. Most variation in species composition is depth related (Axis 1). At upper middle slope depths there is some hor-izontal variation in species composition associated with the dominance of quill worms (low values on Axis 2) and clay outcrops colonized by suspension feeders (high values on Axis 2 and low values on Axis 3). The general pattern is one of continual change with depth and more idiosyncratic variation at shallower depths, suggesting a more heterogeneous environ-ment. From Hecker (1994), with permission of the author and Elsevier.

samples from the northernmost of the seven transects (solid squares), the only site dominated by the quill worm *Hyalinoecia artifex* (see Appendix A). Depth segments with outcrops colonized by sessile epibenthic suspension feeders had low reflections on Axis 3 (open circles). Despite the shift in relative abun-dance of some species, again apparently the consequence of food supply, the basic pattern of beta diversity off Cape Hatteras is similar to that found south of New England. Most change is depth related. The horizontal component of variation is found primarily at upper to upper middle slope depths, with more faunal consistency on the lower slope.

Howell et al. (2002) analyzed beta diversity of asteroids, an important megafaunal component, in the Porcupine Seabight and in the Porcupine Abyssal Plain of the eastern North Atlantic. The Porcupine Seabight is an amphitheater-shaped embayment in the continental margin (Figure 5.7) about 200 km southwest of Ireland (Rice et al. 1991). The slope in the Seabight is fairly gentle to around 3000 m and then descends more steeply to a narrow opening (100 km) to the abyss (Billett and Hansen 1982, Rice et al. 1991). The upper northeastern flanks (600–900 m) support banks and mounds of living and dead cold-water corals (*Lophelia pertusa* and *Madrepora*

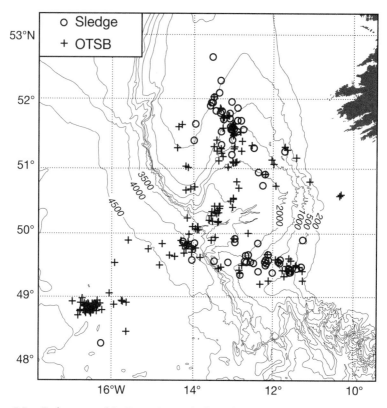

Figure 5.7. Bathymetry of the Porcupine Seabight and Porcupine Abyssal Plain showing the location of epibenthic sled (sledge) and trawl (OTSB) samples used to assess beta diversity in asteroids. From Howell et al. (2002), with permission of the authors and Elsevier.

oculata). At 1000–1300 m there are dense aggregations of the hexactinellid sponge *Pheronema carpenteri* (Rice et al. 1990). Both corals and sponges have patchy distributions and coexist with other fauna. Their restricted depth distributions are perhaps the closest thing to the classic concept of "zones," at least for the megafauna. Both the corals and the sponges are suspension feeders (albeit using quite different mechanisms), and their depth distributions are probably controlled by local hydrography and food supply. The east wall of the Seabight is steep and carved into a more rugged terrain by the Gollum Channel system, which consists of small vertically oriented channels that coalesce downslope into a deep trough extending to the mouth of the Seabight. As we noted in Chapter 1, the lower Seabight and the abyssal plain are the sites of periodic dense concentrations of holothurians related to phytodetritus deposition (Billett and Hansen 1982, Billett et al. 2001).

The depth distributions of the asteroids that Howell et al. (2002) studied were shown in Figure 3.19. Referring back to this figure, we can see that the asteroid fauna changes gradually with considerable overlap of species ranges. Most species have depth ranges of about 1000 m, with their densest populations centered within a narrower range of about 200–300 m. One conspicuous feature is the apparent faunal hiatus at around 3000–3500 m, where there are few co-occurring species. This region is steep and rocky, making trawling difficult, and represents a sampling gap in the dataset. Another sampling gap occurs at 4000–4500 m; this one, however, is not reflected in the depth distributions, possibly because abyssal species are more broadly distributed.

The overall pattern of faunal change with depth is summarized in Figure 5.8. The reflections of stations on the first nonmetric multidimensional scaling axis (a measure of faunal similarity) are plotted against depth. The gradual pattern of change is quite evident. Moreover, the rate of change

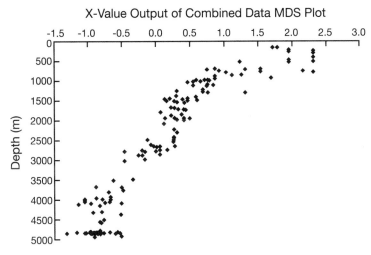

Figure 5.8.   Reflections of the asteroid samples shown in Figure 5.7 on the first axis of nonmetric multidimensional scaling (MDS) plotted against depth. The MDS axis represents species composition in multivariate space. See Figure 3.19 for depth distributions of the species. The plot shows that species turnover is gradual and continuous with depth. The rate of change (difference between samples on the MDS axis) decreases with increasing depth. From Howell et al. (2002), with permission of the authors and Elsevier.

varies with depth, being greatest at upper to middle slope depths (i.e., a larger difference of sample scores per depth difference) and less at greater depths. Again most variation in species makeup is depth related. Surprisingly, no difference in composition was detected between eastern and western sectors of the Seabight, indicating that horizontal variation is minimal despite the differences in topography. Predatory and scavenging species dominate the asteroid assemblages at upper to middle slope depths, but give way to deposit feeders toward the abyss. This suggests that a decrease in food supply with depth has a part in driving beta diversity since carnivores occupy higher trophic levels that only exist upslope, where there is sufficient energy to support them. Howell et al. (2002) also speculated that water mass structure might be implicated in the change in species composition, as it seems to be in the distribution of coral and sponge species noted earlier.

Macrofauna

Olabarria (2005) analyzed the beta diversity of bivalves in the Porcupine Seabight and Abyssal Plain. Depth distributions of species were shown in Figure 3.6. A gradual pattern of species replacement with depth is evident, with most species having very broad overlapping ranges. The apparent discontinuities at 2000–2500 m and 3100–3500 m are actually sampling gaps. The abyssal fauna is largely composed of range extensions of bathyal species living at low density. Nonmetric multidimensional scaling identified five groups of stations based on similarity of species composition (Figure 5.9). The D1–D9 designations for stations in the plot indicate sequentially deeper 500-m depth increments. The larger groups are loosely organized; Group A includes stations from D1 to D9 (763–4510 m), and Group B stations from D5–D9 (2857–4866 m). Most of the variation in species makeup is depth related, as evinced by the roughly left-to-right arrangement of progressively deeper samples. However, horizontal variation is also quite significant in this case, as represented by the displacement of groups C and D, because of geographically restricted ranges in several species.

The spatially complicated pattern of faunal similarity is shown in Figure 5.10. The geographic distribution of the five groups identified by nonmetric multidimensional scaling (Figure 5.9) can be superimposed on the bathymetry of the Seabight shown in Figure 5.7. Though largely arrayed according to depth, some groups occupy a broad horizontal expanse within the Seabight and interface in an irregular jigsaw puzzle fashion. Most of the

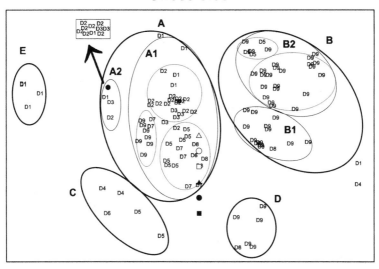

Figure 5.9. A nonmetric multidimensional scaling (MDS) plot of bivalves collected in the Porcupine Seabight and Abyssal Plain. See Figure 3.6 for depth distributions of the species. The plot identifies five groups of samples (A–E). The D1–D9 designations indicate sequentially deeper 500 m depth bins in which the samples were collected. Most variation in species composition is associated with depth, but there is also a complicated pattern of horizontal variation represented by overlap of the groups on the two axes, as shown in Figure 5.10. From Olabarria (2005), with permission of the author and Elsevier.

spatial overlap of groups occurs in the bathyal region, which implies that faunal affinities within the Seabight depend to some extent on environmental factors that are not strictly depth related. Olabarria (2005) suggested that the complicated hydrography and topography of the Seabight and seasonal phytodetrital deposition act to distribute food resources in a spatially heterogeneous fashion.

Blake and Grassle (1994) analyzed macrofaunal beta diversity from box cores collected along four transects off the Carolinas as part of the ACSAR program. The transects ranged in depth from 600 to 3500 m and spanned 500 km along the continental margin. Two of the transects, off Cape Fear and Cape Lookout, are in fairly typical soft-sediment habitats similar to those south of New England. A third, off Cape Hatteras, contains a major depocenter at upper slope depths, as discussed in Chapter 1. Upper slope habitats in the fourth transect, off Charleston, are sandy and swept by the Gulf Stream.

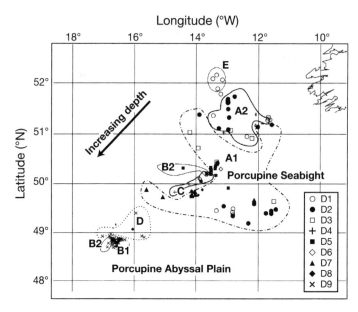

Figure 5.10. The faunal groups of bivalves shown in Figure 5.9 plotted geographically in the Porcupine Seabight and Abyssal Plain (compare spatial distribution to the topography in Figure 5.7). Note that the groups formed by MDS are oriented mostly by depth, but they interface in a spatially complicated way that indicates a horizontal component to variation in species composition. From Olabarria (2005), with permission of the author and Elsevier.

Thus upper slope depths among the transects present a very heterogeneous environment at regional scales.

The patterns of species similarity revealed by reciprocal averaging ordination are shown in Figure 5.11. Axis 3 corresponds to depth. A conspicuous departure from the depth trend is Station 9 at 600 m off Cape Hatteras, with an Axis 1 score that is much higher than those for the other stations at 600–1000 m. Station 9 is directly in the depocenter, where POC flux is unusually high, so it has a much higher density than any other station and depressed alpha diversity; moreover, it is dominated by species more typical of the continental shelf than the slope. Station 10, also off Cape Hatteras, occupies a different position in ordination space than the other stations at 2000 m. It has much higher density and lower diversity than other lower slope stations and probably experiences an increased deposition rate from downslope transport. Station 14A from the upper slope off Charleston is differentiated by its high score on Axis 2. Located in a sandy environment with strong current activity, it has very low density and depressed diversity. Thus,

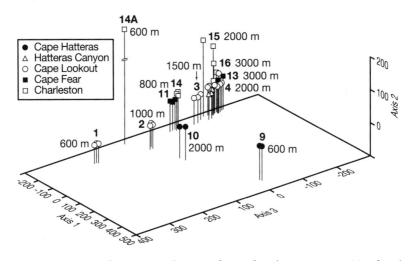

Figure 5.11. Reciprocal averaging ordination of macrofaunal species composition from box cores collected along four transects (and Hatteras Canyon) off the Carolinas. Most of the change in species composition is depth related, as represented in the array of samples along Axis 3. Departures from the main depth trend are largely upper bathyal, including Station 9 at 600 m (high reflection on Axis 1), which is in a major depocenter, and Station 14A at 600 m (high reflection on Axis 2), which occurs in a sandy environment with strong currents. From Blake and Grassle (1994), with permission of the authors and Elsevier.

although there is a continuous and gradual pattern of species replacement at middle to lower slope and upper rise depths, upper slope samples show strong variation in dominant species associated with differences in nutrient loading, currents, and sediment type.

Svavarsson et al. (1990) documented asellote isopod species composition from 25 epibenthic sled samples collected from 800 to 3700 m in the Norwegian Sea. Their results were presented as a cluster analysis based on presence and absence data with similarity estimated by Euclidian distance (Figure 5.12). The authors detected three basic clusters, one composed of stations at about 800 m, a second of those between 1000 and 1500 m, and a third of those between 2000 and 3700 m. The strongest transition (highest distance) is between 800 and 1000 m. Note that none of the linkages among stations approach one, which would indicate no similarity. This is because three species span the entire depth range and several others are very widely distributed, perhaps reflecting a tendency for high-latitude species to be more eurybathic. The authors suggest that the weaker division between the second and third clusters may represent a 524 m sampling gap. Below

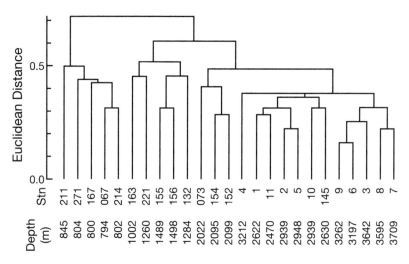

Figure 5.12. Cluster diagram of isopod species composition with depth in the Norwegian Sea. The diagram is based on faunal similarity measured as Euclidean distance applied to presence and absence data and a farthest-neighbor average linking algorithm for clustering. The diagram reveals a major transition between 800 and 1000 m followed by a gradual and continuous change with depth and higher faunal homogeneity at greater depths. From Svavarsson et al. (1990), with permission of the authors and Elsevier.

1000 m there is essentially a gradual increase in similarity among samples, indicated by progressively lower distance couplings between stations.

Isopod alpha-species diversity simply decreases monotonically across the depth gradient from about 20 to fewer than 10 species (Svavarsson et al. 1990). The entire regional pool is only 40 species, fewer than for some individual samples in the Atlantic, which again illustrates the very depressed diversity of the Norwegian Sea. The deeper assemblage is mainly an attenuation of the upper bathyal fauna. Only six new species (15%) are encountered below 1500 m and only one (3%) below 2500 m. Similarity is higher among deeper samples because they share a relatively small subset of upper bathyal (800–1500 m) species. The difference in alpha diversity between the upper and lower bathyal limits calculated similarity. Accepting that the low linkage between 1500 and 2000 m may simply represent a sampling gap, as the authors point out, we see that what the cluster diagram (Figure 5.12) shows is a convincing region of rapid change between 800 and 1000 m followed by a continuous and fairly steady turnover from 1000 to 3700 m, which essentially represents a progressive reduction of the upper bathyal fauna.

Bett's (2001) study of the bathyal macrobenthos north and south of the Greenland-Iceland-Faeroe Ridge revealed very interesting differences in beta diversity on opposite sides of a major topographic feature that strongly influences hydrography. The patterns of alpha diversity were seen in Figure 3.10; diversity increases with depth in the Atlantic south of the ridge, but decreases north of the ridge in the southeastern Norwegian Sea. Figure 5.13 shows an ordination of samples by using nonmetric multidimensional scaling, with the position of the ridge indicated. The sill depth of the ridge in this sector is about 200–600 m. It separates two fundamentally different water masses. Relatively warm Atlantic water lies to the south, flows over the ridge, and extends northward. On the north side, this warm water overlays extremely cold Norwegian Sea Deep Water. The boundary between warm and cold water is dynamic and occurs at around 400–700 m. At shallow depths (<500 m) both sides of the ridge share a common fauna (labeled A in Figure 5.13). Below 500 m the physical barrier of the ridge separates distinctive Atlantic (B, C) and Norwegian Sea (D, E, F) faunas. On both sides, faunal change is largely depth related, although at intermediate depths (500–1200 m) there is some regional distinction of faunas collected west and north of Shetland (D, E).

Bett (2001) suggested that the major factor responsible for the difference between Atlantic and Norwegian Sea fauna is temperature; the similarity in the shallow faunas (<500 m) is associated with a common relatively warm water mass, and the faunal difference below 500 m corresponds to a difference in temperature between southern (warm) and northern (cold) flanks of the GIF Ridge. The clear association of faunal composition with bottom-water temperature on the north side is shown in Figure 5.14. The reflections of stations are plotted against depth for the x- and y-axes of a nonmetric multidimensional scaling analysis. Stations are labeled warm (W, minimum > 0°C), cold (C, maximum $\leq$ 0°C), and intermediate (I). Reflections on the MDSx-axis show the rapid faunal transition across the thermocline at 400–700 m. Reflections on the MDSy-axis show the continuous gradual faunal change with depth within both warm and cold regions. It is difficult to know precisely from the analysis how much of the difference between northern and southern faunal assemblages is due to different species or to the very substantial differences in alpha diversity. There is certainly an endemic element to the Norwegian Sea fauna (Bouchet and Warén 1979).

There are also potential differences in food supply north and south of the GIF Ridge. Thick phytodetritus coverage on the seafloor is commonly ob-

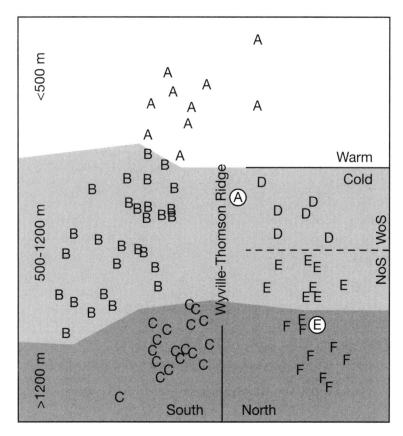

Figure 5.13. Nonmetric multidimensional scaling of samples of macrobenthos north and south of the Wyville-Thomson Ridge (called the Greenland-Iceland-Faeroe Ridge in the text), with the position of the ridge, depth, and temperature of bottom water indicated. NoS and WoS refer to north and south of Shetland, respectively. The letters A–F refer to groups in a cluster analysis of faunal similarity (not shown). Circled sites are misclassified (i.e., samples that fall outside their cluster into a cluster representing a different habitat). From Bett (2001), with permission of the author and Elsevier.

served south of the ridge (>600 m), but never north of the ridge. Standing-stock–depth relationships are somewhat erratic, but on average both abundance and biomass of the megafauna appear to be higher south of the ridge (see Bett 2001, p. 942, Figure 13). Conceivably this difference in food supply explains the divergent alpha-diversity–depth trends to the north and south. However, phytodetritus settlement can be seasonal or episodic and simply could have been missed during the sampling north of the ridge. Bett

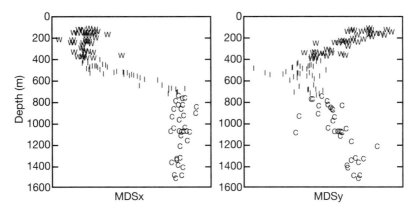

Figure 5.14. Nonmetric multidimensional scaling of samples taken north of the Greenland-Iceland-Faeroe Ridge. Reflections on the *x*-axis (left) and *y*-axis (right) of the ordination are plotted against depth. W, I, and C indicate warm (minimum temperature > 0°C), intermediate, and cold (maximum temperature ≤ 0°C) water masses, respectively. From Bett (2001), with permission of the author and Elsevier.

(2001) did not think that potential differences in food supply, if real, were as significant as the more clear-cut association of beta diversity with water mass and temperature.

Oxygen Minimum Zones

Oxygen minimum zones (OMZs) can dramatically alter patterns of beta diversity over the narrow band where they intersect the slope. The OMZ on the Oman margin of the Arabian Sea provides a good example. Levin et al. (2000) showed that the core of the zone (400–850 m) is dominated almost entirely by surface deposit-feeding polychaetes and a single ampeliscid amphipod. Species diversity in the core is very depressed (see Figure 3.14). Beginning at its lower margin (1000 m), diversity recovers and the fauna becomes a taxonomically and trophically more balanced deep-sea assemblage. Variation in species composition of polychaetes within and below the zone is shown in Figure 5.15. The group within the core is distinctive and occupies a narrow space at the top of the plot. The apparent uniqueness of assemblages at 1250 and 3400 m, which still share some species, reflects the large sampling gap between them. The index of dispersion (below the figure) measures variation among replicate samples at each depth. Increased dispersion below the core of the OMZ may indicate either that more envi-

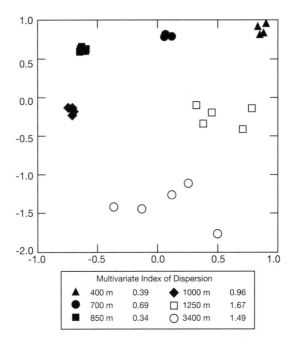

Figure 5.15. Nonmetric multidimensional scaling (MDS) plot of polychaete species composition within (solid symbols) and below (open symbols) the OMZ in the Arabian Sea. The core of the OMZ (400–850 m) harbors a low-diversity community dominated by surface deposit-feeding polychaetes, accounting for its distinct position at the top of the plot. Beginning at 1000 m, diversity increases and the community becomes more trophically complex, continuing to change in species makeup with increasing depth. OMZs can interrupt the typical pattern of more gradual continuous beta diversity at upper bathyal depths. From Levin et al. (2000), with permission of the authors and Elsevier.

ronmental heterogeneity is realized by the assemblage or simply the fact that much lower densities introduce more sampling error. The core assemblage is thought either to have evolved tolerance to low oxygen concentrations and sulfide stress or to have a competitive advantage under conditions of high density; some combination of these factors is also a possibility. This kind of abrupt change in community composition over such a short depth range and horizontal distance unobstructed by topographic barriers in the deep sea requires an extreme environmental shift. (For the important effects of OMZs on standing stock, species diversity, and evolution, see Chapters 1, 3, and 6, respectively.)

Meiofauna

Much less is known about beta diversity in the meiofauna. Foraminiferans in the Norwegian Sea occupy restricted depth ranges, dominant species replace one another along depth gradients, and ranges expand with increasing depth (Mackensen et al. 1985). In the Gulf of Mexico, harpacticoid copepods show strong divergence in species composition with both depth and longitudinal separation, and there is little faunal overlap on spatial scales less than 50 km (Baguley et al. 2006). Harpacticoids also show a rapid turnover with depth in the western North Atlantic (Coull 1972). Most research on beta diversity in nematodes is based on generic rather than species composition, making comparisons with macro- and megafauna difficult (e.g., Soetaert and Heip 1995, Vanaverbeke et al. 1997, Flach et al. 2002, Netto et al. 2005). Abyssal nematode species do appear to be very widely distributed (Lambshead and Boucher 2003). Presumably the same environmental factors that shape beta diversity in larger organisms affect the meiofauna. The TROX model, discussed in Chapter 3, may also apply. If foraminiferan and metazoan meiofaunal species are adapted to exploit different microhabitats in the sediment, then variation in microhabitat variability with depth could drive beta as well as alpha diversity.

Differences among Faunal Groups

Different rates of species turnover with depth are found in different taxa (Sanders 1977, Grassle et al. 1979, Flach and de Bruin 1999, Cartes et al. 2003) and functional groups (Rex 1977, Haedrich et al. 1980, Cartes and Carrassón 2004). By plotting faunal similarity of samples against difference in depth to selected reference samples, Rex (1977) showed that the megabenthos has a higher rate of species replacement than elements of the macrobenthos in the western North Atlantic (Figure 5.16). Among the macrofauna, gastropods have a higher rate of change than polychaetes and bivalves. He suggested that turnover rate might be a function of trophic position. This explanation combined two existing ecological theories: Menge and Sutherland's (1976) model of how the intensity of different biological interactions controls diversity and Terborgh's (1971) theory of species distributions along environmental gradients. Menge and Sutherland reconciled the roles of competition and predation, long felt to be conflicting forces, by showing that their structuring

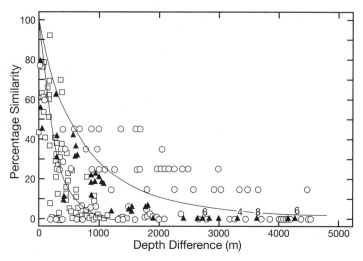

Figure 5.16. Percentage faunal similarity values plotted against depth for selected reference samples south of New England in the western North Atlantic. Three groups are shown: megafauna (open squares), gastropods (solid triangles), and polychaetes and bivalves (open circles). See Rex (1977) for sources of data. Lines are fitted by eye to the megafaunal and polychaete-bivalve data. The highest rate of change is found in the megafauna, followed by gastropods and then polychaetes and bivalves. Groups clearly differ in beta diversity, with megafauna showing a higher rate than macrofaunal elements. From Rex (1977), with permission of the author and Elsevier.

influence varied with trophic level. Competition for prey limits the diversity of predators, and predation pressure alleviates competition among prey. Terborgh showed that the distribution of species along a resource gradient, such as one accompanying depth or elevation, depends on either resource availability or interspecific competition. Where competition is less important, ranges can overlap. If competition is strong, species repulse one another at their boundaries, which compresses ranges and results in a sequence of more mutually exclusive ranges arrayed along the gradient. Uniting the two theories led to the prediction that if members of upper trophic levels are more regulated by competition than those of lower trophic levels, they should show a greater rate of turnover with depth. If megafaunal species rely on macrofauna for prey, which they certainly do to some extent, this could account for their higher rate of change with depth. Among the macrofaunal groups, gastropod predators, which dominate the taxon, consume polychaetes and bivalves as principal prey items, which may explain their higher replacement rate.

Although it is still a plausible hypothesis, subsequent work has shed doubt on the relative trophic positions of the mega- and macrofauna. As we pointed out in Chapter 1, standing stocks of the two groups do not suggest a simple Eltonian food pyramid. Some invertebrate megafaunal taxa may have a direct link to the base of the trophic structure by consuming phytodetritus. Some demersal fish species rely on both benthic and pelagic prey. Indirect evidence on competition from dispersion of size ratios among congeners (Rex et al. 1988) and on predation from the incidence of nonlethal repaired shell damage (Vale and Rex 1988, 1989) in gastropods suggests that deep-sea communities are loosely organized. It is clear from Figure 5.16 that replacement rates are gradual within groups and differ among groups. These different rates tend to average out turnover in the fauna as a whole, again indicating that perceived "zones" and "boundaries" are primarily a matter of moderate differences in the rate of turnover with depth and are evident only in restricted components of the benthos, most conspicuously in the megafauna.

Haedrich et al. (1980) proposed that turnover rate among megafaunal groups is related to adult mobility and breadth of diet. Demersal fishes are highly mobile and exploit a wide range of pelagic and benthic prey, whereas invertebrates are less mobile, often sessile in fact, and are limited to whatever food resources exist in their local ambits. Rex (1981) showed that demersal fishes do have significantly slower rates of turnover compared to megafaunal invertebrates south of New England, which supports the mobility-diet hypothesis of Haedrich et al. (1980). However, Hecker (1990b) in her photographic survey of the megafauna found that sessile filter feeders, slow-moving deposit feeders, and highly mobile carnivores had similar bathymetric ranges.

POTENTIAL CAUSES

Ultimately, beta diversity must devolve from the factors that regulate species' depth ranges. The single most obvious commonality of studies on deep-sea beta diversity is that species do have restricted depth ranges, at least regionally. In general terms, depth ranges represent a balance between the adaptive properties of species and the environmental variables that select for or against them. Relevant adaptive characters that have been proposed include feeding and habitat type (Jumars and Fauchald 1977), metabolism and locomotor capacity (Seibel and Drazen 2007), morphological specialization (Thistle and Wilson 1996), larval dispersal (Young et al. 1997a), adult mobility (Haedrich et al. 1980), body size (Rex and Etter 1998), body shape (Soetaert et al. 2002), and enzymatic activity (Siebenaller and Somero 1978). Environmental parameters include sediment type (Etter and Grassle 1992), hydrographic conditions (Gage 1997), oxygen content (Levin 2003), temperature and pressure (Hochachka and Somero 2002), $CaCO_3$ availability (McClain et al. 2004), water mass structure (Bett 2001), topography (Hecker 1990b), food supply (Rex et al. 2005a), biological interactions (Rex 1977), and microhabitat space in sediments (Jorissen et al. 1995). All of these are potentially involved in determining species depth ranges and, consequently, beta diversity. Many undoubtedly act in concert. For example, bottom topography can affect current velocity, which in turn determines the kind of POC flux that mediates the success of suspension feeders (Hecker 1990b); major topographic barriers can block deep circulation, leading to exceptionally strong thermal gradients that influence faunal makeup (Bett 2001); strong near-bottom currents can erode sediments, favoring a fauna that can sustain disturbance (Blake and Grassle 1994); heavy nutrient loading can cause OMZs, where only species that can tolerate low oxygen levels can exist (Levin et al. 2000); and so on. The interaction of species' traits and environmental influences that position species along depth gradients must be exceedingly complicated and is not known with any precision or certainty.

Another underlying feature of the studies is that the rate of species replacement coincides roughly with the rate of change in depth. The most pronounced turnover occurs at the shelf-slope transition, and beta diversity is higher across the steeper bathyal zone than in the abyss. Physical factors that change continuously with depth, such as temperature and pressure, necessarily change at a higher rate with distance in the steeper bathyal region. Organic carbon flux to the seabed is an exponentially decreasing function of depth. The bathyal zone is also a more heterogeneous environment than the abyss, being the site of major erosional features such as canyons and gullies, mass wasting, rock outcrops, glacial debris, sediment variation, conflicting currents and water masses, and OMZs. This heterogeneity causes horizontal as well as vertical variation in community composition. There are, of course, exceptions to the generalization that continental margins are more heterogeneous than the abyss. Strong deep near-bottom currents (Hollister et al. 1984) and submarine landslides (Talling et al. 2007) do occur at great depths, and phytodetritus accumulation can generate temporary patchiness (see Figure 1.3). However, the bathyal region, overall, is much more biologically and physically complex and dynamic than the abyss and depth-related environmental gradients are stronger, which is why beta diversity is higher in the bathyal zone.

One convincing environmental influence on depth range occupancy is food supply. Suspension-feeding megafaunal species are only abundant where currents provide sufficient POC flux or zooplankton prey (Hecker 1990b, 1994, Rice et al. 1990). Carnivore abundance decreases with depth as prey resources diminish, leaving the abyss to be populated largely by deposit feeders (Rex 1976, Howell et al. 2002). We suggest that the effect of food supply on beta diversity is much more pervasive than just in these well-known examples. The decrease in food supply with depth is the strongest and most universal biological gradient in the deep sea and is responsible for the exponential decrease in standing stock with depth, the effects of which are most pronounced in the bathyal region (see Chapter 1). In terms of size categories, there is a disproportionate decrease in the biomass and density of larger organisms—the macrofauna and megafauna. The megafauna depend on small infaunal prey, organically rich sediments, or food-laden near-bottom currents, all of which decrease with depth. In the macrofauna, there is a shift from broad participation in the food web to primarily deposit feeding. The average size of organisms decreases with depth as a consequence of diminished food supply (see Chapter 1). The decline of larger organisms with depth results in a perforce change in community composition. The abyss is heavily dominated by bacteria, foraminiferans, and minute deposit-feeding meiobenthos.

In Chapter 3 we argued that the decrease in alpha diversity from the bathyal region to the abyss is largely attributable to the dramatic decrease in population densities. In some cases the abyssal community appears to be composed primarily of range extensions for a subset of bathyal species living at densities too low to be reproductively viable. These abyssal populations may be maintained by source-sink dynamics. If this phenomenon is widespread, it helps to explain beta as well as alpha diversity. Abyssal communities may appear to constitute a separate group in multivariate analyses because their low alpha diversity precludes a high calculated level of faunal similarity with bathyal regions. Abyssal clusters are often more coherent because samples include the same few common species that are broadly distributed and adapted to abyssal life and the same subset of rare species whose larval dispersal ability permits sparse abyssal sink populations to exist. To the degree that this might be the case, there are clear exceptions. Holothurians do include a substantial number of endemic abyssal species that are adapted to exploit superficial sediment layers (Billett 1991). Isopods underwent an ancient and very successful radiation in the deep sea and achieve their peak diversity at abyssal depths in the Atlantic (Wilson 1998).

CONCLUSIONS

Beta diversity, the rate of change in species composition, is especially pronounced along depth gradients in the deep sea. The causes of species turnover are still inferential and are almost certainly multivariate. Presumably, beta diversity reflects the interaction of species' adaptive constraints with the depth-related environmental gradients they exploit, mediated by interspecific competition and predation along these gradients. Numerous combinations of adaptive properties and environmental opportunities have been proposed, but there is little consensus about their relative importance. Beta diversity in soft-sediment habitats is continuous and gradual except where interrupted by topography, hard substrate, or extreme environmental circumstances such as OMZs, high-energy bottom currents, nutrient depocenters, and abrupt shifts in water masses. The rate of faunal replacement is roughly proportional to the rate of change in depth, being higher in the bathyal zone and lower in the abyssal plain. In studies that include both horizontal and vertical components of faunal change, most variation in species composition is depth related.

The proximal causes of beta diversity must be complex and taxon specific. However, at large scales, between the continental margin and the abyss, beta diversity ultimately may be driven by organic carbon flux to the seafloor and its effect on the standing stock of different size categories of organisms. Assemblages of larger animals are progressively depleted with increasing depth, resulting in a change in the composition of the whole fauna. The decrease in alpha diversity (sample richness) from the bathyal zone to the abyss is in large part caused by a decrease in population size. Much of the abyssal fauna is a more homogeneous subset of the lower bathyal fauna. The change in alpha diversity limits species similarity along the depth gradient. Standing stock and alpha and beta diversity are not independent aspects of community structure. Standing stock affects alpha diversity, which, in turn, affects beta diversity, and this may be the overall large-scale framework in which a wide variety of other biotic and abiotic factors act on smaller scales to regulate community structure and function in the deep-sea benthos.

6

EVOLUTIONARY PROCESSES IN THE DEEP SEA

Why are there so many kinds of animals?

G. Evelyn Hutchinson (1959)

Why are there so few kinds of animals?

Joseph Felsenstein (1981)

The deep sea supports a rich and highly endemic fauna that varies in diversity on local, regional, and global scales. Our discussion thus far, as well as the majority of work on deep-sea communities, has focused on the question of what ecological circumstances might allow such a surprising diversity of species to coexist. An even more basic question asks where all these species came from. How and where did they evolve? Given the limited ecological opportunity and the lack of obvious mechanisms that would allow population differentiation and speciation, it is unclear how new species form, especially at a rate sufficient to explain the high levels of diversity found in the deep ocean.

A comprehensive understanding of evolution must include the invasion of the deep sea, population divergence, adaptive radiation, and the global spread of higher taxa. Currently, our knowledge of these processes is rudimentary and will certainly expand and change with future exploration. However, a faint picture is already beginning to emerge that allows us to formulate fairly specific hypotheses that can serve to guide future research. In this chapter, we review existing evidence for evolutionary processes and then focus on what we feel is the most basic and tractable problem: the pattern of bathymetric differentiation revealed by population genetic structure.

THE IMPORTANCE OF
EVOLUTIONARY STUDIES

Understanding how evolution unfolds is essential for developing a comprehensive understanding of the ecological forces that shape spatial and temporal patterns of diversity in the deep sea. Recent evidence from terrestrial and shallow-water ecosystems underscores the importance of historical and evolutionary processes for explaining geographic variation in diversity (e.g., Ricklefs 1987, 2007, Jablonski 1993, Ricklefs and Schluter 1993, Cadena et al. 2005, Vellend 2005, Gillooly and Allen 2007, Hawkins et al. 2007a,b, Kelly and Eernisse 2007, McPeek and Brown 2007, Mittelbach et al. 2007, Roy and Goldberg 2007, Wiens 2007). Regional species pools are ultimately the product of evolutionary diversification, and numerous studies have demonstrated a strong influence of regional species pool size on local diversity (e.g., Ricklefs 1987, Cornell and Karlson 1996, Karlson et al. 2004, Witman et al. 2004), including in the deep sea (Stuart and Rex 1994). Moreover, large-scale gradients of diversity in terrestrial and near-shore communities reflect geographic differences in evolutionary rates. For example, paleontological (Stehli and Douglas 1969, Jablonski 1993, Jablonski et al. 2006), molecular (Martin and Palumbi 1993, Martin and McKay 2004, Allen and Gillooly 2006, Gillooly and Allen 2007, Kelly and Eernisse 2007), and phylogenetic (Cardillo 1999, Cardillo et al. 2005, Ricklefs 2007, McPeek and Brown 2007, Wiens 2007) evidence suggests that geographic variation in origination and extinction rates help shape latitudinal diversity gradients. It is likely that bathymetric and geographic gradients in diversity explored so far in this book represent, in part, spatial differences in evolutionary rates.

Evolutionary phenomena in the deep sea may also provide a broader perspective on the general problem of how diversification occurs in nature. It is difficult to see how some features of allopatric speciation could operate in this environment. First, what are the barriers that would isolate gene pools and allow populations to diverge sufficiently to become separate species? The deep sea, although not as homogeneous as initially thought, is still considerably less variable and more continuous than the continental shelf environment. The pronounced changes in currents, temperature, and topography that occur in coastal waters are either reduced or absent below the shelf. Below the permanent thermocline at around 500–1000 m, temperature changes little over vast stretches of the deep ocean; equatorial abyssal temperatures are within a degree or two of those at the poles.

Topographic features of the continents and their margins can have a profound influence on biogeographic and phylogeographic patterns. For example, near-shore populations on either side of Point Conception (Burton 1998, Wares et al. 2001) and Cape Cod (Wares 2002, Jennings et al. 2009) are genetically different due to topographic steering of currents and temperature differences that reduce or preclude larval exchange. The biotas separated by these features are consequently quite different (Engle and Summers 1999, Wares et al. 2001). A more dramatic example is the uplift of Central America, which severed gene flow between Pacific and Caribbean populations and resulted in the formation of sister taxa for many near-shore organisms separated by the isthmus (Bermingham and Lessios 1993, Knowlton et al. 1993, Knowlton and Weigt 1998). Complex and temporally shifting currents during the Pliocene and Pleistocene in Indonesia caused populations to diverge, leading to repeated rounds of allopatric speciation (Barber et al. 2000, 2002, 2006, Crandall et al. 2008). At smaller scales, populations in nearby bays and inlets of coastal waters can diverge genetically because of strong selective gradients despite high gene flow (Hilbish and Koehn 1985, Bertness and Gaines 1993, Wilhelm and Hilbish 1998, Schmidt and Rand 1999, 2001, Schmidt et al. 2000), although hydrodynamic effects may also play a role (Hare et al. 2005, Pringle and Wares 2007).

Second, because of the high dispersal potential of the deep-sea fauna (see Chapter 4), connectivity among populations is likely to be high, reducing the chances for gene pools to become isolated. Although modes of reproduction and larval development vary, the majority of deep-sea species have some form of pelagic larval development (e.g., Rex and Warén 1982, Young and Eckelbarger 1994, Young 2003, Rex et al. 2005a). Frigid temperatures may reduce metabolic demands on larvae, allowing them to disperse in the water column for much longer periods, extending the distances over which they are transported (Shilling and Manahan 1991, Welborn and Manahan 1991, Marsh et al. 1999, Kelly and Eernisse 2007). If larvae can absorb dissolved organic material or survive long periods of starvation, as was recently found among shallow-water invertebrates (Jaeckle and Manahan 1989, Shilling and Manahan 1991, Moran and Manahan 2004), larval durations and dispersal distances would be further augmented.

Third, most deep-sea species are detritus feeders. Detrital food decreases exponentially with depth and distance from productive coastal waters (see Chapter 1). Even if populations differentiate genetically, it is unclear how new species can partition essentially the same limited resources to coexist, particularly at great depths. This has important implications for speciation and adaptive radiation. Classical theory suggests that newly formed species must diverge ecologically to avoid competitive exclusion (e.g., MacArthur and Levins 1967, Abrams 1983, McPeek 2007, Kelly et al. 2008, Phillimore and Price 2008), although some work suggests that ecological equivalency is common (e.g., Sale 1979, Hubbell 2001). Food resources are likely to be among the most important factors controlling diversity within deep-sea communities, as shown by correlative (Rex 1981, Levin et al. 1994, Levin and Gage 1998, Glover et al. 2002, A. B. Smith and Stockley 2005) and experimental (Turner 1973, 1977, C. R. Smith 1986, Snelgrove et al. 1996, C. R. Smith et al. 1998) studies, and by observed natural shifts in community structure that are associated with changes in food supply (Smith 1986, Tietjen et al. 1989, Smith et al. 1998, Ruhl and Smith 2004, Ruhl 2007).

Fourth, as noted in Chapter 4 and taken up again later, global anoxic events appear to have caused mass extinctions among the deep-sea fauna. Major events were most common in the Mesozoic; the most recent occurred in the Late Cretaceous and during the Paleocene epoch of the Cenozoic. Regional-scale events took place as late as the Pleistocene (Hayward 2001, Hayward et al. 2007, O'Neill et al. 2007). Evidence for anoxia-driven extinction comes primarily from marked decreases of foraminiferan diversity in deep seabed cores. Some clades of larger taxa such as echinoids (A. B. Smith and Stockley 2005) and isopods (Wilson 1999) survived. Chemosynthetic communities appear to have remained unscathed (Little and Vrijenhoek 2003, Kiel and Little 2006). The geographic extent and taxonomic breadth of these extinctions are hard to gauge with precision. However, it is possible that much of the remarkable diversity observed today in the deep sea evolved in a relatively short geological time span, which is surprising considering the seemingly limited potential for diversification and resource partitioning.

INVASION OF THE DEEP SEA

Patterns of Invasion through Geological Time

The prevailing view is that the modern deep-sea fauna inhabiting soft sediments originated in coastal systems. Paleontological evidence indicates that during the Phanerozoic there was a recurrent cycle of novel higher taxa arising in near-shore habitats subsequently radiating and spreading across the continental shelves, displacing earlier forms in deep water (Jablonski et al. 1983, Jablonski and Bottjer 1990, 1991, Sepkoski 1991). This onshore-offshore evolutionary pattern was attributed ultimately to greater physical disturbance in the shallow-water environment. Disturbance may create new opportunities for evolutionary innovation followed by replacement of older taxa through the biological interactions of competition and predation. Jacobs and Lindberg (1998) pointed out that this cycle of origin, expansion, and replacement was not continuous throughout the Phanerozoic and may have been strongly influenced by the periodic regional and global anoxic-dysoxic conditions in the deep sea noted in Chapter 4. These events presumably would have retarded or halted invasion. The other implication of deep-sea mass extinctions owing to anoxia is that invasion may have been enabled by recovery from oxygen depletion rather than driven by cycles of onshore origination and displacement. A. B. Smith and Stockley (2005) showed that invasions by detritivore clades of deep-sea echinoids coincided with periods of increased seasonal organic carbon flux, which may have created new opportunities for colonization owing to more food availability. It seems likely that there are environmental factors both pulling and pushing migration into the deep sea.

Despite the extensive anatomical and biochemical adaptations to life in deep-sea vertebrates and invertebrates (Gage and Tyler 1991, Herring 2002, Hochachka and Somero 2002), which might seem to restrict reinvasion of shallow water, colonization evidently has not been a one-way street. After all, the coastal fauna is just as highly adapted to conditions of light, warmer temperatures, lower pressure, greater environmental variation, and a much more abundant food supply; yet it managed to invade the deep sea. A recent molecular genetic analysis of the phylogeny of stylasterid corals suggests that the group originated in deep water and invaded tropical coastal reef environments repeatedly (Lindner et al. 2008).

The origin, phylogeny, and biogeography of isopods are, by far, the best known among the macrofauna. Kussakin (1973) proposed that the deep-sea fauna is the youngest component of the global isopod radiation and arose through submergence of the Antarctic shelf fauna. Many Antarctic species are unusually eurobathic. As we discussed in Chapter 4, global cooling during the post-Paleocene created a cold isothermal water column around Antarctica, which could have served as a colonization conduit to the deep sea. Invasion may have been further facilitated by the depression of the Antarctic shelf to 400–800 m, much deeper than elsewhere. The Antarctic isopod fauna itself has diverse origins consisting of remnants of southern Pangaea, colonizations from other continents in the Southern Hemisphere, and extensive diversification in Antarctic waters (Brandt 1992). Menzies et al.

(1973), like Kussakin (1973), suggested that the modern deep-sea fauna represented Cenozoic polar submergence, but that an earlier fauna may have been extirpated by deep-water cooling.

By contrast, Hessler and Thistle (1975), Hessler et al. (1979), and Wilson (1999) argued convincingly that the deep-sea isopod fauna, which is diverse and evinces a high degree of specialization, represents an ancient in situ radiation. Its invasion of the deep sea may extend back to the early Mesozoic or even Paleozoic era. Members of the early invasion and diversification must have survived Mesozoic and early Cenozoic anoxic events in refugia. Deep-sea isopods are blind, and the presence of eyeless deep-sea forms in Antarctic waters almost certainly represents polar emergence. As we discussed in Chapter 4, isopod invasion of the deep sea is an ongoing process with the more recent flabelliferan colonization centered in the Southern Hemisphere superimposed on the earlier asellotan radiation that now occupies the entire deep World Ocean. It seems likely that invasions of the deep sea by coastal isopods happened independently at different times and places over hundreds of millions of years and that both submergence and emergence have occurred. This scenario has recently been broadly supported by phylogeographic analyses of isopods using molecular genetics (Held 2000, Raupach et al. 2004). Other taxa such as echinoids (A. B. Smith 2004), foraminifera (Kuhnt 1992), and ostracods (Benson 1975) also show evidence of multiple invasions into the deep sea, making it seem likely that the evolutionary-historical development of the deep-sea fauna represents a sequence of invasions, radiations, and extinctions. Much of the fauna must represent colonizations and radiations since the Late Cretaceous and Paleocene global anoxic events, but there are clearly more ancient elements—for example, asellotan isopods, which survived mass extinction events and remain major constituents of deep benthic communities.

Invasion Potential of Dispersing Larvae

One important factor determining whether species are able to invade the deep sea is the physiological tolerance of dispersing larval stages to high pressures and low temperatures. The extensive experimental research on echinoderm development by Craig Young and Paul Tyler and their colleagues has shed considerable light on this problem. Deep-sea echinoderm species have either planktotrophic (swimming-feeding) or pelagic lecithotrophic (nonfeeding) larvae. Species with pelagic lecithotrophic larvae tend to have broader geographic distributions (Young et al. 1997b), possibly because metabolic rates are retarded at low temperatures, which allows a protracted dispersal phase (Shilling and Manahan 1994). The distribution of species with planktotrophic development and ontogenetic vertical migration may be restricted by the availability of high surface production, which varies regionally as shown in Chapter 1. However, although planktotrophic larvae appear to have sufficient energy reserves for vertical migration (Young et al. 1996a), only those of certain species can tolerate the higher temperatures encountered in the euphotic zone (Young et al. 1998), suggesting that others may rely for development on small organic particulates or dissolved organic carbon sources deep in the water column.

Young and Tyler (1993) and Tyler and Young (1998) examined temperature and pressure tolerances in planktotrophic larvae in three species of the sea urchin *Echinus* that occupy different depth ranges in the North Atlantic. *Echinus affinis* inhabits the upper bathyal zone (1600–2400 m). Spawning was induced in adults collected at 2000 m and the fertilized eggs then incubated in vitro through early cleavage at different combinations of temperature and pressure representing coastal and bathyal conditions. Normal development occurred at conditions approximating 1000–2000 m with optimal rates corresponding to 2000 m. Development was strongly inhibited at shallower (0–500 m) conditions. This suggested that the upper depth boundary of *E. affinis* populations may be limited by the physiological tolerances of its larvae and that its larvae are truly barophilic. *E. esculentus* is a subtidal species. Its early embryos can withstand conditions at 1000 m and its larvae to around 2000 m, indicating that dispersal stages of coastal species are capable of invading bathyal depths well below adult ranges. A third species *E. acutus* has both subtidal and upper bathyal populations that extend to depths of 1200 m. As with *E. esculentus,* its embryos and larvae develop normally in conditions found at greater depths than the adult populations can tolerate. Interestingly, developmental stages of adults collected at 900 m were able to grow at lower temperatures and higher pressures than those of shallower populations, suggesting that bathyal populations were adapting to the physical environment of the deep sea. Tyler and Young (1998) proposed that the three species might represent sequential invasion of the bathyal zone. Larvae of the two shallower species were capable of surviving at much greater depths, and both adults and larvae of *E. affinis* appear to have fully evolved to bathyal conditions and are now restricted to those depths.

Similar patterns of larval tolerance have been found in other coastal and deep-sea echinoderms. Adult populations of the asteroid *Plutonaster bifrons* live at 1000–2500 m in the Rockall Trough. The highest incidence of normal embryonic development occurs at a pressure equivalent to 2000 m; pressures above and below the adult depth range severely retarded development (Young et al. 1996b). Three species of subtidal urchins in the Mediterranean were all found to have embryos and larvae that tolerated higher pressures and lower temperatures; tolerance increased for the later larval stages, which would presumably participate in invading greater depths. Embryos and larvae of the coastal Antarctic urchin *Sterechinus neumayeri* are very stenothermal, but can develop normally at bathyal pressures (Tyler et al. 2000). Since both the Mediterranean and Antarctic have fairly isothermal water columns and are sites of deep-water formation, Tyler et al. (2000) suggested that the deep thermohaline conveyor belt may serve as an invasion route to the deep sea. Villalobos et al. (2006) showed that larval stages of two common coastal starfish in the Atlantic, *Asterias rubens* and *Marthasterias glacialis,* survive upper bathyal pressures and that later developmental stages are more tolerant.

In general, it seems clearly established that embryos and especially planktotrophic larvae of coastal echinoderms are capable of invading the upper bathyal zone and that larvae of some bathyal species develop optimally at conditions approximating those across the depth range occupied by adult populations. Tolerance of larvae to higher pressures and lower temperatures is species- and stage-specific, but adult populations seem to reside well within the physical tolerances of dispersing larvae. It is also known that larvae and postlarvae of some echinoderm species do settle outside adult bathymetric ranges and that they can grow to juvenile stages and even begin gametogenesis, but seldom attain reproductive maturity and fail to establish viable populations (e.g., Gage and Tyler 1981a,b). This kind of mass effect may be a common downslope phenomenon in the deep sea (Rex et al. 2005a). As we noted in our discussion of beta diversity in Chapter 5, the physical and biological factors that prevent successful colonization outside adult ranges, and consequently restrict species depth ranges, remain almost completely unknown.

AN OVERVIEW OF POPULATION GENETIC STRUCTURE IN THE DEEP SEA

Given the paucity of fossil evidence for deep-sea taxa, except for foraminiferans and ostracods collected in deep seabed cores, our best hope for deciphering evolutionary trends is to examine genetic variation. Assessing genetic population structure using modern molecular methods can help us infer the scale of within-species differentiation, the potential roles of geographic isolating barriers and gene flow, the phylogeography of higher taxa, and even the time frame of divergence. This approach, especially using actual DNA sequences, has only quite recently been applied to deep-sea organisms. Existing studies are based on sequencing material collected for a variety of other purposes, usually to measure spatial variation in community structure. Later we discuss the difficulties of carrying out genetic studies on deep-sea species and then review what is known about population genetic structure and how this might relate to evolutionary processes. Previous reviews of evolutionary processes in the deep sea have noted the lack of direct quantitative genetic evidence (Wilson and Hessler 1987, Creasey and Rogers 1999, Rogers 2002). Although the evidence is understandably disparate, it does begin to suggest some commonalities. Rather than consider the details of each study, we have tried to synthesize what they tell us about evolution. In taking this approach, we used the same studies and the same basic phylogeographic patterns as evidence for several mechanisms, not only because the mechanisms are often not mutually exclusive but also because we lack critical tests for determining the forces that define the patterns.

There are severe logistical and technical difficulties associated with genetic analyses of the macrofauna and meiofauna because of their small size (usually <2.0 mm), low density, and the fact that they are recovered from great depths in small samples that are bulk sorted to size fractions on board ship and then later sorted to species in the laboratory. Most species do not survive the extreme temperature and pressure changes they are subject to when the samples are brought to the surface. For those that do survive, the time (weeks to months) required to sort and identify each specimen to species makes it impractical to keep them alive. On recovery, samples are typically fixed in 10% buffered formalin for 24–48 h and then transferred to 70% ethanol. Fixing the samples creates significant problems for genetic analysis because formalin degrades proteins and nucleic acids (Karlsen et al. 1994). The effects of fixation coupled with the very small amount of tissue in macro- and meiofaunal individuals had made it extremely difficult to obtain usable DNA, so most genetic studies focused on megafaunal or large macrofaunal species, which can be collected more readily and quickly frozen for DNA analysis. However, in the past decade, several laboratories have independently developed protocols that allow DNA to be amplified and sequenced from formalin-fixed tissues (France and Kocher 1996a, Chase et al. 1998b, Boyle et al. 2004, Bucklin and Allen 2004), and these new techniques opened the door to a new era of evolutionary studies on the deep-sea fauna by sequencing archived material.

A number of mechanisms have been identified that promote divergence in marine organisms with good dispersal, including isolation by distance, extrinsic barriers to gene flow, selection, and historical events (e.g., Palumbi 1994, Grosberg and Cunningham 2001, Riginos and Nachman 2001, Hellberg et al. 2002, Hedgecock et al. 2007). There is extensive evidence concerning these mechanisms operating in shallow-water systems, but as yet little for the deep sea. Although the picture is still very fragmentary, recent work on deep-sea organisms has demonstrated that divergence occurs on a variety of scales and might be influenced by some of the same mechanisms, as well as perhaps by some novel ones.

Potential Isolating Barriers

Distance

The deep sea covers most of the Earth's surface. As all the major oceans are connected through the Southern Ocean, the deep soft-sediment environment is nearly continuous. The sheer size of the deep-water ecosystem suggests that isolation by distance might be an important mechanism for driving the formation of new species. Both theoretical (Irwin 2002, Doebeli and Dickmann 2003, Doebeli et al. 2005, Filin et al. 2008, Leimar et al. 2008) and empirical evidence (T. B. Smith et al. 1997, Irwin et al. 2005) indicate that a reduction in gene flow with distance can lead to population differentiation and eventually speciation. The relevant scales of divergence in the deep sea are unknown, but they are likely to depend on dispersal ability, the magnitude and direction of deep-water currents, and the intensity of geographic and bathymetric selective gradients.

Few studies have directly calculated isolation by distance (IBD), that is, genetic distance as a function of the geographic distance separating samples. In those that have included such calculations, the results are mixed. For example, using allozymes France (1994) found weak IBD among populations of scavenging amphipods (*Abyssorchomene*) from deep-water basins of the California Borderlands. Gene flow estimates were high except for a population in a single deep-sill basin that was thought might represent a separate species. Similarly, IBD was nonsignificant for three species of protobranch bivalves occupying different depth regimes in the western North Atlantic

(Etter et al. 2005). The only exception was *Deminucula atacellana*, which exhibited strong IBD among populations arrayed along a depth gradient from 1100 to 3800 m. The differences largely reflected a pronounced divergence between upper and lower bathyal populations (Chase et al. 1998a, Etter et al. 2005) and, as we argue later, are more likely to be a consequence of environmental changes with depth than geographic distance per se. In the eastern North Atlantic the deep-water coral *Lophelia pertusa* also exhibits no IBD despite strong population structure and distinct offshore and fjord populations (Le Goff-Vitry et al. 2004).

The comparisons just noted involve relatively short distances (hundreds of kilometers), but even at much larger scales IBD can fail to emerge. In a study of thornyhead fishes across the slopes of the eastern and western North Pacific, Stepien et al. (2000) found no IBD in the longspine thornyhead (*Sebastolobus altivelis*), yet significant IBD in its sister species the shortspine thornyhead (*S. alascanus*). They attributed the contrasting patterns to differences in migration as juveniles and adults. No significant IBD occurred at more than 6000 km in the central North Atlantic for the deep-water redfish *Sebastes mentella* (Roques et al. 2002). However, when additional samples were included from more marginal populations on the eastern and western edges of the North Atlantic, IBD was evident. The Dover sole, *Microstomus pacificus*, in the eastern North Pacific also exhibits little IBD despite significant population structure (Stepien 1999). Populations separated by hundreds of kilometers were often more divergent than those separated by thousands of kilometers. A similar pattern was found for the protobranch bivalve *Deminucula atacellana* (Zardus et al. 2006), which shows significant population structure within and among basins of the North and South Atlantic, but no IBD among populations separated by 12,000 km. In the most dramatic example yet, there was virtually no genetic divergence among Arctic and Antarctic populations of three foraminiferan species based on a nuclear ribosomal RNA gene (Pawlowski et al. 2007), but this may reflect an invariant marker within these species rather than extensive gene flow over 17,000 km.

What seems to be typical for many of these studies is that divergence among populations occurs, but the degree of divergence is not directly related to the geographic distance separating populations. Other forces are at work besides a reduction of gene flow with distance. One of the biggest surprises among the patterns that were discovered was that strong divergence can occur over extremely small (10–100 km) scales (e.g., France 1994, France and Kocher 1996b, Creasey et al. 1997, Chase et al. 1998b, Etter et al. 1999, 2005, Stepien et al. 2000, Kojima et al. 2001, Quattro et al. 2001, Howell et al. 2004, Iguchi et al. 2007a,b), although the mechanisms driving such small-scale divergence remain unclear.

Currents

Currents play a key role in the evolution of marine organisms because they often determine the broad-scale movement of larvae and adults and thus influence levels of connectivity among populations (Palumbi 1994, 2004, Mullineaux et al. 2002, Cowen et al. 2006, Pineda et al. 2007). Through time, currents can shift causing dramatic changes in gene flow, which profoundly affect population differentiation and speciation (e.g., Barber et al. 2000, 2002, 2006, Crandall et al. 2008). The role of hydrography in determining gene flow in shallow-water organisms is well established. Deep-water currents are not as well known, especially at scales that might influence larval transport.

The major currents in the deep sea are driven by thermohaline circulation as part of the ocean conveyor belt. In the Atlantic, where the deep circulation is best understood, deep-water currents derive primarily from the North Atlantic Deep Water (NADW) and the Antarctic Bottom Water (AABW). The cooling of surface waters in the Greenland, Labrador, and Norwegian seas causes the cold dense water to sink to depths of 1–5 km and spread southward along the continental margin of the Americas, forming the Western Boundary Undercurrent. The AABW forms by the cooling of surface waters in the Weddell Sea, which sink and spread northward into the Argentine and Brazil basins in the western South Atlantic, but are restricted from moving as far north in the eastern corridor by the Walvis Ridge. Although the thermohaline circulation creates flows that are primarily north-south, the interaction of these flows as well as geostrophic and topographic steering of the currents produce east-west components, reversal of flows, and complex patterns of recirculation on regional scales within and between basins (Figure 6.1; also see Schmitz and McCartney 1993 for detailed descriptions of flows).

Neutrally buoyant passive drifters (Figure 6.2) follow trajectories that are highly variable and considerably different from the average flows of the Eulerian estimates (Richardson 1993, Boebel et al. 1999, Hogg and Owens 1999, Richardson and Fratantoni 1999, Lavender et al. 2000, Speer et al. 2003, Shoosmith et al. 2005). Although these passive drifters are probably poor

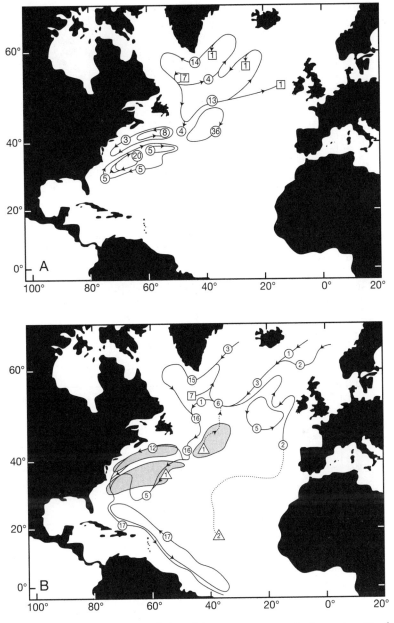

Figure 6.1. Composite diagrams of selected deep-water currents in the western North Atlantic for (A) temperatures below 7°C (800–1000 m) and for (B) 1.8–4°C (2500 m). Trans-
ports are in sverdrups with squares representing sinking and triangles representing upwelling.
Modified from Schmitz and McCartney (1993), with permission of the authors and the
American Geophysical Union.

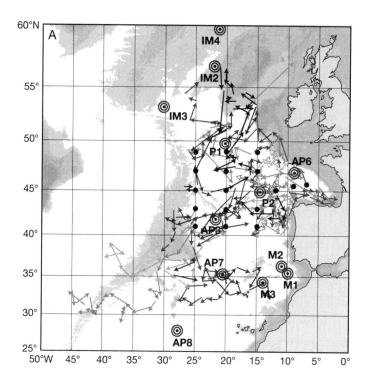

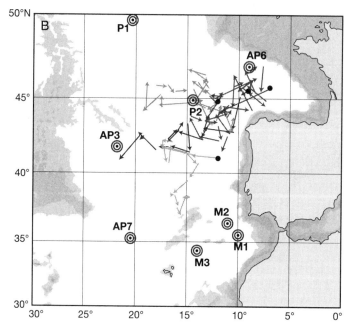

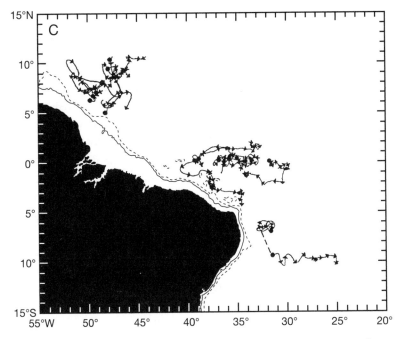

Figure 6.2. Subsurface float trajectories at (A) 1000 m and (B) 1500–2000 m in the eastern North Atlantic from the Eurofloat program. Each vector represents the direction and distance moved by a subsurface float at 3-month intervals from November 1996 to December 2002. A color version of this figure is included in the insert following p. 50. Modified from Speer et al. (2003) and http://www.ifremer.fr/lpo/eurofloat/disptraj.html, with permission of Dahlem University Press. (C) Trajectories at 3000 m in the western Tropical Atlantic from the SOFAR float program. Each vector represents the direction and distance moved by a subsurface float at 1-month intervals. Modified from Richardson et al. (1993), with permission of the authors and Elsevier.

mimics of how passive or behaving larvae might disperse, they provide great insight into the potential for highly stochastic and different dispersal trajectories that complicate interpretations of hydrodynamic effects on connectivity in the deep sea. These complications should not be underestimated. Recent studies of two congeneric mytilids (*Mytilus californianus* and *M. galloprovincialis*) dispersing in the same water masses at the same time yielded very different patterns of dispersal and population connectivity (Becker et al. 2007). Without a much better understanding of the physical oceanography at the scale that might influence larvae and the behavior of larvae during ontogeny, it will be challenging to formulate rigorous hypotheses about how deep-water currents might affect long-term gene flow among populations and explain empirical patterns of genetic structure based on hydrography.

Despite the potential for highly variable hydrographic effects on dispersal, some authors have inferred that deep-water currents affect population structure in bathyal organisms. The lack of genetic structure across the central North Atlantic in the redfish (*Sebastes mentella*) was attributed to the conjunction of three major currents (Irminger, Greenland, and Labrador), which induces larval mixing over the entire central North Atlantic Basin (Roques et al. 2002). In contrast, the divergence on the western and eastern margins of the redfish range was thought to result from regionally localized currents that retain larvae. The retention of larvae in localized currents was similarly invoked to account for the strong geographic divergence of the Dover sole along the west coast of North America (Stepien 1999). Although the Dover sole is benthic as an adult, the larvae have a long pelagic period, which is expected to enhance gene flow and homogenize populations. This is also typical for shallow-water organisms, where actual dispersal distances are far shorter than potential estimates based on larval biology and the duration of the larval phase (Bohonak 1999, Gilg and Hilbish 2003, Palumbi 2004; for deep-sea echinoderms also see Young et al. 1997a). The strong differentiation between fjord and offshore populations of the deep-sea coral *Lophelia pertusa* in the eastern North Atlantic has been attributed to local hydrological effects (Le Goff-Vitry et al. 2004), although the details of how the local currents might limit gene flow were not described.

Topography

The physical structure of the deep ocean varies in complexity from the gentle rolling hills of the abyss to the rugged terrain of continental margins to the largest mountain chain on Earth—the mid-ocean ridge system. Just as in terrestrial systems (Janzen 1967), these topographical features might influence connectivity among populations and potentially lead to population differentiation and speciation. Topographic features such as the mid-ocean ridge could disrupt gene flow if larvae or adults cannot pass over the ridge because vertical migration would exceed their physiological tolerance. Furthermore, topographic steering of deep-water currents would affect larval transport and thus connectivity (Mullineaux et al. 2002, Young et al. 2008). At slope depths, the rough seascape (Figure 6.3) interacts with mesoscale flows to create complex patterns of circulation that could potentially isolate populations in particular sections.

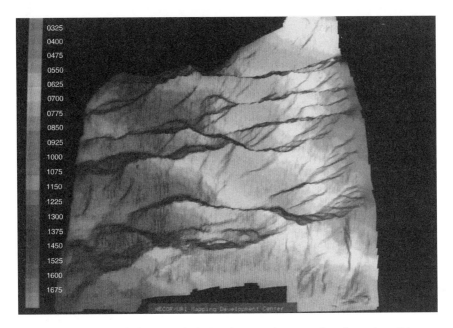

Figure 6.3. Sea beam bathymetry depicting the rugged topography of a section of the upper slope off North Carolina, U.S., measuring approximately 7 × 8 km, looking south and color coded by depth (25-m intervals). A color version of this figure is included in the insert following p. 50. From Mellor and Paull (1994), with permission of the authors and Elsevier.

To what extent do topographic features affect gene flow in the deep sea? There are some interesting examples, but they are highly inferential because they were not originally designed to test the effect of topographic barriers. The Mid-Atlantic Ridge (MAR) is the largest and most distinctive topographic feature in the Atlantic and has the potential to restrict gene flow between the eastern and western basins. Oceanwide phylogeographic studies of abyssal mollusks suggest the MAR may limit transoceanic gene flow at these depths. Both the abyssal protobranch bivalve *Ledella ultima* (Etter et al. in preparation) and the prosobranch gastropod *Benthonella tenella* (Boyle et al. in preparation) exhibit divergence between eastern and western Atlantic populations (Figures 6.4, 6.5), with less variation across similar distances within each basin. The strong divergence for *B. tenella* is unexpected because its planktotrophic larvae migrate to surface waters during development (Bouchet and Fontes 1981, Killingley and Rex 1985) and should be able to cross over the ridge.

Although the ridge may be an effective barrier for lower bathyal and abyssal species, it may be more permeable for upper bathyal organisms or those

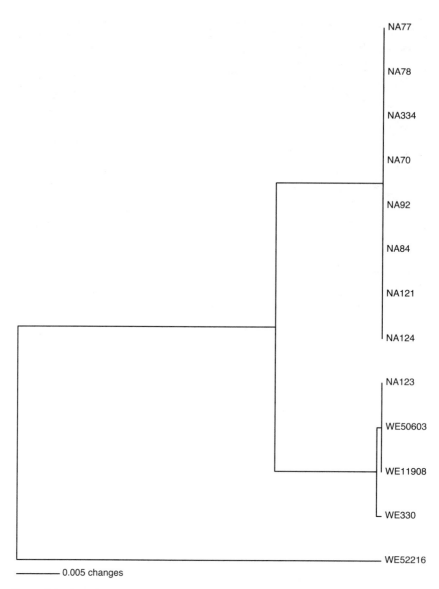

NA77

NA78

NA334

NA70

NA92

NA84

NA121

NA124

NA123

WE50603

WE11908

WE330

WE52216

0.005 changes

Figure 6.4. Population genetic structure of the abyssal protobranch bivalve *Ledella ultima* based on a 198-bp fragment of the 16S mtDNA gene. UPGMA tree derived from sample pairwise Φ_{st}. Station numbers are given at the terminal branches with a prefix indicating the basin (NA, North American Basin; WE, West European Basin). A distinct division occurs between samples from NA and WE basins (AMOVA $P = 0.03$) separated by the MAR.

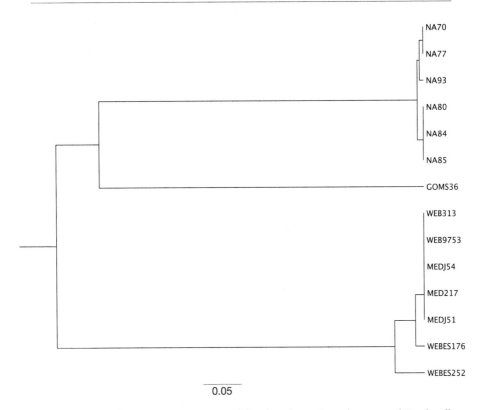

Figure 6.5. Population genetic structure of the abyssal prosobranch gastropod *Benthonella tenella* based on a 198-bp fragment of the COI mtDNA gene. UPGMA tree derived from sample pairwise Φ_{st}. Station numbers are given at the terminal branches with a prefix indicating the basin (GOM, Gulf of Mexico; MED, Mediterranean; NA, North American Basin; WEB, west European basin). A distinct division occurs between samples from NA and WE basins (AMOVA *P* < 0.0001) separated by the MAR. Also note that WEB and MED samples do not differ, whereas NA basin samples differ from those in the GOM. Courtesy of Elizabeth Boyle.

with highly mobile adults. In a broad-scale study of the widely distributed amphipod *Eurythenes gryllus,* France and Kocher (1996b) found no difference between eastern and western North Atlantic populations, despite strong population divergence within each basin (Figure 6.6). The bathyal protobranch bivalve *Deminucula atacellana* exhibited a very similar pattern (Zardus et al. 2006). Despite strong bathymetric divergence in the western North Atlantic, populations on either side of the ridge were genetically homogeneous (Figure 6.7). There was also little divergence among conspecifics from eastern and western equatorial Atlantic sites for two cold-seep *Bathymodiolus* mussels (Roy

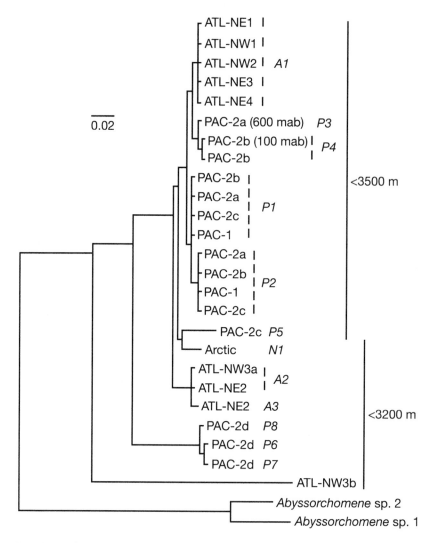

Figure 6.6. Phylogenetic structure in the deep-sea amphipod *Eurythenes gryllus* based on the 16S mtDNA gene. Unrooted neighbor-joining tree derived from gamma distances among haplotypes with *Abyssorchromene* sp. outgroups. Identical haplotypes found in multiple locations are indicated by vertical hatched bar. All haplotypes collected from abyssal depths are in a single clade (indicated by solid vertical bar labeled >3500 m) except P5. Sample abbreviations: ATL-NE, eastern North Atlantic; ATL-NW, western North Atlantic; PAC, Pacific. Haplotype abbreviations: A, Atlantic; P, Pacific. Modified from France and Kocher (1996b), with permission of the authors and Springer Science.

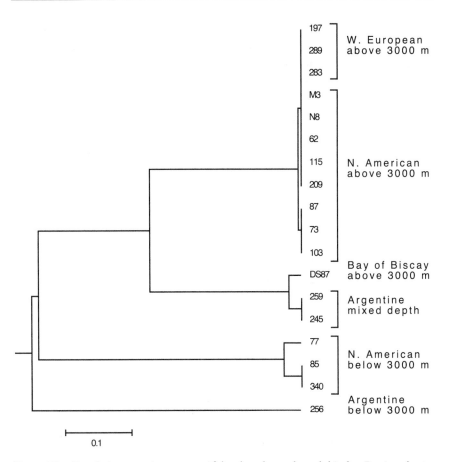

Figure 6.7. Population genetic structure of the abyssal protobranch bivalve *Deminucula atacellana* based on a 198-bp fragment of the 16S mtDNA gene. Neighbor-joining tree derived from sample pairwise Φ_{st}. Station numbers are given at the terminal branches and brackets depict different geographic and bathymetric divisions of the samples. Above 3000 m there is no difference between samples from the eastern and western North Atlantic but a clear difference between those from different depths (above and below 3300 m) in both the North and the South Atlantic. From Zardus et al. (2006), with permission of the authors and Blackwell Publishing.

et al. 2007). The upper bathyal bluemouth fish, *Helicolenus dactylopterus,* provides an interesting contrast. Found throughout the North Atlantic (200–1000 m), it exhibits much greater divergence across the Atlantic than within the eastern North Atlantic (Aboim et al. 2005). In this case, overall distance might be important because the MAR is unlikely to be an effective barrier to gene flow above 1000 m and the smaller but still significant divergence within the eastern North Atlantic suggests that dispersal is limited.

Of course there are numerous possible explanations for why conspecifics might, or might not, diverge over such vast distances. If the depth-related divergence patterns reflect the efficacy of the MAR as a barrier, changes in permeability with depth might be due to topographic steering of currents. In the North Atlantic, the ridge is topographically complex and can rise to depths above 1000 m. However, there are numerous deep-water corridors and fractures that may allow movement between the eastern and western basins, especially for bathyal species. Deep-water currents influenced by the ridge can be highly complex with both along-axis and across-axis flows (Bower and Hunt 2000a,b, Lavender et al. 2000, Bower et al. 2002, Ollitrault et al. 2002, Speer et al. 2003). Subsurface drifters dispersing at upper bathyal depths (400–1750 m) typically moved north or south along the ridge axis, but several were swept across the ridge through deep-water corridors (Bower et al. 2002, Sparrow et al. 2002, Speer et al. 2003, Lavender et al. 2005). These corridors may provide important conduits for the movement of mobile adults and larvae between the eastern and western basins of the North Atlantic (Mullineaux et al. 2002, Speer et al. 2003, Young et al. 2008).

The shallow sills that delineate semi-isolated peripheral seas, such as the Mediterranean and the Norwegian, also represent potential topographic barriers to gene flow. The Mediterranean Sea is separated from the eastern North Atlantic by a sill that rises to 280 m in the Strait of Gibraltar (Bouchet and Taviani 1992). The sill coupled with restricted flow through the strait should limit dispersal between the Atlantic and Mediterranean, especially for deep-sea species. Both biogeographic patterns of the deep-sea fauna (Bouchet and Taviani 1992, Carney 2005) and population genetic studies of shallow-water organisms (Patarnello et al. 2007) are consistent with this notion, although there are considerable differences in levels of divergence among shallow-water taxa.

The only specific genetic test of this hypothesis for the deep sea comes from Elizabeth Boyle's Ph.D. work (Boyle et al. in preparation). She analyzed the phylogeographic structure of the pan-Atlantic prosobranch gastropod *B. tenella* (see Appendix A) and found Mediterranean populations to be similar to those in the eastern North Atlantic (Figure 6.5), which were considerably different from conspecifics in the western North Atlantic. The lack of divergence may result from ontogenetic vertical migration of the larvae. If larvae migrate to surface waters during development (Bouchet and Fontes 1981, Killingley and Rex 1985), the sill may not act as a barrier, even if the adults are restricted to depths well below 280 m. However, the genetic similarity may also reflect a recent colonization of the Mediterranean, as has been inferred for many shallow-water species that exhibit little divergence (Patarnello et al. 2007).

In contrast, populations of *B. tenella* in the Gulf of Mexico differ from those in the adjacent western North Atlantic (Figure 6.5). Several other deep-water species exhibit concordant patterns with clear differences between the Gulf of Mexico and the western North Atlantic (Weinberg et al. 2003, Roy et al. 2007), which parallel similar geographic patterns of divergence in taxa from the continental shelves (Avise 2004). Although topographic isolation is a plausible explanation for such broad taxonomic congruence, other explanations such as historical environmental shifts or differences in selective regimes are equally plausible and not mutually exclusive. Much more work is required before we can unequivocally identify specific mechanisms.

Topography can also influence gene flow if populations are separated by deep basins or submarine canyons that species cannot cross because of physiological or environmental tolerances. This is more likely for species with limited dispersal ability that are unable to disperse across the gaps and thus experience required habitat as isolated patches. For example, two whelks, *Neptunea constricta* and *Buccinum tsubai*, in the Sea of Japan (Figure 6.8) exhibit clear population divergence corresponding to four geographic regions separated by deep basins or submarine canyons (Iguchi et al. 2007a,b). The whelks have direct development and appear restricted to depths of about 200–1800 m. Because of their weak dispersal abilities they cannot traverse the deeper regions, thus limiting gene flow among disjunct patches of the slope. The geographic divergence is less clear for *N. constricta,* presumably because its range within the Sea of Japan has recently expanded and there has been less time for lineage sorting.

The role of topographic features in isolating populations depends on the nature, duration, and depth of dispersal, as well as on the reproductive biology (e.g., number of propagules and the timing and frequency of reproduction) and the bathymetric range of each species.

Oxygen Levels

The levels of dissolved oxygen within the oceans vary geographically, bathymetrically, and temporally and have the potential to act as an isolating mechanism if they drop below the physiological tolerance of organisms to hypoxic conditions (White 1987, Wilson and Hessler 1987, Levin and Gage

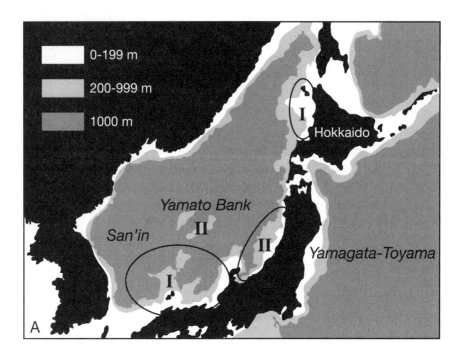

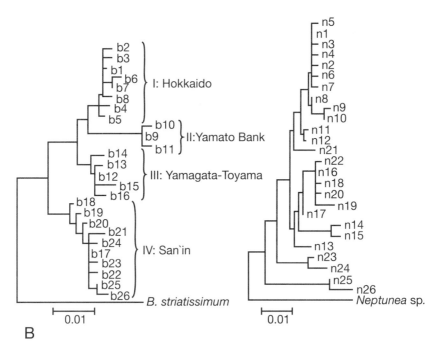

1998, Rogers 2000, Helly and Levin 2004). Enormous areas of the World Ocean are impacted by oxygen minimum zones (OMZs), where dissolved oxygen concentrations drop below 0.2 ml l^{-1} (Figure 6.9). Contemporary OMZs are widespread in the eastern South Atlantic, eastern Pacific, and especially the northern Indian Ocean, with recent estimates suggesting that permanently hypoxic conditions exist over more than 1×10^6 km^2 of seafloor at continental shelf to bathyal depths (Helly and Levin 2004). OMZs, which typically develop in regions of high surface productivity and weak circulation, may be spreading due to anthropogenic influences (Diaz and Rosenberg 2008). Decomposition of sinking organic matter under highly productive regions depletes the mid-waters of dissolved oxygen, creating subsurface OMZs that impinge the seafloor from the continental shelf to more than 1500 m (Helly and Levin 2004).

Tolerance to hypoxic conditions varies among organisms (Diaz and Rosenberg 1995), but OMZs are well known to affect the abundance, distribution, and diversity of the benthos, as we described in earlier chapters. Recent evidence indicates that expansion of hypoxic waters to the continental shelf, where few organisms are adapted to low oxygen, causes widespread loss of typically abundant shelf species (Chan et al. 2008). Because conditions can persist for thousands of years (Reichart et al. 1998), OMZs could easily fragment populations. On longer time scales, the geographic and bathymetric extent of OMZs shift, expanding during periods of global warming and retracting as global temperatures decline. The expansion and contraction of these hypoxic regions would provide shifting barriers to gene flow that might promote population differentiation and speciation (White 1987, Rogers 2000). Although it seems possible that OMZs could interrupt gene flow, no studies have actually quantified genetic variation among populations (or sister taxa) separated by an OMZ. Given the wide distribution of OMZs and their potential relevance to evolution, this is an important direction for future research.

In addition to forming a barrier to gene flow, regions of hypoxia might also cause strong changes in the genetic makeup of organisms that are tol-

---

Figure 6.8. (A) Sampling locations within the Sea of Japan for two whelks. (B) Phylogenetic (neighbor-joining) relationships among mitochondrial COI haplotypes for (left) *Buccinum tsubai* and (right) *Neptunea constricta*. Clades in *B. tsubai* corresponded to the different geographic regions sampled (indicated by brackets), whereas clades in *N. constricta* were not as geographically distinct. From Iguchi et al. (2007b), with permission of the author and Springer Science.

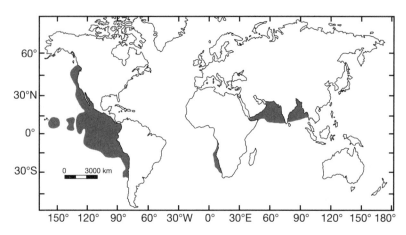

Figure 6.9. Geographic distribution of contemporary OMZs (gray) in the World Ocean. From Levin (2003), with permission of the author and CRC Press.

erant of low oxygen. For example, populations of the spider crab within the OMZ on the Oman slope in the Indian Ocean differ genetically from samples just below (650 m) in presumably more oxygenated water and from samples above (150 m) and outside the OMZ (Creasey et al. 2000). The strong genetic divergence at such a small spatial scale may reflect a selective gradient mediated by changes in levels of hypoxia and other concomitant environmental changes associated with the OMZ, although differences in sex ratio, age structure, and population size might also account for the observed genetic differences. In a similar study, there were genetic differences in squat lobsters on the Oman slope just below the OMZ between samples from 900 and 1000 m, possibly in response to a gradient in hypoxia. Whether the selective gradients near OMZs can lead to the formation of new species is unclear, but recent theoretical and empirical work indicates that strong environmental gradients can produce adaptive speciation (Doebeli and Dickmann 2003, Doebeli et al. 2005, Irwin et al. 2005, Filin et al. 2008, Leimar et al. 2008).

Vicariance

Historical events can leave a long-lasting distinct signal in DNA. One of the greatest advantages of using DNA combined with a phylogenetic approach is that we can estimate the timing of divergence by using molecular clocks. In the deep sea, dramatic environmental changes during the Pleistocene have been invoked to account for contemporary structure in several bathyal fish. Phylogeographic analysis of mtDNA in the black scabbardfish, *Aphanopus carbo,* from the eastern North Atlantic identified two deeply divided clades (Figure 6.10) corresponding to a geographic separation between the central Azores and various other sites within the North Atlantic (Stefanni and Knutsen 2007). However, numerous haplotypes from each clade were shared among the regions. The deep phylogenetic division coupled with haplotype sympatry suggested vicariance followed by secondary contact. Molecular clock estimates placed the division at about 400,000 years ago, when the climate cooled and sea levels dropped, possibly isolating populations within the central Azores. Subsequent climate and sea level changes presumably allowed the once isolated populations to come back into contact.

The Pleistocene is also the backdrop for explaining trans-Atlantic population structure in the bluemouth fish, *Helicolenus dactylopterus* (Aboim et al. 2005). Strong divergence existed between eastern and western North Atlantic populations, with weaker, but still significant, structure within the eastern North Atlantic. Genetic analyses (pairwise mismatch distributions) indicated that population bottlenecks followed by expansions took place about 1 million years ago in both the Azores and the western North Atlantic. The authors argue that the observed structure probably reflected Pleistocene climate changes, which more heavily impacted the western North Atlantic. Glacial advances around 1 million years ago could have caused extinction of populations in the western North Atlantic and severely depleted those in the eastern North Atlantic. Genetic evidence suggests that the western North Atlantic was recolonized by an unusual long-distance dispersal event (approximately 0.64–1.2 million years ago) from refugia in the eastern North Atlantic and that the newly founded populations expanded rapidly.

Even more recent vicariant events are thought to have influenced the phylogeographic structure of an abundant demersal fish, *Bothrocara hollandi,* in the Sea of Japan. Kojima et al. (2001) sampled 17 sites ranging in depth from 375 to 1675 m along the western margin of Japan and found pronounced genetic structure forming two major phylogroups or clades (Figure 6.11). Clade A was restricted to relatively shallow sites (above 1100 m) in the western part of the Sea of Japan, whereas Clade B was widespread but genetically impoverished. The divergence between the two clades was estimated to be about 34,000 years ago, at the beginning of the last glacial period when the deeper portions of the Sea of Japan became anoxic. Deep-water anoxia may have led to the depth–related divergence and reduced genetic diversity in Clade B.

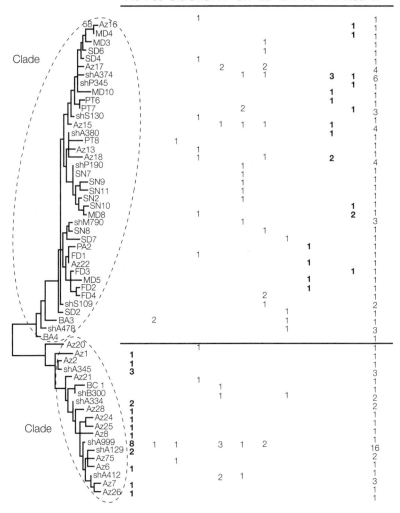

Figure 6.10. Phylogenetic (neighbor-joining) relationship among haplotypes of the mitochondrial control region for the black scabbardfish, *Aphanopus carbo*. The tree is rooted at the midpoint. The table at the right shows the geographical distribution of the haplotypes. AZ, Azores; BA, Azores Bank; FA★, Fayal Island and Condor Bank; Fco, Flores and Corvo islands; FD, Faraday seamount; Gra, Graciosa Island; MAD, Madeira; PA, Princess Alice Banks; Pic, Pico Islands; PT, mainland Portugal; Sma, Santa Maria Island; SN, Seine seamount. N, total haplotypes; Sh, Shared haplotypes. Bold numbers indicate samples encountered only in one of the two clades. The deep phylogenetic break between the two clades with several haplotypes shared among geographic regions is indicative of vicariance followed by secondary contact. From Stefanni and Knutsen (2007), with permission of the authors and Elsevier.

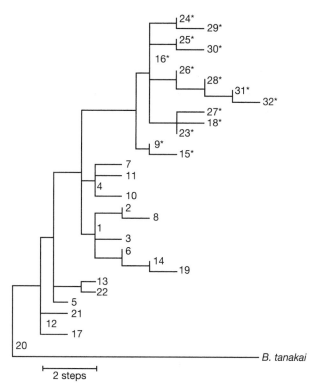

Figure 6.11. Phylogenetic (maximum parsimony) relationship among haplotypes of the mitochondrial control region in the deep-sea fish *Bothrocara hollandi* from the Sea of Japan. The haplotypes form two phylogroups (clades) indicated by Clade A (asterisks) and Clade B. From Kojima et al. (2001), with permission of the authors and Inter-Research.

Environmental Differences

Strong environmental differences can also impede gene flow among populations by selecting against immigration of larvae or adults. As we describe throughout this book, the deep sea exhibits spatial and temporal environmental heterogeneity on a range of scales, which could create different selective regimes that limit or preclude gene flow. Few studies have directly related patterns of genetic variation to specific selective forces, in part because selective pressures are poorly known in the deep sea, but also because most population genetic studies are conducted on an ad hoc basis and many factors co-vary among sampling locations.

Depth-Related Trends

The strongest environmental gradients in the deep sea are those associated with changes in depth. Temperature, pressure, the sedimentary regime, food resources, and community structure all change rapidly along depth gradients (see Chapters 1–5). If environmental-driven selection pressures counteract gene flow, the effects should be most apparent where the gradients are most intense. Numerous species exhibit pronounced genetic differences among populations from different depth regimes (France 1994, France and Kocher 1996b, Chase et al. 1998b, Etter et al. 1999, 2005, Kojima et al. 2001, Quattro et al. 2001, Howell et al. 2004). A common finding in genetic studies of the deep-sea fauna is that divergence is considerably greater between populations separated vertically than those separated horizontally (Bucklin et al. 1987, France and Kocher 1996b, Kojima et al. 2001, Etter et al. 2005, Zardus et al. 2006), probably in response to the intense vertical environmental gradients. For example, Atlantic and Pacific populations of *E. gryllus* are genetically homogeneous at abyssal depths, yet there is pronounced divergence (based on 16S mtDNA) among populations above and below 3200 m in individual basins (France and Kocher 1996b; Figure 6.6). The bathymetric divergence within ocean basins far exceeds that found between the Atlantic and Pacific populations at similar depths. The strong bathymetric divergence with little divergence horizontally over much larger scales was also observed in an earlier study of *E. gryllus* using allozymes (Bucklin et al. 1987). Similarly, bathymetric divergence among populations within the North American Basin is larger than between the North and South Atlantic in a protobranch bivalve (Figure 6.7). Levels of divergence among populations separated by a 12,000-km distance were equivalent to those separated by a 1.5-km depth in the same basin (Figure 6.12). Clearly, small changes in depth appear to be much more significant for inducing population divergence than enormous changes in geographic distance.

The forces that have shaped depth–related patterns of population structure are unknown, but the strong environmental gradients and extremely small scales involved suggest that environmentally driven selection might be important. If so, the selective forces are likely to be similar to those thought to be important in shallow water (e.g., temperature, salinity, oxygen, sediment characteristics, food resources, competition, predation). As there is no direct evidence for these forces operating in the deep sea and many have been discussed elsewhere (reviewed in Palumbi 1994, 2004, Grosberg and

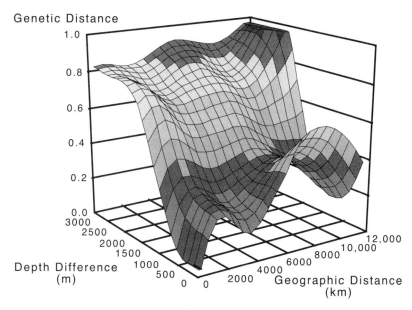

Figure 6.12. Relationship among genetic distance (Φ_{st}), geographic distance, and depth difference among samples of the protobranch bivalve *Deminucula atacellana* in the Atlantic Ocean. Genetic distances are based on 16S mtDNA haplotypes. Surface smoothing by linear interpolation. From Zardus et al. (2006), with permission of the authors and Blackwell Publishing.

Cunningham 2001, Avise 2004), we only consider hydrostatic pressure, which is exclusive to the deep sea. Our focus on pressure is not meant to imply that it is more important or likely, only that it is often ignored and has received much less attention.

Hydrostatic Pressure

Pressure increases as a simple linear function of depth (1 atm for every 10 m depth) and may play an important and novel role in the evolution of deep-sea organisms. Both temperature and pressure influence biochemical reactions (Somero 1990). Temperature varies widely over the surface of the planet and is thought to influence large-scale biogeographic patterns of diversity owing to its ubiquitous effect on physiology and biochemistry (Allen et al. 2002, Hawkins et al. 2003a, 2007a,b, Gillooly and Allen 2007). Organisms are adapted to particular temperature regimes, and although physiological tolerances may vary among taxa, biogeographic boundaries are often associated with rapid shifts in temperature (e.g., Point Conception and Cape

Cod, United States). In the deep sea below the permanent thermocline, temperature changes are minor, so physiological limits and biogeographic boundaries are more likely to be set by changes in hydrostatic pressure. Just as terrestrial and coastal organisms exhibit adaptations to specific temperature regimes, denizens of the deep sea appear to be adapted to specific pressure regimes. Pressure changes of 50 atm can significantly affect the structure, function, and kinetics of proteins and enzymes (Somero 1990, 1992). Numerous studies have demonstrated that closely related species from different depth regimes have adapted by producing different enzymes and/or structural proteins (e.g., Siebenaller and Somero 1978, 1979, 1982, 1984, Somero 1990, 1992). Thus, in terms of physical factors, hydrostatic pressure represents a potentially important selective agent, like temperature in terrestrial and coastal environments, that may isolate gene pools and ultimately lead to population differentiation and speciation.

Several lines of evidence support the notion that hydrostatic pressure might be involved in the origin of deep-sea biodiversity. Populations from different depth regimes can exhibit pronounced genetic divergence despite any obvious selective or nonselective explanation (Siebenaller 1978, France 1994, France and Kocher 1996b, Chase et al. 1998b, Creasey and Rogers 1999, Creasey et al. 2000, Kojima et al. 2001, Howell et al. 2004, Etter et al. 2005). For example, populations of the protobranch bivalve *Deminucula atacellana* (see Appendix A) from upper and lower bathyal depths (above or below 3300 m) in the western North Atlantic possess highly divergent haplotypes (Figure 6.13), forming distinct populations (Figure 6.14). There are no conspicuous oceanographic or topographic features that would limit gene flow between these regions. Deep-water currents within this area of the western North Atlantic are complex (Richardson 1993, Schmitz and McCartney 1993) with basinwide recirculation and eastward components that could move larvae up- or downslope. Moreover, the distance separating the two closest samples showing divergent genetic compositions is less than 140 km—very likely within the dispersal window of the pelagic demersal larvae of *D. atacellana*. The lack of differentiation among samples above or below 3300 m, despite much greater bathymetric and geographic separation, suggests that dispersal can homogenize populations over much larger spatial scales. A similar break occurs at 3300 m in the South Atlantic (Zardus et al. 2006), although the data are more limited (Figure 6.7).

Interestingly, *D. atacellana* is not the only species to show a sharp genetic break at 3300 m. In a global analysis of population structure in the cosmo-

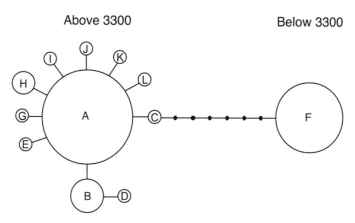

Figure 6.13. Statistical-parsimony network indicating interrelationships among haplotypes for 16S mtDNA for *Deminucula atacellana* from a series of stations along a depth gradient in the North American Basin known as the Gay-Head Bermuda transect. Letters within circles represent different haplotypes. The area of the circle is directly proportional to the number of individuals possessing that haplotype. The line connecting haplotypes represents a single mutational step. Haplotypes that were not found but are necessary intermediates are shown as small solid circles. The population tree based on these haplotypes is shown as part of Figure 6.14. From Etter et al. (2005), with permission of the authors and Blackwell Publishing.

politan amphipod *E. gryllus,* France and Kocher (1996b) found a similar divergence at virtually the same depth (Figure 6.6). These two species have extremely different lifestyles, natural histories, phylogenetic affinities, and geographic distributions. Their congruent genetic divergence at 3300 m in different regions of the World Ocean may suggest that 3300 m represents an unrecognized phylogeographic barrier that isolates organisms inhabiting different depths. On the other hand, because there is nothing obvious that would isolate gene pools between these depth regimes, the divergence might result from disruptive selection favoring different pressure-sensitive biochemistries. If true, the break at 3300 m may simply be coincidental— 330 atm exceeds the barophysiological tolerance of these two species, and selection has favored modified enzymes and/or proteins below this depth. Not all species are expected to exhibit a genetic break exactly at 3300 m, just as all species do not exhibit genetic or distributional breaks at the same temperature. Tolerance to hydrostatic pressure varies among species (Young et al. 1997a).

Bathymetric divergence is often sufficient to suggest that different populations represent unrecognized cryptic species (e.g., France 1994, France and Kocher 1996b, Chase et al. 1998b, Etter et al. 1999, 2005, Kojima et al. 2001, Howell et al. 2004). If this pattern is widespread, many species defined by morphological criteria that are thought to have broad bathymetric distributions may instead be composed of sibling species separated bathymetrically. Most species that have been analyzed genetically across large (>2000 m) vertical ranges exhibit significant depth-related genetic divergence. The only known exception concerns two species of foraminifera (Gooday et al. 2004, Pawlowski et al. 2007) that appear to have broad bathymetric distributions and little genetic divergence, although this may be more a consequence of invariant markers rather than extensive gene flow. Of course many environmental factors change with depth and might affect observed levels of divergence, but so far we lack compelling evidence of adaptive responses to other environmental changes.

There are several possible mechanisms that might account for how pressure differentials might reduce gene flow. One way might be by affecting larval development. As noted earlier in this chapter, some echinoid larvae develop only at particular ranges of hydrostatic pressure and temperature (Young and Tyler 1993, Tyler and Young 1998), and the bathymetric ranges of some echinoderms correspond to the upper and lower pressure limits of their developing larvae (Young et al. 1996a,b, Tyler and Young 1998). Water viscosity also varies with pressure and temperature (Macdonald 1975) and can influence the mechanics of ciliary motion in larvae (Podolsky 1994). Thus, mechanical or physiological capabilities of larvae may be adapted to specific depth regimes, which could limit gene flow among populations at different depths, at least for sessile species.

Another mechanism by which pressure gradients might limit gene flow is through gamete recognition systems, which have been implicated in speciation of free-spawning marine invertebrates (Palumbi and Metz 1991, Palumbi 1992, 1999, Metz and Palumbi 1996, Geyer and Palumbi 2003, Riginos et al. 2004, 2006). Successful fertilization is dependent on compatibility of recognition proteins on the surface of the sperm and egg. These proteins are controlled by a few loci, and small changes, often affecting their three-dimensional structure, can lead to reproductive isolation and possibly speciation (Palumbi and Metz 1991, Palumbi 1992, 1999, Metz and Palumbi 1996, Gavrilets and Waxman 2002). If the three-dimensional structure of the gamete recognition proteins is pressure sensitive, depth-related changes in the proteins could affect fertilization success and lead to reproductive isolation in a manner similar to mutations in bindin and lysin genes that influence protein conformation.

One prediction from both of these potential mechanisms is that we should expect to find sibling species adjacent to one another but displaced bathymetrically along the depth gradient. The strong bathymetric divergence in many species and the putative cryptic species separated bathymetrically match these expectations precisely. Although we have focused on hydrostatic pressure, many environmental factors change with depth and have to be considered as possible selective forces.

THE DEPTH-DIFFERENTIATION HYPOTHESIS

The deep sea, once considered to be a monotonous ecosystem, is now known to be highly complex with strong geographic and bathymetric gradients that are likely to influence the location, scales, and dynamics of evolution (Etter and Rex 1990, Etter et al. 2005, Rex et al. 2005a). The rate of environmental change is a function of the rate of change in depth and proximity to coastal production and is steepest at upper bathyal depths, decreasing below the continental slope and with distance from land. In addition, both biotic and abiotic environmental heterogeneity are considerably greater on the continental margins than in the abyss (see later). As evolutionary dynamics can be influenced by levels of heterogeneity and the intensity of environmental gradients (Doebeli and Dickmann 2003, Doebeli et al. 2005, Filin et al. 2008, Leimar et al. 2008), we have argued that the narrow bathyal environment is much more conducive to population differentiation and speciation than the more extensive abyss and is likely to be the primary physiogeographic region for the evolution and adaptive radiation of the deep-sea fauna (Etter and Rex 1990, Etter et al. 2005). In what follows we summarize the rationale and evidence for this conjecture.

Vertical, horizontal, and temporal environmental heterogeneity is greater at bathyal depths than below the continental margins. The intensity of environmental gradients in the deep sea parallels changes in depth and is steepest in the upper bathyal zone, where pressure increases and temperature, currents, and POC flux decrease rapidly along the slope face (reviewed in Gage and Tyler 1991). Across the same depths, and probably in response, the biotic environment is also rapidly shifting in composition, diversity, and trophic complexity (Carney 2005; see Chapter 5), creating strong differences in ecological and evolutionary pressures over relatively small scales. Below the continental margins, changes in depth and associated biotic and abiotic environmental gradients are highly attenuated, providing more uniform conditions and limiting the potential for divergence. Depth-related differences in the magnitude of morphological (Etter and Rex 1990, Rex and Etter 1990) and genetic (France and Kocher 1996b, Etter et al. 2005) divergence suggest that organisms respond in a direct quantitative way to the intensity of these vertical gradients.

The bathyal zone also exhibits more pronounced horizontal heterogeneity than the abyss. The continental slope is incised by deep submarine canyons at intervals of tens of kilometers. Slope face and canyon environments are quite different (e.g., Hecker et al. 1983, Cooper et al. 1987). Canyons experience strong vertical currents and contain a wide variety of substrate types, including rock and clay outcrops, boulder fields, gravel, sand, and mud. The slope face is more physically stable and is characterized by more uniform fine-grained sediments (MacIlvaine and Ross 1979). Strong descending currents scour the canyon floor and their V-shaped topography can accelerate oscillatory tidal flows to resuspend sediments (Gardner 1989a,b). Not surprisingly, the megafaunal communities of canyon and slope-face habitats are quite different. Canyons support a higher density of megafauna that cause erosion and slumping through their burrowing activity (Hecker et al. 1983, Vetter and Dayton 1998). The different substrate types, high-energy environment, and unique communities of canyons might effectively fragment the continental slope and isolate slope-face habitats.

On even smaller scales (kilometers) the bathyal environment can be topographically complex (Mellor and Paull 1994; Figure 6.3), affecting current velocities, sedimentation, and nutrient flux on very small scales (Rhoads and Hecker 1994) and creating rapidly shifting environments with distinct "microclimates." Moreover, larvae are likely to be entrained in small-scale topographically driven currents. Populations living on or within certain topographic features (e.g., gullies, canyons, valleys, leeward cliffs perpendicular to flow) might become geographically isolated, allowing population differentiation to occur on much smaller scales than previously thought. In contrast, the gentle topography of the abyss affords less opportunity for isolation or for the formation of distinct habitats. Although there is some degree of heterogeneity at abyssal depths (e.g., phytodetritous patches, large-scale catastrophic events; see Chapter 1), it is generally considerably less than that found on the continental margins.

Bathyal depths are also frequently impacted by OMZs, which, as noted earlier, can disrupt gene flow, thus fragmenting populations and potentially leading to population differentiation and speciation through selective or nonselective processes. Over geologic time scales, OMZs change in geographic and bathymetric extent, location, and intensity, creating a shifting barrier to gene flow and promoting repeated rounds of divergence. Although the abyss can experience anoxic conditions (Jacobs and Lindberg 1998, Hayward 2001, Hayward et al. 2007), the frequency is much less.

The bathyal region must have experienced strong environmental swings during past episodes of climate change, particularly during Pliocene and Pleistocene glaciation. Glacial advances lowered sea level by as much as 120 m (Lambeck and Chappell 2001), exposing continental shelves and promoting canyon formation (Cooper et al. 1987, Mellor and Paull 1994) and weakened the deep thermohaline circulation (Adkins et al. 1998, Raymo et al. 1998). Climate change during glaciation could be surprisingly abrupt with major shifts in temperature and ice volume occurring on time scales of only tens to hundreds of years (Severinghaus et al. 1998). Glacial cycles were associated with geographic and depth range shifts of bathyal species (Kurihara and Kennett 1988, Cronin and Raymo 1997) and must have affected population sizes (Slowey and Curry 1995). The isolating effects of active canyon formation and the shifting landscape of species distributions created a dynamic environment that was very likely conducive to population differentiation and speciation.

In addition to the more pronounced spatial and temporal environmental heterogeneity at bathyal depths, several lines of evidence support the notion that this region promotes population differentiation and speciation. First, intraspecific population genetic divergence is greater at bathyal than abyssal depths for both mollusks (Chase et al. 1998b, Etter et al. 1999, 2005, Quattro et al. 2001, Zardus et al. 2006) and crustaceans (France and Kocher 1996b). The cosmopolitan amphipod *E. gryllus* was essentially genetically homogeneous at abyssal depths across the Atlantic and Pacific oceans (Figure 6.6), whereas at bathyal depths, there was pronounced genetic divergence among basins in both oceans (France and Kocher 1996b). At smaller within-basin scales, population divergence among conspecifics was greater for clams and snails at bathyal depths than for similar species in the abyss (Etter et al. 2005). Patterns of genetic divergence in four common bivalves, *Nuculoma similis, Deminucula atacellana, Malletia abyssorum,* and *Ledella ultima,* distributed along a depth gradient in the western North Atlantic indicate strong population structure in upper bathyal species but little divergence in lower bathyal and abyssal species (Figure 6.14). A similar pattern is found in gastropods. The upper bathyal *Frigidoalvania brychia* shows extreme divergence in mitochrondrial DNA on small spatial scales (Quattro et al. 2001), whereas two abyssal prosobranch gastropods, *Benthonella tenella* (Boyle et al. in preparation) and *Xyloskenea naticiformis* (unpublished data), exhibited virtually no variation. In many cases, the genetic divergence at bathyal depths is sufficient to suggest cryptic species (e.g., France 1994, France and Kocher 1996b, Chase et al. 1998b, Etter et al. 1999, 2005, Kojima et al. 2001, Quattro et al. 2001, Howell et al. 2004), indicating that the greater environmental heterogeneity and potential isolating mechanisms at bathyal depths have culminated in the formation of new or incipient species.

Second, depth-related changes in morphological divergence of gastropod shell architecture parallel the genetic patterns we described earlier. Etter and Rex (1990) measured a series of conchological characters on eight species of prosobranch gastropods arrayed along a depth gradient in the western North Atlantic to test whether phenotypic variation differed with depth. Levels of intraspecific morphological variation within each species were quantified with a multivariate measure of phenotypic distance (Mahalanobis' distance). Similar to the genetic patterns, bathyal species exhibit considerably more intraspecific phenotypic divergence than abyssal species, with divergence decreasing exponentially with depth (Figure 6.15). Phenotypic divergence was highly negatively correlated with faunal similarity among the same locations (Figure 6.16), indicating bathymetric variation in morphological divergence mirrors and is possibly responding to changes in community structure. The highly significant correlation between phenotypic divergence and faunal similarity and the congruent patterns of genetic divergence suggest that processes at the genetic, population, and community levels are intertwined and are responding to similar environmental changes with depth. The reductions in genetic, morphological, and faunal divergence with depth all suggest that the abyss is much less variable than bathyal depths and has little potential for divergence.

Third, the extraordinarily low organic flux to the abyssal regions of the World Ocean severely constrains the ability of organisms to grow, reproduce, and sustain viable populations, limiting their ability to evolve. Perhaps the single most important gradient influencing ecological and evolutionary opportunity in the deep sea is the rate of nutrient input from sinking phytodetritus, which decreases exponentially with depth (see Chapter 1).

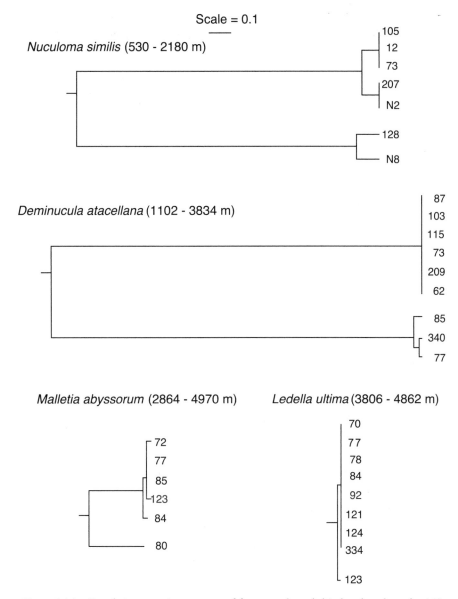

Figure 6.14. Population genetic structure of four protobranch bivalves based on the 16S mtDNA gene and derived from pairwise-modified coancestry coefficients among stations. Station numbers are shown at branch tips and are from a depth gradient in the North American Basin known as the Gay-Head Bermuda transect. The depth range of the samples used for each species is given. Note that the two bathyal species exhibit pronounced genetic structure, whereas the two abyssal forms show little divergence among samples. From Etter et al. (2005), with permission of the authors and Blackwell Publishing.

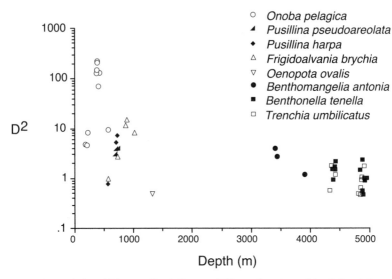

Figure 6.15. Mahalonobis' generalized distance (D^2) as a function of depth for eight species of prosobranch gastropods arrayed across a depth gradient in the North American Basin. Standardized morphological characters (e.g., shell length, width, shape) were measured and used to calculate the Mahalonobis' distance. Depth was determined as the midpoint in depth between each pair of conspecific populations. From Etter and Rex (1990), with permission of the authors and Elsevier.

Consequently, benthic standing stock decreases two to three orders of magnitude with depth across the bathyal zone. It reaches very low levels of around 10–100 individuals m^{-2} and 1 g m^{-2} or less at the base of the bathyal zone and then continues to gradually decline with increased distance seaward in the abyss (Rex et al. 2005a). For many species this translates into densities of much less than 1 individual m^{-2}. Given such low densities, it is difficult to imagine how many species, especially minute gonochoristic species with low adult mobility, could be reproductively self-sustainable. As we outlined in Chapter 4, this led Rex et al. (2005a) to argue that the abyss may function as a sink habitat with much of the abyssal fauna existing as nonsustainable populations maintained by continued immigration from more successful bathyal sources. If true, abyssal environments would provide little chance for populations to diverge or respond to selective pressures and would lack sufficient ecological opportunity to support new species.

Finally, a rapidly expanding body of evidence from paleontology (Jablonski 1993, Flessa and Jablonski 1996, Sepkoski 1998, Buzas et al. 2004, Jablonski et al. 2006), comparative phylogenetics (Cardillo 1999, Davies et al. 2004, Cardillo et al. 2005, Ricklefs 2007, McPeek and Brown 2007, Wiens

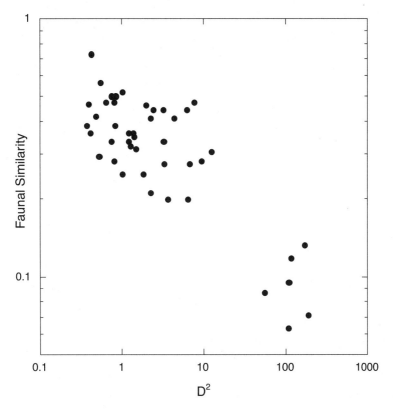

Figure 6.16. Percentage faunal similarity (based on presence/absence) of prosobranch gastropods in the North American Basin as a function of phenotypic divergence (Mahalonobis' D^2). The same stations were used for calculating both faunal similarity and D^2. From Etter and Rex (1990).

2007, Wiens et al. 2007), and molecular evolution (Martin and Palumbi 1993, Bromham and Cardillo 2003, Wright et al. 2003, Martin and McKay 2004, Williams and Reid 2004, Xiang et al. 2004, Gillooly et al. 2005, Vellend 2005, Allen and Gillooly 2006, Gillooly and Allen 2007, Kelly and Eernisse 2007, McPeek and Brown 2007, Ricklefs 2007, Wiens 2007) suggests that geographic variation in evolutionary rates may play an important role in producing large-scale gradients in diversity. Similarly, depth-related variation in evolutionary rates is very likely a factor in species diversity being greatest at bathyal depths.

The patterns of genetic and morphological variation within and among species, the extraordinarily low food supply to the abyss, and the correlation between evolutionary rates and species diversity all suggest that the bathyal environment may be the center for the adaptive radiation and the origin of the endemic deep-sea fauna. The potential for isolation is greater at bathyal depths and once gene pools are fragmented, the greater geographic, bathymetric, and temporal heterogeneity will favor rapid selective divergence. Even without specific isolating mechanisms, the strong environmental gradients and greater biotic and abiotic heterogeneity at bathyal depths might impose different selective regimes that increase the probability of population differentiation and speciation (Doebeli and Dickmann 2003, Hare et al. 2005, Irwin et al. 2005, Polechova and Barton 2005).

CONCLUSIONS

Biogeographic variation in species diversity represents the culmination of ecological and evolutionary processes operating across a wide variety of spatial and temporal scales. Attempts to explain existing geographic and bathymetric patterns of diversity in the deep sea have focused primarily on the ecological causes. It is becoming apparent that spatiotemporal differences in evolution may play a fundamental role in creating and maintaining large-scale gradients in diversity in shallow-water and terrestrial ecosystems. Thus it is clear that to develop a more comprehensive understanding of the forces that shape macroecological patterns in the deep sea we have to incorporate the effects of geographic, bathymetric, and temporal variation in the processes that generate new species.

Our knowledge of evolution in this vast and remote ecosystem is at a very formative stage and we have just begun to document the scales and geography of population differentiation. Yet, even at this early stage hints about potentially important patterns have emerged. Research over the past two decades has demonstrated that populations of deep-sea species can possess considerable genetic structure despite the absence of obvious isolating barriers. Population divergence occurs on a range of scales and appears to be associated with depth, distance, deep-water circulation, topographic features, various selective gradients, and historical environmental fluctuations, although the exact mechanisms in each case remain speculative. Depth differences have a much greater effect on levels of population differentiation than geographic separation within isobaths, probably because of the intense vertical environmental gradients. Finally, we propose that the continental margins, which form a narrow ribbon around the edges of the World Ocean, may be the primary site of evolution and adaptive radiation of the deep-sea fauna.

7

SUMMARY

The local community is an epiphenomenon that has relatively little explanatory power in ecology and evolutionary biology.... Ecologists are moving toward a community concept based on interactions between populations over a continuum of spatial and temporal scales within entire regions, including the population and evolutionary processes that produce new species.

Robert E. Ricklefs (2008)

In preceding chapters, we reviewed known patterns of biodiversity in the deep-sea benthos inhabiting soft sediments and attempted to integrate their potential causes across scales of time and space. Although knowledge of deep-sea ecosystem structure and function remains very incomplete, this synthesis was designed to provide a broader perspective, and we hope it will inform a future research agenda. Certainly, discoveries of the last two decades have fundamentally changed the way scientists look at deep-sea ecology. In this summary, we avoid long lists of references and instead refer the reader to the relevant chapters, where detailed case studies and citations can be found.

ENERGY AND STANDING STOCK

Compared to most terrestrial and coastal marine systems, the deep sea is an extremely energy-deprived environment, and energy constraints become more severe with increasing depth. The most conspicuous feature of the soft-sediment community is the exponential decrease in standing stock with depth. This is caused by the reduction of particulate organic carbon flux from surface production with increased depth and distance from nutrient-rich coastal waters. The complex process of pelagic-benthic coupling that conveys food to the deep sea is still imperfectly known, but its ultimate expression at the seafloor is benthic standing stock—the raw material that is shaped by ecological and evolutionary forces into biodiversity. As we showed in Chapter 1, all major size categories of the benthos (bacteria, meiofauna, macrofauna, and megafauna) decline significantly with depth. Not surprisingly the sequence of relative abundance levels across all depths is size related: bacteria > meiofauna > macrofauna > megafauna. As their populations are more vulnerable to low energy availability, larger animals (macrofauna and megafauna) decrease in abundance with depth at a higher rate than the minute meiofauna, which results in a decrease in average organism size with depth. Similarly, the biomass of all four groups diminishes with depth, animals declining much more rapidly than bacteria. There appears to be a shift from the upper bathyal zone, where macrofaunal biomass dominates the community, to lower bathyal and abyssal depths, where bacterial and meiofaunal biomass prevails.

Bathymetric patterns of standing stock are modulated by basin-to-basin variation in surface production, oxygen minimum zones, topographic focusing, oceanographic conditions that extend high production seaward, strong bottom currents that expose reactive sediments, proximity to sea ice, and lateral advection. However, the decrease in standing stock from the continental shelf to the abyss appears to be a general defining characteristic of the deep-sea environment. Since there are currently so few long-term studies of pelagic-benthic coupling, benthic standing stock remains the best indication that we have of spatial variation in the food supply that supports seafloor life. It serves as a proxy variable for productivity in this system that enables us to assess its ecological connections to both alpha and beta species diversity and to the evolutionary potential for diversification. Moreover, the changing relationship among depth trends in standing stocks of the major size categories of organisms is, in itself, a dimension of biodiversity. The diversity of body size and of consumers in the community decreases with depth from a complex assemblage in the bathyal zone to a simplified assemblage dominated by bacteria and meiofauna in the abyss.

Although geographic variation in standing stock is the best-documented aspect of benthic community structure, much more needs to be learned. A glance at Figure 1.11 shows that vast regions of the deep ocean have never been sampled, and doing so will certainly expand knowledge of patterns of standing stock and the complicated processes that regulate them. More important, it is necessary to continue to develop sampling programs that simultaneously combine measuring standing stock with ecologically related variables, including alpha and beta diversity, spatial dispersion, body size, natural history information, and environmental properties of sediment habitats. Relatively few studies have done this, so associations among standing stock, other facets of community structure, and environmental variables are mostly inferential.

BIODIVERSITY

The causes of biodiversity are scale dependent. The task of deep-sea ecologists is to integrate potential causes across scales of time and space to account for observed patterns of biodiversity. This may seem premature given the current state of knowledge, but a conservative interpretation of existing information begins to suggest how deep-sea biodiversity arose and is now distributed.

At the largest scales that we have examined, oceanic dimensions and geological time, patterns of species diversity reflect both evolutionary-historical processes and contemporary ecological factors that vary along geographic gradients. There are still no reliable estimates of total deep-sea biodiversity, but it is clear that the impressive levels of local diversity (see Chapter 2) cannot be extrapolated in a simple way to larger scales because many common and rare species appear to be very broadly distributed (see Chapter 4). New data on population genetic structure (Chapter 6) caution that some coherent morphospecies may actually be constellations of cryptic species, further complicating any attempt to determine the contribution of the deep-sea fauna to biodiversity of the global biosphere.

It is now obvious that there is considerable basin-to-basin variation in species diversity (see Chapters 3 and 4). A question of very general interest to ecologists and evolutionary biologists is whether there are oceanwide latitudinal gradients of diversity like those found in terrestrial and coastal marine systems. If so, this could shed light on the underlying causes of these major biogeographic phenomena because the deep sea is a distinctive environment where only a subset of the many proposed causal mechanisms could operate. There is intriguing evidence for poleward declines in diversity in some macrofaunal taxa (gastropods, bivalves, isopods, cumaceans) in the North Atlantic, but patterns in the South Atlantic are less convincing and exhibit strong regional variation. Depressed species richness at higher latitudes is accompanied by a decrease in evenness of the relative abundance distribution, suggesting that it may be caused by higher or more seasonal surface productivity translated down through the water column.

Deep-sea foraminiferans show latitudinal gradients in both hemispheres. Assemblages at high latitudes are dominated by opportunistic species that exploit sinking phytodetritus. The historical development of these latitudinal gradients during global cooling since the Eocene reveals the emergence of these same opportunists at high latitudes. In the North Atlantic, the diversity of deep-sea foraminiferans can be predicted by the seasonality of surface production. If productivity is implicated in contemporary large-scale patterns of diversity, it is difficult to tease out the potential roles of rate and seasonality. As we pointed out in Chapters 3 and 4, deep-sea diversity is depressed in all known circumstances of nutrient loading (topographic focusing in canyons and trenches, lateral advection and deposition of organically rich sediments, fluvial input, stimulation of bacterial growth through sediment erosion and deposition by strong currents, proximity to oxygen minimum zones), but the rate and seasonal variation of food supply nearly always covary. We stress that the association between productivity and diversity at large scales is tenuous and nothing definite is known about the proximal mechanisms through which productivity regulates community structure.

If diversity at higher latitudes is depressed by high and seasonal organic carbon flux, then perhaps the formation of latitudinal gradients during the Cenozoic was driven partly by a poleward increase in extinction rates. It is difficult to explain why the tropical deep sea would have higher origination rates or lower extinction rates, leading to the higher diversity found there. One possible evolutionary cause of higher tropical diversity in the deep sea is that the coastal tropical species pool available for invasion and subsequent speciation is simply much higher than those at temperate and polar latitudes.

In complete contrast to long-held assumptions, the deep sea has not been an environmentally stable environment on geological time scales. There is ample evidence for regional catastrophic events from massive submarine landslides, volcanic eruptions, and asteroid impacts. Global-scale anoxia has periodically extirpated much of the deep-sea foraminiferan biota. These episodes must also have limited the accessibility of the deep ocean for invasion and possibly accelerated extinction rates and retarded origination rates in animal taxa. The geological record of foraminiferans and ostracods in seabed cores shows that species diversity fluctuated in a highly regular way with glacial cycles during the Pliocene and Pleistocene, suggesting links to food supply and changes in the deep thermohaline circulation. The deep sea is not an isolated environment, but an integral part of the global biosphere.

Scientists have only just begun unraveling the evolutionary origin of the deep-sea fauna (see Chapter 6). The most basic question is: How and where do species form? This is an especially puzzling problem because the relative homogeneity and continuity of the soft-sediment environment and the long-range dispersal ability, similar body sizes, and feeding modes of many species would seem to violate basic tenets of the canonical allopatric speciation model. Another paradox is that periodic mass extinction events from global anoxia imply that much of the biodiversity of the deep-sea fauna must have arisen fairly recently during the Cenozoic. The deep-sea fauna, by and large, appears to have evolved from shallow-water progenitors, but phenotypic and genetic evidence suggests that emergence also occurred. We do not know the extent to which the fauna evolved through in situ radiation or secular migration (independent evolutionary change occurring during invasion). Research on physiological tolerance during development indicates that larvae of coastal species can withstand upper bathyal conditions enabling them to invade. Some bathyal species have evolved barophilic larvae whose physiological tolerances correspond to those in the depth range of adult populations, but ontogenetic vertical migration, even in abyssal species, also occurs.

Since there is very little fossil evidence, except for foraminiferans and ostracods in deep seabed cores, we must rely on molecular genetic evidence to reconstruct how evolution has unfolded in the deep sea. Patterns of differentiation can be inferred from genetic population structure. A review of existing trends (see Chapter 6) suggests a number of possible mechanisms that might affect divergence; but the evidence is inconclusive, primarily because it is so hard to relate genetic change to specific environmental features in the deep sea and to the dispersal potential of species. In such a vast and continuous ecosystem, isolation by distance seems like an obvious possibility, but the degree of divergence detected so far is not related to distance in a consistent way, and a surprising amount of differentiation occurs at relatively small within-basin scales. Evidence for the isolating influence of currents, topographic features, oxygen minimum zones, and vicariant events is mixed. One fairly consistent trend to emerge is that more genetic difference is associated with depth difference than with horizontal distance—hundreds of meters in depth can produce more differentiation than thousands of kilometers in horizontal separation.

The narrow more steeply descending bathyal zones that rim the oceans appear to be more conducive to population differentiation and speciation than the vast abyssal plains that stretch between the continental margins and mid-oceanic ridges (see Chapter 6). Across the bathyal zone, biological variables (such as the rate of food supply, biomass, abundance, body size, trophic structure, bioturbation, and species makeup) and physical variables (such as temperature, pressure, oxygen concentration, currents, and sediment composition) can change rapidly with depth over short distances. The slope face is also deeply incised by numerous canyons and gullies, creating a heterogeneous and fragmented environment. Potential selective gradients and opportunities for geographic isolation are more pronounced in the bathyal zone. The abyss is considerably more topographically and environmentally homogeneous. With low energy and extremely low population densities, the abyss affords more limited opportunity for coexistence through resource partitioning among the products of speciation.

The environmental differences between the bathyal zone and the abyss are reflected in patterns of population differentiation. In the western North Atlantic, mollusks show significant phenotypic and genetic clinal variation along depth gradients in the bathyal zone. Clines are generally steepest in the upper bathyal and become less pronounced with increased depth. Abyssal populations show remarkably little variation over huge areas (see Chapter 6). These clinal effects parallel rates of faunal replacement with depth (see Chapter 5), suggesting that both are responding to the same environmental gradients—strong in the bathyal zone and weaker in the abyss. We hypothesize that the bathyal zone is the center of evolution in the deep sea, with the abyss playing only a minor role. If this is correct, then the higher species diversity observed at bathyal depths may result from high origination rates and the passive accumulation of species, as well as the potential ecological causes we discussed in Chapter 3. It will be interesting to see whether these geographic patterns of differentiation are borne out in other taxa and in other regions of the World Ocean.

Understanding evolution at larger interbasin and global scales and how this has affected contemporary large-scale patterns of biodiversity, such as latitudinal gradients, requires determining the phylogeography of whole clades. Accomplishing this in a meaningful way presents huge logistical challenges. First, it necessitates more sampling to establish the geographic ranges of individual species as, except for some megafaunal taxa, very little is known about them. Second, a major initiative on basic alpha taxonomy of deep-sea groups is needed to ascertain prospective clades, which can then be analyzed by molecular genetic methods.

More is known about diversity-depth trends at regional scales than any other aspect of deep-sea biodiversity (see Chapter 3), but most of this research has centered on the macrofauna. These show a variety of patterns in different regions, but where there are typical offshore gradients of surface production and where the level of production is high to moderate, the predominant pattern is unimodal—diversity increases from the abyss to a peak at mid-bathyal depths and then decreases toward upper bathyal depths. The unimodal pattern cannot be explained as a null mid-domain effect caused by the coast and abyssal seafloor acting as physical boundary constraints. If the exponential decline in standing stock with depth can be accepted as a proxy for the rate of food supply, this diversity pattern is broadly consistent with the productivity-diversity theory. According to this theory, the increase in diversity from the abyss to mid-bathyal depths represents the accumulation of species with increased population size. Depressed diversity at upper bathyal depths may result from rapid population growth and intensified biological interactions driven by high and variable nutrient input. We stress, again, that the association between standing stock and productivity is inferential and that the proximal mechanisms of the productivity-diversity theory remain poorly characterized in the deep sea.

Much of the basin-to-basin variation in diversity-depth trends for the macrofauna can be explained by shifting the scale of productivity toward the lower range of realized values in circumstances where surface production, and consequently food supply to the benthos, is reduced. This is associated with a lowering of the overall level of diversity and a shifting of the mode to shallower depths until, under severely impoverished conditions, diversity simply decreases with depth, representing only the ascending limb of the unimodal productivity-diversity relationship (see Chapter 3). Diversity-depth relationships in the meiofauna and megafauna conform reasonably well to this explanation, although fewer data on both diversity and standing stock are available. In general, meiofaunal diversity is less sensitive to the decline in productivity and megafaunal diversity is more sensitive, as might be expected from their relative average body size and energy requirements. Meiofaunal diversity remains higher at abyssal depths, and megafaunal diversity becomes very low. While productivity-diversity relationships appear to apply to large functional groups, there are exceptions at the level of individual taxa. For example, macrofaunal isopods remain diverse at abyssal depths, possibly because they represent an ancient radiation in which species have evolved adaptations to exist at very low density. Megafaunal holothurians remain diverse in the abyssal eastern North Atlantic, where some species are apparently adapted to monopolize pulses of sinking phytodetritus.

The productivity-diversity theory is one that can be tested indirectly by using comparative data under a wide range of circumstances in the deep sea. However, it is clearly not the sole, or perhaps even the primary, ecological explanation for diversity-depth trends. There is more limited, but convincing, evidence for the effects of disturbance from strong currents, sediment grain-size diversity, and oxygen depletion as well. Undoubtedly, a complete explanation is multivariate and must await more experimental manipulations that vary these parameters in a controlled way. Enough comparative information now exists to design such an experiment.

Less is known about patterns of beta diversity (species turnover along depth gradients) than about alpha diversity (see Chapter 5). Measuring beta diversity requires careful taxonomic assessment of the fauna to determine how species are shared among sampling sites. The wide variety of multivariate methods used to analyze beta diversity makes it difficult to compare studies. Two generalizations to emerge from existing studies are that more turnover is associated with depth difference than horizontal separation along isobaths and that the rate of change is higher at bathyal depths than in the abyss. Both trends can be attributed to steeper environmental gradients with depth in the bathyal zone.

Unlike the patterns of alpha diversity, there is no coherent theory or set of theories that can account for beta diversity. In other ecosystems, it is usually assumed that species replacement along an environmental gradient represents a correspondence between the adaptive properties of species and resources arrayed along the gradient. The very fact that deep-sea species do have restricted depth ranges and an optimal depth where abundance is maximized suggests that adaptation is a factor, although ranges vary from basin to basin. A large number of potentially relevant adaptive features (feeding and habitat types, locomotion, metabolic rates, larval dispersal, body size, enzymatic activity) and environmental characteristics (sediment type, hydrographic conditions, oxygen content, physical oceanographic conditions, food supply) have been suggested, but very few specific interactions between species' adaptations and the environment are known (see Chapter 5). We point out that beta diversity is not an independent aspect of community structure as it is generally treated in deep-sea ecology, but rather is closely related to standing stock and to alpha diversity. Differences in alpha diversity limit similarity among samples and thus strongly affect the rate of change in species composition. Studies of standing stock (see Chapter 1) and trophic makeup (see Chapter 5) show that part of species replacement with depth can be attributed to the progressive loss of larger animals and consumers with decreased food supply. If source-sink dynamics between bathyal and abyssal populations are widespread (see Chapter 3), then lower rates of species replacement found at great depths simply reflect the fact that a small coherent subset of bathyal species with high dispersal ability inhabits the abyss. Food supply may strongly affect both alpha and beta diversity. For beta diversity, it may provide the overall context within which species-specific links between adaptations and environmental resources can operate.

Unfortunately, very little is known about beta diversity on very large basin-to-basin and oceanic scales. As with evolutionary studies, the most serious obstacles to more fully understanding variation in species composition at large scales are the paucity of taxonomic studies and the lack of information on the biogeographic distributions of individual species.

Local species coexistence in the deep sea remains an enigma (see Chapter 2). Individual box-core samples taken at upper bathyal depths in the western North Atlantic yield an impressive figure of some 100 macrofaunal species, about twice the number found on the adjacent continental shelf. Upper bathyal assemblages are characterized by more rare species and more even relative abundance distributions. It is important to stress, however, that not all regions of the deep sea support high local diversity (see Chapter 3). Diversity is much higher on the continental margin than in the abyss and varies among basins (see Chapter 4). Thus, it is more meaningful to consider local diversity in a specific geographic context. Much of the variation in local diversity may represent ecological and evolutionary processes that operate at much larger scales of time and space. These processes ultimately form the regional species pools that contribute species to local communities and favor the life-history traits that permit them to disperse among local habitats.

The ecological causes that might mediate coexistence at local scales have been the object of much debate (see Chapter 2). Most field research on community structure at local scales has been driven by the spatial mosaic theory—the idea that high diversity is fostered by small physical and biological disturbances at the seafloor that create patches of successional sequences that are temporally out of phase. Predictions of the model have been tested by precision-sampling studies designed to detect patchiness at the ex-

pected scales and by controlled manipulative experiments that monitor colonization of defaunated or enriched sediments. Some patchiness does exist,
but at small scales most species appear to be randomly distributed. Succession does occur as a response to disturbance, but the number of affected
species is small and typically includes opportunists that are rare or not even
present in background communities. Collectively, these findings do not suggest that the spatial mosaic theory accounts for high diversity. However, there
is a clear need for much more work to test the significance of patchiness in
maintaining diversity. The technology and numerical analytical methods to
conduct this research are now well developed, but progress has been hin-
dered by the vagaries of funding and the severe logistical difficulties of doing fine-scale detailed research in the deep-sea sediment environment.

We also urge that new experiments be designed to test alternative theories. A promising hypothesis with strong comparative support is that increased sediment grain-size diversity enhances species diversity by providing
more food resources or microhabitat heterogeneity to partition. Experimental tests would provide insight into the potential role of equilibrial
processes, as opposed to the spatial mosaic theory, which relies on a nonequilibrial mechanism.

There is also evidence suggesting that not all deep-sea communities are
structured by local ecological factors. For mollusks, abyssal assemblages may
be composed largely of noninteracting sink populations (see Chapter 3).
Many abyssal populations appear to be too sparsely distributed to be reproductively viable. They are primarily deeper range extensions for a subset of
bathyal species with larval dispersal ability. This suggests that abyssal populations experience chronic extinction from vulnerabilities to Allee effects and
are maintained by continued immigration from bathyal sources. If this is true,
abyssal communities may be ecologically unstructured. This hypothesis can
be tested in other taxa by documenting depth range distributions, by
population genetic analyses for the asymmetrical gene flow and haplotype
distributions expected in source-sink dynamics, and by examining the reproductive state of conspecific bathyal and abyssal populations.

Finally, we should consider the possibility, implied in Robert Ricklefs's
quote at the beginning of this chapter, that "local communities" are illusory
as meaningful discrete entities that are primarily regulated by biological interactions at small scales in ecological time. This may be true even of the
highly diverse communities found at bathyal depths. Our use of the terms
local, regional, and global is a heuristic convenience dictated by the very frag-

mented picture available of community structure at different scales in the deep sea. But space and time are continua and the processes that shape community structure operate at all scales. Local communities in the deep-sea soft-sediment environment are not physically bounded units. At local scales, species tend to have random dispersion patterns and interspecific interactions appear to be weak (see Chapter 2). Most deep-sea species are small deposit feeders with dispersal ability. Local diversity is a positive function of regional diversity and dispersal potential, at least in gastropods, which is the one group in which these relationships have been analyzed. This suggests that regional enrichment influences local diversity. Species diversity at any particular site may simply represent the overlap of species generated by evolutionary diversification whose biogeographic distributions are determined by interactions with the environment and other species at larger scales. Currently, it is difficult to assess the relative impact of large- and small-scale processes on local community structure, but it is now clear that our understanding of patterns and causes of biodiversity in the deep sea would benefit from more macroecological research.

Meiofauna. The deep-sea meiofauna is composed of minute organisms that are retained on the 32 μm mesh sieves that are used to separate organisms from surrounding sediment, although it is often defined as a specific subset of taxa. The meiofauna includes two components, the protozoan foraminiferans and the metazoan (animal) meiofauna. Shown here are representatives of the most common groups.

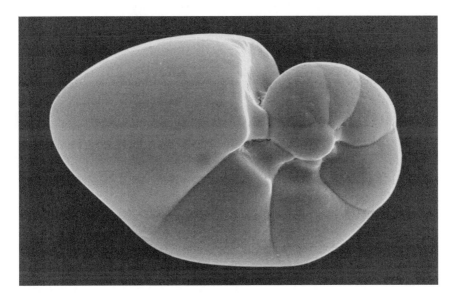

Figure A.1. The foraminiferan *Nonionella iridea* with a calcium carbonate test, 170 μm in width, collected at 1960 m in the Rockall Trough of the eastern North Atlantic. For the giant agglutinated foraminiferan *Bathysiphon filiformis* see Figure A.32. Courtesy of Andrew Gooday, from Gooday and Hughes (2002), with permission of the authors and Elsevier.

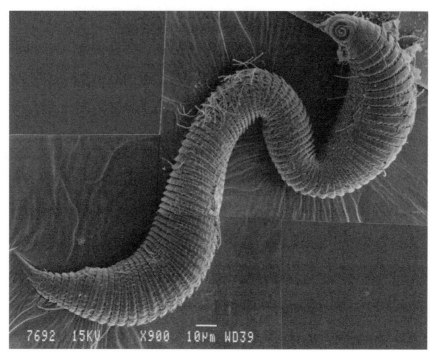

Figure A.2.   The nematode *Bathyepsilonema lopheliae* with an annulated cuticle collected at 1005 m in the Porcupine Seabight of the eastern North Atlantic, associated with the coral *Lophelia*. Courtesy of Ann Vanreusel, photograph by Maarten Raes, from Raes et al. (2003), with permission of the authors and Springer.

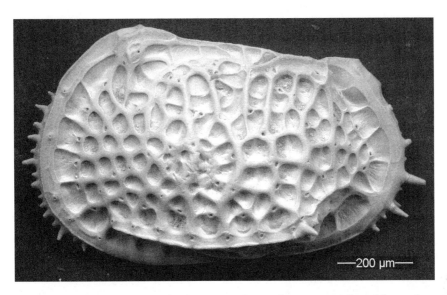

Figure A.3. Left valve of the shell of the ostracode *Poseidonamicus* sp. collected from a deep seabed core in the western equatorial Atlantic at 3040 m, Quaternary period, 490,000 ka. Courtesy of Moriaki Yasuhara.

Figure A.4. A harpacticoid copepod of the family Canuellidae collected at 2893 m in the Scotia Sea off Antarctica, length 400 μm. Courtesy of Pedro Martinez Arbizu.

Macrofauna. Macrofaunal species are larger than meiofaunal species and represent most marine phyla. They typically range in size from about a millimeter to several centimeters and are defined as those organisms retained on 250 to 300 μm sieves used to separate animals from surrounding sediment. The three most dominant groups are the polychaetes, the peracarid crustaceans, and the mollusks in declining order of abundance and species diversity.

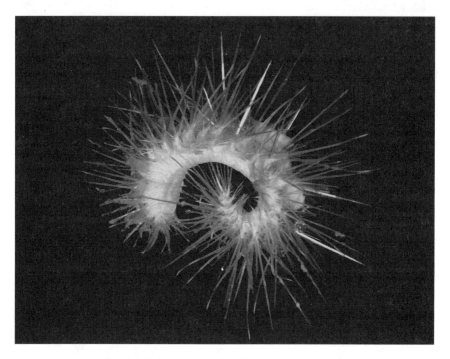

Figure A.5.   Macrofaunal polychaete. The epibenthic carnivore *Eunoe spica,* 25 mm in length, collected at 595 m in the Weddell Sea off Antarctica. A color version of this figure is included in the insert following p. 50. Courtesy of Brigitte Ebbe, photograph by Dieter Fiege.

Figure A.6. Macrofaunal polychaete. The epibenthic carnivore *Macellicephala laubieri,* 7 mm in length, collected at 1249 m in the Aegean Sea of the Mediterranean. Courtesy of Brigitte Ebbe, photograph by Dieter Fiege.

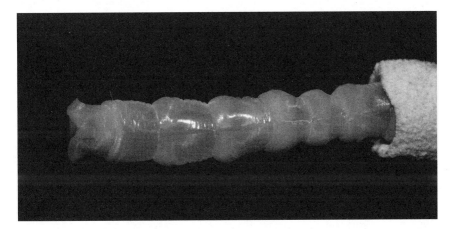

Figure A.7. Macrofaunal polychaete. A subsurface deposit feeding tube dweller of the family Maldanidae, 10 mm in length, collected at 595 m in the Weddell Sea off Antarctica. A color version of this figure is included in the insert following p. 50. Courtesy of Brigitte Ebbe, photograph by Dieter Fiege.

Figure A.8. Macrofaunal polychaete. The burrowing subsurface deposit feeder *Oligobregma* sp., 30 mm in length, collected at 5467 m in the Angola Basin, eastern South Atlantic. Courtesy of Brigitte Ebbe, photograph by Dieter Fiege.

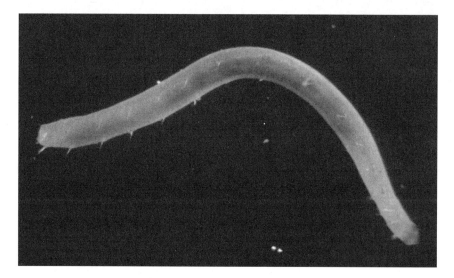

Figure A.9. Macrofaunal polychaete. A burrowing deposit feeder at the sediment-water interface, *Fauveliopsis brevis,* 7 mm in length, collected at 4739 m in the Weddell Sea off Antarctica. Courtesy of James Blake.

Figure A.10. Macrofaunal polychaete. A burrow-dwelling surface deposit feeder, *Aphelochaeta* sp., image of body 6 mm across, collected at 1000 m in Monterey Bay off California. A color version of this figure is included in the insert following p. 50. See also Figure A.29 for the quill worm *Hyalinoecia tubicola.* Courtesy of James Barry and Linda Kuhnz, 2003 MBARI.

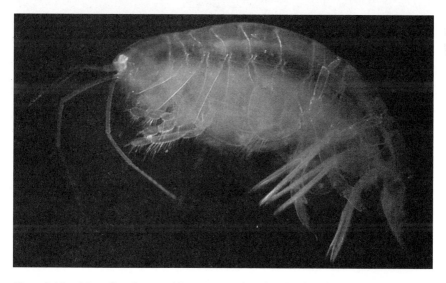

Figure A.11. Macrofaunal peracarid crustacean. The tube-dwelling filter-feeding amphipod *Ampelisca unsocalae,* length 4 mm, collected at 1040 m in Monterey Canyon off California. A color version of this figure is included in the insert following p. 50. Courtesy of James Barry and Linda Kuhnz, 2003 MBARI.

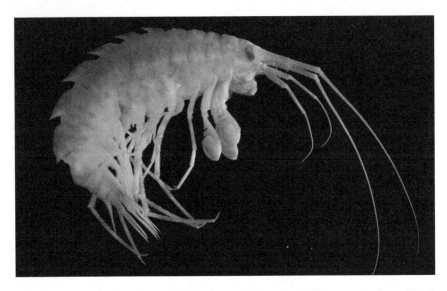

Figure A.12. Macrofaunal peracarid crustacean. A large (probably scavenging) amphipod, *Eusirus giganteus,* 8 cm in length, collected at 4000 m off King George Island (Antarctica) in the Southern Ocean. Although belonging to a largely macrofaunal taxon, scavenging amphipods often attain megafaunal size. Courtesy of Angelika Brandt.

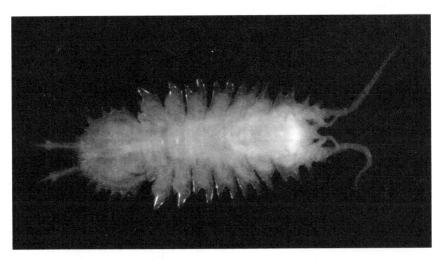

Figure A.13. Macrofaunal peracarid crustacean. A juvenile deposit-feeding isopod, *Acanthaspidia* sp., 4 mm in length, collected at 2700 m in the Weddell Sea off Antarctica. A color version of this figure is included in the insert following p. 50. Courtesy of Angelika Brandt.

Figure A.14.   Macrofaunal peracarid crustacean. A deposit feeding cumacean, *Cyclaspis* sp., length 3 mm, collected at 3000 m in the Lazarev Sea off Antarctica. A color version of this figure is included in the insert following p. 50. Courtesy of Angelika Brandt.

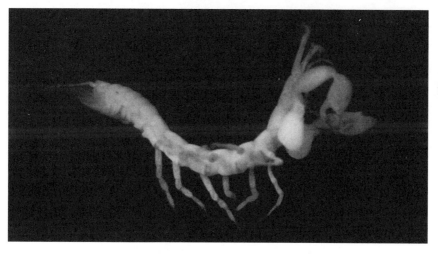

Figure A.15.   Macrofaunal peracarid crustacean. A deposit-feeding tanaid, *Neotanais antarcticus,* length 10 mm, collected at 4000 m off Elephant Island (Antarctica) in the Southern Ocean. A color version of this figure is included in the insert following p. 50. Courtesy of Angelika Brandt.

Figure A.16. Macrofaunal mollusk. The deposit-feeding protobranch bivalve *Deminucula atacellana*, 2.67 mm in length, collected at 2040 m in the western North Atlantic. A color version of this figure is included in the insert following p. 50. Courtesy of Elizabeth Boyle and Selina Våge.

Figure A.17. Macrofaunal mollusk. The deposit-feeding gastropod *Benthonella tenella*, 3.84 mm in height, collected at 3800 m in the western North Atlantic. A color version of this figure is included in the insert following p. 50. Photograph by Michael Rex, from Rex and Etter (1990), with permission of Elsevier.

Figure A.18. Macrofaunal mollusk. The predatory gastropod *Benthomangelia antonia,* 11.53 mm in height, collected at 3800 m in the western North Atlantic. Photograph by Michael Rex, from Rex and Etter (1990), with permission of Elsevier.

Megafauna. The megafauna is composed of larger organisms that are readily visible in bottom photographs or collected in large trawls with mesh sizes larger (1–3 cm) than those used for macrofauna. The megafauna includes both invertebrates and fishes. We use bottom photographs as much as possible because they help convey an impression of bottom topography, sediment surface texture, and the epifaunal community.

Figure A.19.   Megafaunal fish. The sit-and-wait predator *Bathypterois* sp. at 1600 m on the Yakutat Seamount in the Corner Rise Seamounts of the western North Atlantic. A color version of this figure is included in the insert following p. 50. Courtesy of Peter Auster and the Deep Atlantic Stepping Stones Science team / NOAA / IFE / URI-IAO.

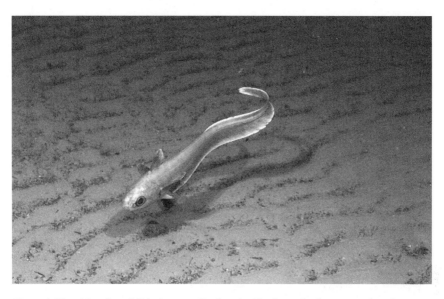

Figure A.20. Megafaunal fish. A generalist foraging both on the bottom and in the water column: *Coryphaenoides rupestris,* at 1340 m on the Manning Seamount of the New England Seamounts of the western North Atlantic. A color version of this figure is included in the insert following p. 50. Courtesy of Peter Auster and the Deep Atlantic Stepping Stones Science team / NOAA/IFE/URI-IAO.

Figure A.21. Megafaunal fish. A forager on bottom invertebrates: *Aldrovandia affinis,* at 1900 m in the western North Atlantic. Note the textured sediment with biogenic mounds, pits, and tracks, along with typical mid-bathyal ophiuroids and anemones. Courtesy of Barbara Hecker, ACSAR project.

Figure A.22. Megafaunal fish. A forager on bottom invertebrates: unidentified halosaur at 1900 m in the western North Atlantic. Note the textured sediment with biogenic mounds, pits, and tracks. Courtesy of Barbara Hecker, ACSAR project.

Figure A.23. Megafaunal fish. A forager on bottom invertebrates: *Lycodes atlanticus,* at 1470 m in Lydonia Canyon on the U.S. continental margin of the western North Atlantic. The globular objects on the sediment surface are xenophyophores, which are megafaunal protozoans, possibly related to foraminiferans. Courtesy of Barbara Hecker, ACSAR project.

Figure A.24. Megafaunal fish. A forager on bottom invertebrates: *Hydrolagus affinis,* at 2300 m off Newfoundland. A color version of this figure is included in the insert following p. 50. Courtesy of Richard Haedrich, Krista Baker, Chevron Canada Ltd., and the SERPENT project.

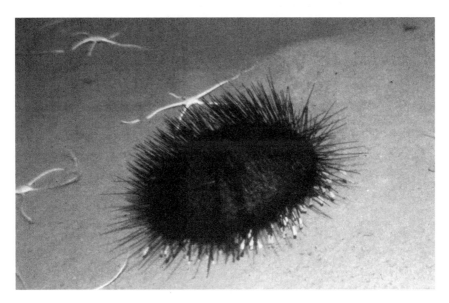

Figure A.25. Megafaunal invertebrate. The deposit-feeding sea urchin *Hygrosoma petersii* and the omnivorous ophiuroid *Ophiomusium lymani* at 1800 m in the western North Atlantic south of New England. A color version of this figure is included in the insert following p. 50. Courtesy of Ruth Turner, from Rex et al. (2006), with permission of Inter-Research.

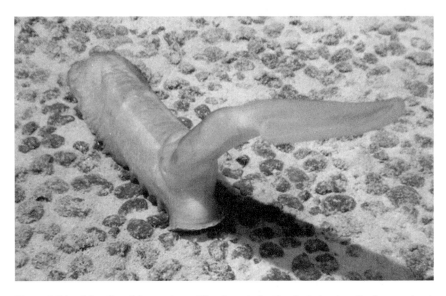

Figure A.26. Megafaunal invertebrate. The deposit feeding holothurian *Psychropotes longicauda* at 5000 m in a manganese nodule field in the tropical East Pacific. The animal is 35 cm in length. Copyright Ifremer-Nautile / Cruise Nodinaut 2004 to study biodiversity of the French Claim Nodule Area. A color version of this figure is included in the insert following p. 50. Courtesy of Joëlle Galéron and Stefanie Keller.

Figure A.27. Megafaunal invertebrate. A filter-feeding brisingid seastar at 5600 m in the central North Pacific, perched on a manganese nodule. Manganese nodule beds cover much of the abyssal seafloor, particularly in the Indo-Pacific. A color version of this figure is included in the insert following p. 50. Courtesy of Robert Hessler.

Figure A.28. Megafaunal invertebrate. The suspension-feeding sea pens *Pennatula aculeata* at 432 m in Lydonia Canyon on the U.S. continental margin of the western North Atlantic. A color version of this figure is included in the insert following p. 50. Courtesy of Barbara Hecker, ACSAR project.

Figure A.29. Megafaunal invertebrate. The suspension-feeding anemone *Bolocera* at 347 m in Lydonia Canyon on the U.S. continental margin of the western North Atlantic. The horizontal elongated strawlike structures next to the anemone and in the background are the tubes of the surface-dwelling motile carnivore *Hyalinoecia tubicola*, a megafaunal polychaete. Courtesy of Barbara Hecker, ACSAR project.

Figure A.30. Megafaunal invertebrate. The omnivorous anemone *Actinoscyphia aurelia* at 1947 m in Atwater Canyon in the eastern Gulf of Mexico. A color version of this figure is included in the insert following p. 50. Courtesy of Ian R. MacDonald, photo by Will Sager.

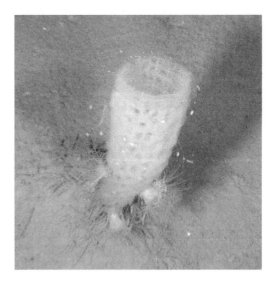

Figure A.31. Megafaunal invertebrate. A suspension-feeding glass sponge with anemones at 2876 m in the Gulf of Mexico off Florida. A color version of this figure is included in the insert following p. 50. Courtesy of Ian R. Mac-Donald.

Figure A.32. Megafaunal protozoan. A dense bed of the agglutinated deposit-feeding megafaunal foraminiferan *Bathysiphon filiformis* at 1897 m off North Carolina on the U.S. continental slope of the western North Atlantic. The tubes project about 10 cm above the sediment, with about 1 cm embedded in the sediment. Courtesy of Barbara Hecker, ACSAR project.

Figure A.33. Megafaunal invertebrate. The giant sea spider *Colossendeis* sp., whose legs can span 50 cm, at 1735 m in the Baltimore Canyon on the U.S. continental slope of the western North Atlantic. It appears to be a predator on cnidarians and possibly other benthos. Courtesy of Barbara Hecker, ACSAR project.

APPENDIX B:

DEEP-SEA SAMPLING GEAR

Figure B.1.   Epibenthic sled. Epibenthic sleds are designed to obtain large qualitative samples of the macrobenthic infauna living near the sediment-water interface. There are numerous designs, all of which have broad sled runners to prevent sinking into the soft sediment and a net to capture organisms mounted in a protective frame. The mesh size of the net (1.0 mm) allows sediments to escape while retaining the animals. Epibenthic sled samples first revealed the surprisingly high species diversity in the deep-sea macrobenthos (Hessler and Sanders 1967). The model shown being deployed here (Brandt and Barthel 1995, Brenke 2005) has two nets, the lower one to collect infauna and the upper one to capture suprabenthic organisms and lighter infaunal elements displaced up into the water column through turbulence caused by the sled. The lever below the sled opens doors to the nets while on the bottom and closes them on recovery to contain the sample. Samples are sieved (mesh size 300 μm) using an elutriation method (Sanders et al. 1965) to separate out the macrofauna and handpick for the megafauna. Most meiofauna are lost. Photograph courtesy of Angelika Brandt.

Figure B.2. Box corer. Box corers are the most widely used gear to obtain quantitative samples of the macrobenthos (Hessler and Jumars 1974). They are often subsampled with small cores for meiofauna and bacteria as well. The sample box in the center of the frame measures 50 × 50 cm and penetrates the sediment to a depth of around 50 cm. When the frame hits bottom, the central column forces the box down into the sediment. The pivoting gate, shown closed, then swings under the box to prevent loss of the sample. Box cores collect a block of relatively undisturbed sediment with a surface area of 0.25 m². The box can be partitioned into subcores to measure spatial dispersion of species, sediment characteristics, or biological and chemical processes (Jumars 1975a). Sediments sampled are sieved using an elutriation method (Sanders et al. 1965) to separate out the macrofauna (mesh size 300 μm) and meiofauna (32 μm). Photograph courtesy of Angelika Brandt.

Figure B.3.   Multicorer. Multiple-coring devices are the most effective way to collect quantitative samples of the meiofauna (Barnett et al. 1984, Bett et al. 1994). Box corers produce a bow wave on descent that can blow away superficial sediment layers and the minute organisms that live there. When the lighter frame of the multicorer lands on the bottom, hydraulic dampening is used to slowly insert the plastic core tubes (25 cm^2) into the sediment. The resulting samples retain even the most flocculent surface material and yield meiofaunal densities about twice those obtained from box cores. Multicorers are also more effective at collecting macrofauna (Hughes and Gage 2004), but the samples are much smaller than box-core samples. Several laboratory methods have been developed to extract meiofauna from sediments in sieve residues, but the most accurate approach is the tedious process of hand sorting samples under a microscope (Soltwedel 2000). Photograph courtesy of Angelika Brandt.

Figure B.4. Camera sled. Photographic surveys provide the most accurate estimates of megafaunal abundance, provided they cover an adequate area or a sufficient period of time. The towed camera shown here rides along the bottom on skids, illuminates an area of about 5 m² with a strobe light, and takes an exposure every 15 s over about 13 h of bottom time (Hecker 1990a). Surveys are also possible from cameras mounted on submersibles and re-motely operated vehicles, and by stationary time-lapse cameras. Density and body-size data can be converted to biomass by using dredged material to calculate a size-weight conversion formula (Ohta 1983). Trawl samples, the other common way to collect the megabenthos, can be used to estimate density and biomass directly as long as bottom time can be determined accurately by an odometer or by acoustic telemetry (Lampitt et al. 1986, Billett et al. 2001). Neither photographic surveys nor trawls adequately census the deep-burrowing megafauna, and actively swimming demersal species can evade both. Photograph courtesy of Barbara Hecker.

Bacterial data from Aller et al. 2002 (western North Atlantic, western South Pacific), Boetius et al. 1996 (Mediterranean), Duineveld et al. 2000 (Mediterranean), Kröncke et al. 1994 (Arctic), Kröncke et al. 2000 (Arctic), Levin et al. 1991b (eastern Tropical Pacific), Relexans et al. 1996 (eastern Tropical Atlantic), Schaff and Levin 1994 (western North Atlantic), Soltwedel and Vopel 2001(Norwegian Sea), Soltwedel et al. 2000 (Arctic), Tietjen et al. 1989 (western North Atlantic, western Tropical Atlantic–Puerto Rico Trench), Vanreusel et al. 1995 (eastern North Atlantic), Yingst and Rhoads 1985 (Gulf of Mexico).

Meiofaunal data from Aller et al. 2002 (western North Atlantic), Alongi and Pichon 1988 (western South Pacific), Alongi 1992 (western Tropical Pacific), Ansari et al. 1980 (Arabian Sea), Coull et al. 1977 (western North Atlantic), Danovaro et al. 1995 (Mediterranean), Danovaro et al. 2000 (Mediterranean), Danovaro et al. 2002 (eastern South Pacific–Atacama Trench), de Bovée et al. 1990 (Mediterranean), Dinet and Vivier 1977 (eastern North Atlantic), Dinet 1973 (eastern South Atlantic), Dinet 1974 in Soltwedel 2000 (Norwegian Sea), Dinet 1976 (Mediterranean), Dinet in Vincx et al. 1994 (eastern Tropical Atlantic), Duineveld et al. 1997 (western Tropical Indian), Escobar et al. 1997 (Gulf of Mexico), Fabiano and Danovaro

1999 (Antarctic), Ferrero in Vincx et al. 1994 (eastern North Atlantic), Flach et al. 2002 (eastern North Atlantic), Galéron et al. 2000 (eastern Tropical Atlantic), Galéron et al. 2001 (eastern North Atlantic), Gooday in Vincx et al. 1994 (eastern North Atlantic), Heip et al. 2001 (eastern North Atlantic), Herman and Dahms 1992 (Antarctic), Kröncke et al. 2000 (Arctic), Lambshead and Ferrero in Vincx et al. 1994 (eastern North Atlantic), Levin and Thomas 1989 (central Tropical Pacific), Levin et al. 1991b (eastern Tropical Pacific), Parulekar et al. 1982 in Soltwedel 2000 (Antarctic), Pequegnat et al. 1990 (Gulf of Mexico), Pfannkuche and Thiel 1987 (Arctic), Pfannkuche et al. 1983 (eastern North Atlantic), Pfannkuche et al. 1990 in Vincx et al. 1994 (eastern North Atlantic), Pfannkuche 1985 (eastern North Atlantic), Relexans et al. 1996 (eastern Tropical Atlantic), Romano and Dinet 1981 in Soltwedel 2000 (Arabian Sea), Rutgers van der Loeff and Lavaleye 1986 (eastern North Atlantic), Shirayama and Kojima 1994 (western North Pacific), Shirayama 1983 (western Tropical, western North Pacific), Sibuet et al. 1984 (western Tropical Atlantic), Sibuet et al. 1989 (eastern North, Tropical, eastern South Atlantic), Snider et al. 1984 (central North Pacific), Soetaert et al. 1991 (Mediterranean), Soltwedel and Thiel 1995 (eastern Tropical Atlantic), Soltwedel et al. 2000 (Arctic), Soltwedel 1997a (eastern Tropical Atlantic), Sommer and Pfannkuche 2000 (Arabian Sea), Tahey et al. 1994 (Mediterranean), Thiel 1966, 1975 (western Tropical Indian), Thiel 1975 (Norwegian Sea, eastern North Atlantic), Thiel 1979a (Red Sea), Thiel 1982 (eastern North Atlantic), Thistle et al. 1985 (western North Atlantic), Tietjen 1971 (western North Atlantic), Tietjen et al. 1989 (western Tropical Atlantic–Puerto Rico Trench), Vanaverbeke et al. 1997 (eastern North Atlantic), Vanhove et al. 1995 (Antarctic), Vanreusel and Vincx in Vincx et al. 1994 (eastern North Atlantic), Vanreusel et al. 1992 (eastern North Atlantic), Vanreusel et al. 1995 (eastern North Atlantic), Vivier 1978 (Mediterranean), Wigley and McIntyre 1964 (western North Atlantic).

Macrofauna data from Aller et al. 2002 (western North Atlantic), Alongi 1992 (western Tropical Pacific), Blake and Grassle 1994 (western North Atlantic), Blake and Hilbig 1994 (western North Atlantic), Carey and Ruff 1974 (Arctic), Carey 1981 (eastern North Pacific), Clough et al. 1997 (Arctic), Cosson et al. 1997 (eastern Tropical Atlantic), Dahl et al. 1976 (Norwegian Sea), Dauwe et al. 1998 (North Sea), Desbruyères et al. 1980 (eastern North Atlantic), Duineveld et al. 2000 (Mediterranean), Flach and Heip 1996 (eastern North Atlantic), Flach et al. 2002 (eastern North Atlantic), Frankenberg and Menzies 1968 (eastern Tropical Pacific), Gage 1977 (eastern North Atlantic), Gage 1979 (eastern North Atlantic), Galéron et al. 2000 (eastern Tropical Atlantic), Galéron et al. 2001 (eastern North Atlantic), Grassle and Morse-Porteous 1987 (western North Atlantic), Grassle 1977 (western North Atlantic), Griggs et al. 1969 (eastern North Pacific), Hecker and Paul 1979 (eastern Tropical Pacific), Hessler and Jumars 1974 (central North Pacific), Houston and Haedrich 1984 (western North Atlantic), Hyland et al. 1991 (eastern North Pacific), Jaźdźewski et al. 1986 (Antarctic), Jumars and Hessler 1976 (central North Pacific–Aleutian Trench), Khripounoff et al. 1980 (western Tropical Atlantic), Kröncke 1998 (Arctic), Kröncke et al. 2000 (Arctic), Kröncke et al. 2003 (Mediterranean), Laubier and Sibuet 1979 (eastern North Atlantic), Levin and Thomas 1989 (central Tropical Pacific), Levin et al. 1991a (eastern Tropical Pacific), Levin et al. 2000 (Arabian Sea), Maciolek and Grassle 1987 (western North Atlantic), Maciolek et al. 1987a,b (western North Atlantic), Nichols and Rowe 1977 (eastern Tropical Atlantic), Pfannkuche et al. 1983 (eastern North Atlantic), Rhoads et al. 1985 (western North Pacific), Richardson et al. 1985 (Caribbean), Richardson et al. 1995 (western Tropical Atlantic–Puerto Rico Trench), Romero-Wetzel and Gerlach 1991 (Norwegian Sea), Rowe and Menzel 1971 (Gulf of Mexico), Rowe 1971b (eastern Tropical Pacific), Rowe et al. 1974 (western North Atlantic, Gulf of Mexico), Rowe et al. 1975 (western North Atlantic), Rowe et al. 1982 (western North Atlantic), Sanders 1969 (eastern South Atlantic), Sanders et al. 1965 (western North Atlantic), Schaff et al. 1992 (western North Atlantic), Shirayama 1983 (western Tropical, western North Pacific), Sibuet et al. 1984 (western Tropical Atlantic), Sibuet et al. 1989 (eastern North, Tropical, eastern South Atlantic), K. L. Smith 1978 (western North Atlantic), K. L Smith 1987 (eastern North, central North Pacific), Spiess et al. 1987 (eastern Tropical Pacific), Tselepides and Eleftheriou 1992 (Mediterranean), Tselepides et al. 2000a (Mediterranean), Witte 2000 (Arabian Sea).

    Megafaunal data from Bett et al. 1995 (eastern North Atlantic), Christiansen and Thiel 1992 (eastern North Atlantic), Grassle et al. 1975 (western North Atlantic), Hecker 1990b (western North Atlantic), Heip et al. 2001 (eastern North Atlantic), Kallianiotis et al. 2000 (Mediterranean), Kaufmann et al. 1989 (central Tropical Pacific), Kröncke and Türkay 2003 (eastern Tropical, eastern South Atlantic), Levin et al. 1991a (eastern Tropical Pacific), Ohta 1983 (western North Pacific), Richardson and Young 1987 (Caribbean), Sardà et al. 1994 (Mediterranean), Sibuet et al. 1984 (western Tropical Atlantic), Sibuet et al. 1989 (eastern North, Tropical, eastern South Atlantic).

Bacterial data from Aller et al. 2002 (western North Atlantic), Boetius et al. 1996 (Mediterranean), Danovaro et al. 1995 (Mediterranean), Deming and Yager 1992 (western North Atlantic, Norwegian Sea, eastern North Atlantic, western Tropical Atlantic), Kröncke et al. 1994 (Arctic), Rowe et al. 1997 (Norwegian Sea), Soltwedel and Vopel 2001 (Norwegian Sea), Tietjen et al. 1989 (western Tropical Atlantic–Puerto Rico Trench), Vanhove et al. 1995 (Antarctic).

Meiofaunal data are from Aller et al. 2002 (western North Atlantic), Danovaro et al. 1995 (Mediterranean), Danovaro et al. 2000 (Mediterranean), Danovaro et al. 2002 (eastern South Pacific–Atacama Trench), Fabiano and Danovaro 1999 (Antarctic), Flach et al. 2002 (eastern North Atlantic), Galéron et al. 2000 (eastern Tropical Atlantic), Heip et al. 2001 (eastern North Atlantic), Jensen et al. 1992 (Norwegian Sea), Pfannkuche and Thiel 1987 (Arctic), Pfannkuche et al. 1983 (eastern North Atlantic), Pfannkuche 1985 (eastern North Atlantic), Relexans et al. 1996 (eastern Tropical Atlantic), Rowe et al. 1997 (Norwegian Sea), Shirayama 1983 (western Tropical, western North Pacific), Sibuet et al. 1984 (western Tropical Atlantic), Sibuet et al. 1989 (eastern Tropical, eastern North Atlantic), Soltwedel 1997a (eastern Tropical Atlantic), Tietjen et al. 1989 (western Tropical Atlantic–

Puerto Rico Trench),Vanhove et al. 1995 (Antarctic),Vanreusel et al. 1992 (eastern North Atlantic),Wigley and McIntyre 1964 (western North Atlantic).

Macrofaunal data from Aller et al. 2002 (western North Atlantic), Alongi 1992 (western Tropical Pacific), Carey and Ruff 1974 (Arctic), Carey 1981 (eastern North Pacific), Clough et al. 1997 (Arctic), Dauwe et al. 1998 (North Sea), Duineveld et al. 2000 (Mediterranean), Flach and Heip 1996 (eastern North Atlantic), Flach et al. 2002 (eastern North Atlantic), Frankenberg and Menzies 1968 (eastern Tropical Pacific), Galéron et al. 2000 (eastern Tropical Atlantic), Gerdes et al. 1992 (Antarctic), Griggs et al. 1969 (eastern North Pacific), Hecker and Paul 1979 (eastern Tropical Pacific), Houston and Haedrich 1984 (western North Atlantic), Jaźdźewski et al. 1986 (Antarctic), Khripounoff et al. 1980 (western Tropical Atlantic), Kröncke 1998 (Arctic), Kröncke et al. 2000 (Arctic), Kröncke et al. 2003 (Mediterranean), Levin et al. 2000 (Arabian Sea), Maciolek et al. 1987a,b (western North Atlantic), Nichols and Rowe 1977 (eastern Tropical Atlantic), Pfannkuche et al. 1983 (eastern North Atlantic), Rhoads et al. 1985 (western North Pacific), Richardson et al. 1985 (Caribbean), Richardson et al. 1995 (western Tropical Atlantic–Puerto Rico Trench), Romero-Wetzel and Gerlach 1991 (Norwegian Sea), Rowe 1971b (eastern Tropical Pacific), Rowe and Menzel 1971 (Gulf of Mexico),Rowe et al. 1974 (western North Atlantic, Gulf of Mexico),Rowe et al. 1975 (western North Atlantic),Rowe et al. 1982 (Atlantic), Rowe et al. 1997 (Norwegian Sea), Schaff et al. 1992 (western North Atlantic), Shirayama 1983 (western Tropical, western North Pacific), Sibuet et al. 1984 (western Tropical Atlantic), Sibuet et al. 1989 (eastern Tropical, eastern North Atlantic), K. L. Smith 1978 (western North Atlantic), K. L. Smith 1987 (eastern North, central North Pacific),Tselepides and Eleftheriou 1992 (Mediterranean), Tselepides et al. 2000a (Mediterranean), Witte 2000 (Arabian Sea).

Megafaunal data from Christiansen and Thiel 1992 (eastern North Atlantic), Haedrich and Rowe 1977 (western North Atlantic), Heip et al. 2001 (eastern North Atlantic), Kröncke and Türkay 2003 (eastern Tropical, eastern South Atlantic), Ohta 1983 (western North Pacific), Richardson and Young 1987 (Caribbean), Sardà et al. 1994 (Mediterranean), Sibuet et al. 1984 (western Tropical Atlantic).

REFERENCES

Aboim, M. A., G. M. Menezes, T. Schlitt, and A. D. Rogers. 2005. Genetic structure and history of populations of the deep-sea fish *Helicolenus dactylopterus* (Delaroche, 1809) inferred from mtDNA sequence analysis. Molecular Ecology **14:**1343–1354.

Abrams, P. A. 1983. The theory of limiting similarity. Annual Review of Ecology and Systematics **14:**359–376.

Adkins, J. F., H. Cheng, E. A. Boyle, E. R. M. Druffel, and R. L. Edwards. 1998. Deep-sea coral evidence for rapid change in ventilation of the deep North Atlantic 15,400 years ago. Science **280:**725–728.

Agassiz, A. 1868. Letter to Fritz Müller. Pp. 91–94, *in* G. R. Agassiz. 1913. Letters and Recollections of Alexander Agassiz with a Sketch of His Life and Work. Houghton Mifflin, Boston.

Agassiz, A. 1888. Three Cruises of the United States Coast and Geodetic Survey Steamer "Blake." Houghton Mifflin, Boston.

Allen, A. P., and J. F. Gillooly. 2006. Assessing latitudinal gradients in speciation rates and biodiversity at the global scale. Ecology Letters **9:**947–954.

Allen, A. P., J. F. Gillooly, V. M. Savage, and J. H. Brown. 2006. Kinetic effects of temperature on rates of genetic divergence and speciation. Proceedings of the National Academy of Sciences, USA **103:**9130–9135.

Allen, J. A., and H. L. Sanders. 1996. The zoogeography, diversity and origin of the deep-sea protobranch bivalves of the Atlantic: The epilogue. Progress in Oceanography **38:**95–153.

Allen, J. P., J. H. Brown, and J. F. Gillooly. 2002. Global biodiversity, biochemical kinetics, and the energy-equivalence rule. Science **297:**1545–1548.

Aller, J.Y. 1989. Quantifying sediment disturbance by bottom currents and its effect on benthic communities in a deep-sea western boundary zone. Deep-Sea Research **36:**901–934.

Aller, J.Y. 1997. Benthic community response to temporal and spatial gradients in physical disturbance within a deep-sea western boundary region. Deep-Sea Research I **44:**39–69.

Aller, J.Y., and R. C. Aller. 1986. Evidence for localized enhancement of biological activity associated with tube and burrow structures in deep-sea sediments at the HEBBLE site, western North Atlantic. Deep-Sea Research **33:**755–790.

Aller, J.Y., R. C. Aller, and M. A. Green. 2002. Benthic faunal assemblages and carbon supply along the continental shelf/shelf break-slope off Cape Hatteras, North Carolina. Deep-Sea Research II **49:**4599–4625.

Aller, R. C. 1982. The effects of macrobenthos on chemical properties of marine sediment and overlying water. Pp. 53–102, *in* P. L. McCall and M. J. S. Tevesz, eds. Animal-Sediment Relations. Plenum, New York.

Alongi, D. M. 1992. Bathymetric patterns of deep-sea benthic communities from bathyal to abyssal depths in the western South Pacific (Solomon and Coral Seas). Deep-Sea Research **39:**549–565.

Alongi, D. M., and M. Pichon. 1988. Bathyal meiobenthos of the western Coral Sea: Distribution and abundance in relation to microbial standing stocks and environmental factors. Deep-Sea Research **35:**491–503.

Alve, E., and S. T. Goldstein. 2003. Propagule transport as a key method of dispersal in benthic foraminifera (Protista). Limnology and Oceanography **48:**2163–2170.

Anderson, O. G. N. 1989. Primary production, chlorophyll, light and nutrients beneath the Arctic sea ice. Pp. 147–191, *in* Y. Hermann, ed. The Arctic Seas: Climatology, Oceanography, Geology and Biology. Van Nostrand, New York.

Angel, M.V. 1997. Pelagic biodiversity. Pp. 35–68, *in* R. F. G. Ormond, J. D. Gage, and M.V. Angel, eds. Marine Biodiversity: Patterns and Processes. Cambridge University Press, Cambridge, UK.

Ansari, Z. A., A. H. Parulekar, and T. G. Jagtap. 1980. Distribution of sub-littoral meiobenthos off Goa coast, India. Hydrobiologia **74:**209–214.

Antoine, D., J.-M. André, and A. Morel. 1996. Oceanic primary production 2. Estimation at global scale from satellite (coastal zone color scanner) chlorophyll. Global Biogeochemical Cycles **10:**57–69.

Avise, J. C. 2004. Molecular Markers, Natural History and Evolution. Chapman and Hall, New York.

Baguley, J. G., P. A. Montagna, W. Lee, L. J. Hyde, and G. T. Rowe. 2006. Spatial and bathymetric trends in Harpacticoida (Copepoda) community structure in the

Northern Gulf of Mexico deep sea. Journal of Experimental Marine Biology and Ecology **330:**327–341.

Bailey, D. M., H. A. Ruhl, and K. L. Smith Jr. 2006. Long-term changes in benthopelagic fish abundance in the abyssal northeast Pacific Ocean. Ecology **87:**549–555.

Baird, B. H., D. E. Nivens, J. H. Parker, and D. C. White. 1985. The biomass, community structure, and spatial distribution of the sedimentary microbiota from a high-energy area of the deep sea. Deep-Sea Research **32:**1089–1099.

Barber, P. H., M. V. Erdmann, and S. R. Palumbi. 2006. Comparative phylogeography of three codistributed stomatopods: Origins and timing of regional lineage diversification in the coral triangle. Evolution **60:**1825–1839.

Barber, P. H., S. R. Palumbi, M. V. Erdmann, and M. K. Moosa. 2000. A marine Wallace's line? Nature **406:**692–693.

Barber, P. H., S. R. Palumbi, M. V. Erdmann, and M. K. Moosa. 2002. Sharp genetic breaks among populations of *Haptosquilla pulchella* (Stomatopoda) indicate limits to larval transport: Patterns, causes, and consequences. Molecular Ecology **11:**659–674.

Barnett, P. R. O., J. Watson, and D. Connelly. 1984. A multiple corer for taking virtually undisturbed samples from shelf, bathyal and abyssal sediments. Oceanologica Acta **7:**399–408.

Beaulieu, S. E. 2002. Accumulation and fate of phytodetritus on the sea floor. Oceanography and Marine Biology: An Annual Review **40:**171–232.

Becker, B. J., L. A. Levin, F. J. Fordrie, and P. A. McMillan. 2007. Complex larval connectivity patterns among marine invertebrate populations. Proceedings of the National Academy of Sciences, USA **104:**3267–3272.

Behrenfeld, M. J., and P. G. Falkowski. 1997. Photosynthetic rates derived from satellite-based chlorophyll concentration. Limnology and Oceanography **42:**1–20.

Belicka, L. L., H. R. Macdonald, and H. R. Harvey. 2002. Sources and transport of organic carbon to shelf slope, and basin surface sediments of the Arctic Ocean. Deep-Sea Research I **49:**1463–1483.

Belyaev, G. M. 1966. Hadal bottom fauna of the World Ocean. Akademiya Nauk SSSR, Institute Okeanologii (Israel Program for Scientific Translations, Jerusalem 1972).

Benson, R. H. 1975. The origin of the psychrosphere as recorded in changes of deep-sea ostracode assemblages. Lethaia **8:**69–83.

Berelson, W. M. 2001. The flux of particulate organic carbon into the ocean interior: A comparison of four U.S. JGOFS regional studies. Oceanography **14:**59–67.

Berelson, W. M. 2002. Particle settling rates increase with depth in the ocean. Deep-Sea Research II **49:**237–251.

Berger, W. H., and G. Wefer. 1990. Export production: Seasonality and intermittency, and paleoceanographic implications. Palaeogeography, Palaeoclimatology, Palaeoecology **89:**245–254.

Bermingham, E., and H. A. Lessios. 1993. Rate variation of protein and mitochondrial DNA evolution as revealed by sea-urchins separated by the Isthmus of Panama. Proceedings of the National Academy of Sciences, USA **90**:2734–2738.

Bernhard, J. M. 1992. Benthic foraminiferal distribution and biomass related to pore-water oxygen content: Central California continental slope and rise. Deep-Sea Research **39**:585–605.

Bernhard, J. M., B. K. Sen Gupta, and J. G. Baguley. 2009. Benthic foraminifera living in Gulf of Mexico bathyal and abyssal sediments: Community analysis and comparison to metazoan meiofaunal biomass and density. Deep-Sea Research II **55**:2617–2626.

Bernhard, J. M., B. K. Sen Gupta, and P. F. Borne. 1997. Benthic foraminiferal proxy to estimate dysoxic bottom-water oxygen concentrations: Santa Barbara Basin, U.S. Pacific continental margin. Journal of Foraminiferal Research **27**:301–310.

Bernstein, B. B., R. R. Hessler, R. Smith, and P. A. Jumars. 1978. Spatial dispersion of benthic Foraminifera in the abyssal central North Pacific. Limnology and Oceanography **23**:401–416.

Bernstein, B. B., and J. P. Meador. 1979. Temporal persistence of biological patch structure in an abyssal benthic community. Marine Biology **51**:179–183.

Bertness, M., and S. Gaines. 1993. Larval dispersal and local adaptation in acorn barnacles. Evolution **47**:316–320.

Bertness, M. D., S. D. Gaines, and M. E. Hay, eds. 2001. Marine Community Ecology. Sinauer, Sunderland, MA.

Bett, B. J. 2001. UK Atlantic Margin Environmental Survey: Introduction and overview of bathyal benthic ecology. Continental Shelf Research **21**:917–956.

Bett, B. J., M. G. Malzone, B. E. Narayanaswamy, and B. D. Wigham. 2001. Temporal variability in phytodetritus and megabenthic activity at the seabed in the deep Northeast Atlantic. Progress in Oceanography **50**:349–368.

Bett, B. J., and A. L. Rice. 1992. The influence of hexactinellid sponge (*Pheronema carpenteri*) spicules on the patchy distribution of macrobenthos in the Porcupine Seabight (bathyal NE Atlantic). Ophelia **36**:217–226.

Bett, B. J., and A. L. Rice. 1993. The feeding behaviour of an abyssal echiuran revealed by in situ time-lapse photography. Deep-Sea Research I **40**:1767–1779.

Bett, B. J., A. L. Rice, and M. Thurston. 1995. A quantitative photographic survey of "spoke-burrow" type Lebensspuren on the Cape Verde Abyssal Plain. Internationale Revue der gesamten Hydrobiologie **80**:1–18.

Bett, B. J., A. Vanreusel, M. Vincx, T. Soltwedel, O. Pfannkuche, P. J. D. Lambshead, A. J. Gooday, T. Ferrero, and A. Dinet. 1994. Sampler bias in the quantitative study of deep-sea meiobenthos. Marine Ecology Progress Series **104**:197–203.

Biggs, D. C., C. Hu, and F. E. Müller-Karger. 2009. Remotely sensed sea-surface chlorophyll and POC flux at Deep Gulf of Mexico Benthos sampling stations. Deep-Sea Research II **55**:2555–2562.

Billett, D. S. M. 1991. Deep-sea holothurians. Oceanography and Marine Biology: An Annual Review **29**:259–317.

Billett, D. S. M., B. J. Bett, C. L. Jacobs, I. P. Rouse, and B. D. Wigham. 2006. Mass deposition of jellyfish in the deep Arabian Sea. Limnology and Oceanography **51**:2077–2083.

Billett, D. S. M., B. J. Bett, A. L. Rice, M. H. Thurston, J. Galéron, M. Sibuet, and G. A. Wolff. 2001. Long-term change in the megafauna of the Porcupine Abyssal Plain (NE Atlantic). Progress in Oceanography **50**:325–348.

Billett, D. S. M., and B. Hansen. 1982. Abyssal aggregations of *Kolga hyalina* Danielssen and Koren (Echinodermata: Holothurioidea) in the northeast Atlantic Ocean: A preliminary report. Deep-Sea Research **29**:799–818.

Billett, D. S. M., R. S. Lampitt, A. L. Rice, and R. F. C. Mantoura. 1983. Seasonal sedimentation of phytoplankton to the deep-sea benthos. Nature **302**:520–522.

Billett, D. S. M., and A. L. Rice. 2001. The BENGAL programme: Introduction and overview. Progress in Oceanography **50**:13–25.

Bintanja, R., R. S. W. van de Wal, and J. Oerlemans. 2005. Modelled atmospheric temperatures and global sea levels over the past million years. Nature **437**:125–128.

Blackburn, T. M., and K. J. Gaston. 1994. Animal body size distributions: Patterns, mechanisms and implications. Trends in Ecology and Evolution **9**:471–474.

Blake, J. A., and R. J. Diaz. 1994. Input, accumulation and cycling of materials on the continental slope off Cape Hatteras: An introduction. Deep-Sea Research II **41**:707–710.

Blake, J. A., and J. F. Grassle. 1994. Benthic community structure on the U.S. South Atlantic slope off the Carolinas: Spatial heterogeneity in a current-dominated system. Deep-Sea Research II **41**:835–874.

Blake, J. A., and B. Hilbig. 1994. Dense infaunal assemblages on the continental slope off Cape Hatteras, North Carolina. Deep-Sea Research II **41**:875–899.

Boebel, O., R. E. Davis, M. Ollitrault, R. G. Peterson, P. L. Richardson, C. Schmid, and W. Zenk. 1999. The intermediate depth circulation of the western South Atlantic. Geophysical Research Letters **26**:3329–3332.

Boetius, A., and K. Lochte. 1996. Effect of organic enrichments on hydrolytic potentials and growth of bacteria in deep-sea sediments. Marine Ecological Progress Series **140**:239–250.

Boetius, A., and K. Lochte. 2000. Regional variation of total microbial biomass in sediments of the deep Arabian Sea. Deep-Sea Research II **47**:149–168.

Boetius, A., S. Scheibe, A. Tselepides, and H. Thiel. 1996. Microbial biomass and activities in deep-sea sediments of the Eastern Mediterranean: Trenches are benthic hotspots. Deep-Sea Research I **43**:1439–1460.

Bohonak, A. J. 1999. Dispersal, gene flow, and population structure. Quarterly Review of Biology **74**:21–45.

Boucher, G., and P. J. D. Lambshead. 1995. Ecological biodiversity of marine nematodes in samples from temperate, tropical and deep-sea regions. Conservation Biology **9**:1594–1604.

Bouchet, P. 1976a. Mise en évidence de stades larvaires planctoniques chez des Gastéropodes Prosobranches des étages bathyal et abyssal. Bulletin du Muséum national d'Histoire naturelle **400**:947–971.

Bouchet, P. 1976b. Mise en évidence d'une migration de larves véligères entre l'étage abyssal et la surface. Comptes Rendus hebdomadaires des séances de l'Académie des Sciences, Paris, Série D **283**:821–824.

Bouchet, P., and J.-C. Fontes. 1981. Migrations verticales des larves de Gastéropodes abyssaux: arguments nouveaux dûs à l'analyse isotopique de le coquille larvaire et postlarvaire. Comptes Rendus hebdomadairs des séances de l'Académie des Sciences, Paris, Série III **292**:1005–1008.

Bouchet, P., P. Lozouet, P. Maestrati, and V. Heros. 2002. Assessing the magnitude of species richness in tropical marine environments: Exceptionally high numbers of molluscs at a New Caledonia site. Biological Journal of the Linnean Society **75**:421–436.

Bouchet, P., and M. Taviani. 1992. The Mediterranean deep-sea fauna: Pseudo-populations of Atlantic species? Deep-Sea Research **39**:169–184.

Bouchet, P., and A. Warén. 1979. The abyssal molluscan fauna of the Norwegian Sea and its relation to other faunas. Sarsia **64**:211–243.

Bouchet, P., and A. Warén. 1993. Revision of the Northeast Atlantic bathyal and abyssal Mesogastropoda. Bollettino Malacologico Supplemento **3**:579–840.

Bouchet, P., and A. Warén. 1994. Ontogenetic migration and dispersal of deep-sea gastropod larvae. Pp. 98–117, *in* C. M. Young and K. J. Eckelbarger, eds. Reproduction, Larval Biology, and Recruitment of the Deep-Sea Benthos. Columbia University Press, New York.

Boudreau, B. P., and B. B. Jørgensen, eds. 2001. The Benthic Boundary Layer: Transport Processes and Biogeochemistry. Oxford University Press, Oxford, UK.

Bower, A. S., and H. D. Hunt. 2000a. Lagrangian observation of the deep western boundary current in North Atlantic Ocean I: Large-scale pathways and spreading rates. Journal of Physical Oceanography **30**:764–783.

Bower, A. S., and H. G. Hunt. 2000b. Lagrangian observations of the deep western boundary current in the North Atlantic Ocean II: The Gulf Stream-deep western boundary current crossover. Journal of Physical Oceanography **30**:784–804.

Bower, A. S., B. LeCann, T. Rossby, W. Zenk, J. Gould, K. Speer, P. L. Richardson, M. D. Prater, and H.-M. Zhang. 2002. Directly measured mid-depth circulation in the northeastern North Atlantic Ocean. Science **419**:603–607.

Boyd, P. W., and T. W. Trull. 2007. Understanding the export of biogenic particles in oceanic waters: Is there consensus? Progress in Oceanography **72**:276–312.

Boyle, E. E., J. D. Zardus, M. R. Chase, R. J. Etter, and M. A. Rex. 2004. Strategies for molecular genetic studies of preserved deep-sea macrofauna. Deep Sea Research I **51**:1319–1336.

Brandt, A. 1992. Origin of Antarctic Isopoda (Crustacea, Malacostraca). Marine Biology **113**:415–423.

Brandt, A., and D. Barthel. 1995. An improved supra- and epibenthic sledge for catching Peracarida (Crustacea, Malacostraca). Ophelia **43**:15–23.

Brandt, A., and B. Ebbe, eds. 2007. ANtarctic benthic DEEP-sea biodiversity: Colonization history and recent community patterns (ANDEEP-III). Deep-Sea Research II **54**:1645–1904.

Brandt, A., A. J. Gooday, S. N. Brandão, S. Brix, W. Brökeland, T. Cedhagen, M. Choudhury, N. Cornelius, B. Danis, I. De Mesel, R. J. Diaz, D. C. Gillan, B. Ebbe, J. A. Howe, D. Janussen, S. Kaiser, K. Linse, M. Malyutina, J. Pawlowski, M. Raupach, and A. Vanreusel. 2007. First insights into the biodiversity and biogeography of the Southern Ocean deep sea. Nature **477**:307–311.

Brandt, A., and B. Hilbig, eds. 2004. ANDEEP (ANtarctic benthic DEEP-sea biodiversity: Colonization history and recent community patterns): A tribute to Howard L. Sanders. Deep-Sea Research II **51**:1457–1919.

Brenke, N. 2005. An epibenthic sledge for operations on marine soft bottom and bedrock. Marine Technology Society Journal **39**:10–21.

Brey, T., M. Klages, C. Dahm, M. Gorny, J. Gutt, S. Hain, M. Stiller, and W. E. Arntz. 1994. Antarctic benthic diversity. Nature **368**:297.

Briggs, J. C. 1994. Species diversity: Land and sea compared. Systematic Biology **43**:130–135.

Bromham, L., and M. Cardillo. 2003. Testing the link between the latitudinal gradient in species richness and rates of molecular evolution. Journal of Evolutionary Biology **16**:200–207.

Brown, J. H. 1995. Macroecology. University of Chicago Press, Chicago.

Brown, J. H., J. F. Gillooly, A. P. Allen, V. M. Savage, and G. B. West. 2004. Toward a metabolic theory of ecology. Ecology **85**:1771–1789.

Bucklin, A., and L. D. Allen. 2004. MtDNA sequencing from zooplankton after long-term preservation in buffered formalin. Molecular Phylogenetics **30**:879–882.

Bucklin, A., R. R. Wilson, and K. L. Smith Jr. 1987. Genetic differentiation of seamount and basin populations of the deep-sea amphipod *Eurythenes gryllus*. Deep-Sea Research **34**:1795–1810.

Buesseler, K. O., C. H. Lamborg, P. W. Boyd, P. J. Lam, T. W. Trull, R. R. Bidigare, J. K. B. Bishop, K. L. Casciotti, F. Dehairs, M. Elskens, M. Honda, D. M. Karl, D. A. Siegel, M. W. Silver, D. K. Steinberg, J. Valdes, B. Van Mooy, and S. Wilson.

2007. Revisiting carbon flux through the ocean's twilight zone. Science **316**:567–570.

Buesseler, K. O., and R. S. Lampitt, eds. 2008. Understanding the ocean's biological pump: Results from VERTIGO. Deep-Sea Research II **55**:1519–1695.

Bugge, T., R. H. Belderson, and N. H. Kenyon. 1988. The Storegga Slide. Philosophical Transactions of the Royal Society of London A **325**:357–388.

Bühring, S. I., N. Lampadariou, L. Moodley, A. Tselepides, and U. Witte. 2006. Benthic microbial and whole-community responses to different amounts of ^{13}C-enriched algae: In situ experiments in the deep Cretan Sea (Eastern Mediterranean). Limnology and Oceanography **51**:157–165.

Burnett, B. R. 1973. Observation of the microfauna of the deep-sea benthos using light and scanning electron microscopy. Deep-Sea Research **20**:413–417.

Burnett, B. R. 1981. Quantitative sampling of nanobiota (microbiota) of the deep-sea benthos-III: The bathyal San Diego Trough. Deep-Sea Research **28**:649–663.

Burton, R. S. 1998. Intraspecific phylogeography across the Point Conception biogeographic boundary. Evolution **52**:734–745.

Buzas, M. A., and T. G. Gibson. 1969. Species diversity: Benthonic Foraminifera in the western North Atlantic. Science **163**:72–75.

Buzas, M. A., L.-A. C. Hayek, B. W. Hayward, H. R. Grenfell, and A. T. Sabaa. 2007. Biodiversity and community structure of deep-sea foraminifera around New Zealand. Deep-Sea Research I **54**:1641–1654.

Cacchione, D. A., and L. F. Pratson. 2004. Internal tides and the continental slope. American Scientist **92**:130–137.

Cadena, C. D., R. E. Ricklefs, I. Jimenez, and E. Bermingham. 2005. Ecology: Is speciation driven by species diversity? Nature **438**:E1–E2.

Campbell, J. W., and T. Aarup. 1992. New production in the North Atlantic derived from seasonal patterns of surface chlorophyll. Deep-Sea Research **39**:1669–1694.

Cannariato, K. G., J. P. Kennett, and R. J. Behl. 1999. Biotic response to late Quaternary rapid climate switches in Santa Barbara Basin: Ecological and evolutionary implications. Geology **27**:63–66.

Cardillo, M. 1999. Latitude and rates of diversification in birds and butterflies. Proceedings of the Research Society of London B **266**:1221–1225.

Cardillo, M., C. D. L. Orme, and I. P. F. Owens. 2005. Testing for latitudinal bias in diversification rates: An example using New World birds. Ecology **86**:2278–2287.

Carey Jr., A. G. 1981. A comparison of benthic infaunal abundance on two abyssal plains in the northeast Pacific Ocean. Deep-Sea Research **28**:467–479.

Carey Jr., A. G., and R. E. Ruff. 1974. Ecological studies of the benthos in the western Beaufort Sea with special reference to bivalve molluscs. Pp. 505–530, *in* M. J. Dunbar, ed. Polar Oceans. Arctic Institute of North America, Calgary.

Carney, R. S. 1995. On the Adequacy and Improvement of Marine Benthic Pre-Impact Surveys: Examples from the Gulf of Mexico Continental Shelf. Academic, New York.

Carney, R. S. 1997. Basing conservation policies for the deep-sea floor on current-diversity concepts: A consideration of rarity. Biodiversity and Conservation 6:1463–1485.

Carney, R. S. 2005. Zonation of deep biota on continental margins. Oceanography and Marine Biology: An Annual Review 43:211–278.

Carney, R. S., and A. G. Carey Jr. 1976. Distribution pattern of holothurians on the Northeastern Pacific (Oregon, U.S.A.) continental shelf, slope, and abyssal plain. Thalassia Jugoslavica 12:67–74.

Carney, R. S., and A. G. Carey Jr. 1982. Distribution and diversity of holothuroids (Echinodermata) on Cascasdia Basin and Tufts Abyssal Plain. Deep-Sea Research 29:597–607.

Carney, R. S., R. L. Haedrich, and G. T. Rowe. 1983. Zonation of fauna in the deep sea. Pp. 371–398, in G. T. Rowe, ed. The Sea, Vol. 8: Deep-Sea Biology. Wiley, New York.

Cartes, J. E., and M. Carrassón. 2004. Influence of trophic variables on the depth-range distributions and zonation rates of deep-sea megafauna: The case of the Western Mediterranean assemblages. Deep-Sea Research I 51:263–279.

Cartes, J. E., C. Huguet, S. Parra, and F. Sanchez. 2007. Trophic relationships in deep-water decapods of Le Danois bank (Cantabrian Sea, NE Atlantic): Trends related with depth and seasonal changes in food quality and availability. Deep-Sea Research I 54:1091–1110.

Cartes, J. E., D. Jaume, and T. Madurell. 2003. Local changes in the composition and community structure of suprabenthic peracarid crustaceans of the bathyal Mediterranean: Influence of environmental factors. Marine Biology 143:745–758.

Caswell, H. 1976. Community structure: A neutral model analysis. Ecological Monographs 46:327–354.

Chan, F., J. A. Barth, J. Lubchenco, A. Kirincich, H. Weeks, W. T. Paterson, and B. A. Menge. 2008. Emergence of anoxia in the California current large marine ecosystem. Science 319:920.

Chase, M. R., R. J. Etter, M. A. Rex, and J. M. Quattro. 1998a. Bathymetric patterns of genetic variation in a deep-sea protobranch bivalve, Deminucula atacellana. Marine Biology 131:301–308.

Chase, M. R., R. J. Etter, M. A. Rex, and J. M. Quattro. 1998b. Extraction and amplification of mitochondrial DNA from formalin-fixed deep-sea mollusks. BioTechniques 24:243–247.

Chown, S. L., and K. J. Gaston. 1999. Patterns in procellariiform diversity as a test of species-energy theory in marine systems. Evolutionary Ecology Research 1:365–373.

Christiansen, B., and H. Thiel. 1992. Deep-sea epibenthic megafauna of the North-east Atlantic: Abundance and biomass at three mid-oceanic locations estimated from photographic transects. Pp. 125–138, *in* G. T. Rowe and V. Pariente, eds. Deep-Sea Food Chains and the Global Carbon Cycle. Kluwer, Dordrecht, The Netherlands.

Clarke, A. 1992. Is there a latitudinal diversity cline in the sea? Trends in Ecology and Evolution **7**:286–287.

Clarke, A. 2003. The deep polar seas. Pp. 239–260, *in* P. A. Tyler, ed. Ecosystems of the World 28: Ecosystems of the Deep Oceans. Elsevier, Amsterdam.

Clarke, A., and K. J. Gaston. 2006. Climate, energy and diversity. Proceedings of the Royal Society of London B **273**:2257–2266.

Clarke, J. A., and A. Crame. 1997. Diversity, latitude and time: Patterns in the shallow sea. Pp. 122–147, *in* R. F. G. Ormond, J. D. Gage, and M. V. Angel, eds. Marine Biodiversity: Patterns and Processes. Cambridge University Press, Cambridge, UK.

Clough, L. M., W. G. Ambrose Jr., J. K. Cochran, C. Barnes, P. E. Renaud, and R. C. Aller. 1997. Infaunal density, biomass and bioturbation in the sediments of the Arctic Ocean. Deep-Sea Research II **44**:1683–1704.

Colwell, R. K., and G. C. Hurtt. 1994. Nonbiological gradients in species richness and a spurious Rapoport effect. American Naturalist **144**:570–595.

Colwell, R. K., and D. C. Lees. 2000. The mid-domain effect: Geometric constraints on the geography of species richness. Trends in Ecology and Evolution **15**:70–76.

Connell, J. H. 1978. Diversity in tropical rain forests and coral reefs. Science **199**: 1302–1310.

Conte, M. H., T. D. Dickey, J. C. Weber, R. J. Johnson, and A. H. Knap. 2003. Transient physical forcing of pulsed export of bioreactive material to the deep Sargasso Sea. Deep-Sea Research I **50**:1157–1187.

Cooper, R. A., P. Valentine, J. R. Uzmann, and R. A. Slater. 1987. Submarine canyons. Pp. 52–63, *in* R. H. Backus, ed. Georges Bank. MIT Press, Cambridge, MA.

Corliss, B. H., C. W. Brown, X. Sun, and W. J. Showers. 2009. Deep-sea benthic diversity linked to seasonality of pelagic productivity. Deep-Sea Research I **56**:835–841.

Cornelius, N., and A. J. Gooday. 2004. "Live" (stained) deep-sea benthic foraminiferans in the western Weddell Sea: Trends in abundance, diversity and taxonomic composition along a depth transect. Deep-Sea Research II **51**:1571–1602.

Cornell, H. V., and R. H. Karlson. 1996. Species richness of reef-building corals determined by local and regional processes. Journal of Animal Ecology **65**:233–241.

Cosson, N., M. Sibuet, and J. Galéron. 1997. Community structure and spatial heterogeneity of the deep-sea macrofauna at three contrasting stations in the tropical northeast Atlantic. Deep-Sea Research I **44**:247–269.

Cosson-Sarradin, N., M. Sibuet, G. L. J. Paterson, and A. Vangriesheim. 1998. Polychaete diversity at tropical Atlantic sites: Environmental effects. Marine Ecology Progress Series **165**:173–185.

Coull, B. C. 1972. Species diversity and faunal affinities of meiobenthic Copepoda in the deep sea. Marine Biology **14**:48–51.

Coull, B. C., R. L. Ellison, J. W. Fleeger, R. P. Higgins, W. D. Hope, W. D. Hummon, R. M. Rieger, W. E. Sterrer, H. Thiel, and J. H. Tietjen. 1977. Quantitative estimates of the meiofauna from the deep sea off North Carolina, USA. Marine Biology **39**:233–240.

Cowen, R. K., C. B. Paris, and A. Srinivasan. 2006. Scaling of connectivity in marine populations. Science **311**:522–527.

Crandall, E. D., M. A. Frey, R. K. Grosberg, and P. H. Barber. 2008. Contrasting demographic history and phylogeographical patterns in two Indo-Pacific gastropods. Molecular Ecology **17**:611–626.

Crawford, W. R., P. J. Brickley, T. D. Peterson, and A. C. Thomas. 2005. Impact of Haida Eddies on chlorophyll distribution in the Eastern Gulf of Alaska. Deep-Sea Research II **52**:975–989.

Creasey, S., A. D. Rogers, P. A. Tyler, J. D. Gage, and D. Jollivet. 2000. Genetic and morphometric comparisons of squat lobster, *Munidopsis scobina* (Decapoda: Anomura: Galatheidae) populations, with notes on the phylogeny of the genus *Munidopsis*. Deep-Sea Research II **47**:87–118.

Creasey, S., A. D. Rogers, P. A. Tyler, C. M. Young, and J. D. Gage. 1997. The population biology and genetics of the deep-sea spider crab, *Encephaloides armstrongi* Wood-Mason 1891 (Decapoda: Majidae). Philosophical Transactions of the Royal Society of London B **352**:365–379.

Creasey, S. S., and A. D. Rogers. 1999. Population genetics of bathyal and abyssal organisms. Advances in Marine Biology **35**:1–151.

Cronin, T. M., D. M. DeMartino, G. S. Dwyer, and J. Rodriguez-Lazaro. 1999. Deep-sea ostracode species diversity: Response to late Quaternary climate change. Marine Micropaleontology **37**:231–249.

Cronin, T. M., and M. E. Raymo. 1997. Orbital forcing of deep-sea benthic species diversity. Nature **385**:624–627.

Culver, S. J., and M. A. Buzas. 2000. Global latitudinal species diversity gradient in deep-sea foraminifera. Deep-Sea Research I **47**:259–275.

Cunningham, J. R., and J. F. Ustach. 1992. Protozoan numbers and biomass in the sediments of the Blake Outer Ridge. Deep-Sea Research **39**:789–794.

Currie, D. J., and J. T. Kerr. 2008. Tests of the mid-domain hypothesis: A review of the evidence. Ecological Monographs **78**:3–18.

Currie, D. J., G. G. Mittelbach, H. V. Cornell, R. Field, J.-F. Guégan, B. A. Hawkins, D. M. Kaufman, J. T. Kerr, T. Oberdorff, E. O'Brien, and J. R. G. Turner. 2004.

Predictions and tests of climate-based hypotheses of broad-scale variation in taxonomic richness. Ecology Letters **7**:1121–1134.

Cutter Jr., G. R., R. J. Diaz, and J. Lee. 1994. Foraminifera from the continental slope off Cape Hatteras, North Carolina. Deep-Sea Research II **41**:951–963.

Dahl, E. 1979. Amphipoda Gammaridea from the deep Norwegian Sea: A preliminary report. Sarsia **64**:57–59.

Dahl, E., L. Laubier, M. Sibuet, and J.-O. Strömberg. 1976. Some quantitative results on benthic communities of the deep Norwegian Sea. Astarte **9**:61–79.

Danovaro, R., A. Dell'Anno, C. Corinaldesi, M. Magagnini, R. Noble, C. Tamburini, and M. Weinbauer. 2008a. Major viral impact on the functioning of benthic deep-sea ecosystems. Nature **454**:1084–1087.

Danovaro, R., A. Dell'Anno, and M. Fabiano. 2001b. Bioavailability of organic matter in the sediments of the Porcupine Abyssal Plain, northeastern Atlantic. Marine Ecology Progress Series **220**:25–32.

Danovaro, R., A. Dell'Anno, M. Fabiano, A. Pusceddu, and A. Tselepides. 2001a. Deep-sea ecosystem response to climate changes: The eastern Mediterranean case study. Trends in Ecology and Evolution **16**:505–510.

Danovaro, R., N. Della Croce, A. Dell'Anno, and A. Pusceddu. 2003. A depocenter of organic matter at 7800 m depth in the SE Pacific Ocean. Deep-Sea Research I **50**:1411–1420.

Danovaro, R., N. Della Croce, A. Eleftheriou, M. Fabiano, N. Papadopoulou, C. Smith, and A. Tselepides. 1995. Meiofauna of the deep Eastern Mediterranean Sea: Distribution and abundance in relation to bacterial biomass, organic matter composition and other environmental factors. Progress in Oceanography **36**:329–341.

Danovaro, R., A. Dinet, G. Duineveld, and A. Tselepides. 1999. Benthic response to particulate fluxes in different trophic environments: A comparison between the Gulf of Lions–Catalan Sea (western-Mediterranean) and the Cretan Sea (eastern-Mediterranean). Progress in Oceanography **44**:287–312.

Danovaro, R., C. Gambi, A. Dell'Anno, C. Corinaldesi, S. Fraschetti, A. Vanreusel, M. Vincx, and A. J. Gooday. 2008b. Exponential decline of deep-sea ecosystem functioning linked to benthic biodiversity loss. Current Biology **18**:1–8.

Danovaro, R., C. Gambi, and N. Della Croce. 2002. Meiofauna hotspot in the Atacama Trench, eastern South Pacific Ocean. Deep-Sea Research I **49**:843–857.

Danovaro, R., C. Gambi, N. Lampadariou, and A. Tselepides. 2008. Deep-sea nematode biodiversity in the Mediterranean Basin: Testing for longitudinal, bathymetric and energetic gradients. Ecography **31**:231–244.

Danovaro, R., A. Tselepides, A. Otegui, and N. Della Croce. 2000. Dynamics of meiofaunal assemblages on the continental shelf and deep-sea sediments of the Cretan Sea (NE Mediterranean): Relationships with seasonal changes in food supply. Progress in Oceanography **46**:367–400.

Dauwe, B., P. M. J. Herman, and C. H. R. Heip. 1998. Community structure and bioturbation potential of macrofauna at four North Sea stations with contrasting food supply. Marine Ecology Progress Series **173**:67–83.

Davies, T. J., V. Savaolainen, M. W. Chase, J. Moat, and T. G. Barraclough. 2004. Environmental energy and evolutionary rates in flowering plants. Proceedings of the Royal Society of London B **271**:2195–2200.

Dayton, P. K., and R. R. Hessler. 1972. Role of biological disturbance in maintaining diversity in the deep sea. Deep-Sea Research **19**:199–208.

DeMaster, D. J., R. H. Pope, L. A. Levin, and N. E. Blair. 1994. Biological mixing intensity and rates of organic carbon accumulation in North Carolina slope sediments. Deep-Sea Research II **41**:735–753.

Deming, J. W. 1985. Bacterial growth in deep-sea sediment trap and boxcore samples. Marine Ecology Progress Series **25**:305–312.

Deming, J. W., and J. A. Baross. 1993. The early diagenesis of organic matter: Bacterial activity. Pp. 119–144, *in* S. A. Macko, ed. Organic Geochemistry. Plenum, New York.

Deming, J. W., and J. A. Baross. 2000. Survival, dormancy, and nonculturable cells in extreme deep-sea environments. Pp. 147–197, *in* R. R. Colwell and D. J. Grimes, eds. Nonculturable Microorganisms in the Environment. AMS, Washington, DC.

Deming, J. W., and P. L. Yager. 1992. Natural bacterial assemblages in deep-sea sediments: Toward a global view. Pp. 11–27, *in* G. T. Rowe and V. Pariente, eds. Deep-Sea Food Chains and the Global Carbon Cycle. Kluwer, Dordrecht, The Netherlands.

Desbruyères, D., J. Y. Bervas, and A. Khripounoff. 1980. Un cas de colonisation rapide d'un sédiment profond. Oceanologica Acta **3**:285–291.

Deuser, W. G., F. E. Muller-Karger, R. H. Evans, O. B. Brown, W. E. Esaias, and G. C. Feldman. 1990. Surface-ocean color and deep-ocean carbon flux: How close a connection? Deep-Sea Research **37**:1331–1343.

Deuser, W. G., F. E. Muller-Karger, and C. Hemleben. 1988. Temporal variations of particle fluxes in the deep subtropical and tropical North Atlantic: Eulerian and Lagrangian effects. Journal of Geophysical Research **93**:6857–6862.

Deuser, W. G., and E. H. Ross. 1980. Seasonal change in the flux of organic carbon to the deep Sargasso Sea. Nature **283**:364–365.

D'Hondt, S., P. Donaghay, J. C. Zachos, D. Luttenberg, and M. Lindinger. 1998. Organic carbon fluxes and ecological recovery from the Cretaceous-Tertiary mass extinction. Science **282**:276–279.

Diaz, R. J., J. A. Blake, and G. R. Cutter Jr., eds. 1994. Input, accumulation and cycling of materials on the continental slope off Cape Hatteras. Deep-Sea Research II **41**:705–982.

Diaz, R. J., and G. Rosenberg. 1995. Marine benthic hypoxia: A review of its ecological effects and the behavioural responses of benthic macrofauna. Oceanography and Marine Biology: An Annual Review **33**:245–303.

Diaz, R. J., and R. Rosenberg. 2008. Spreading dead zones and consequences for
marine ecosystems. Science **321**:926–929.

Dinet, A. 1973. Distribution quantitative du méiobenthos profond dans la région de
la dorsale de Walvis (Sud-Ouest Africain). Marine Biology **20**:20–26.

Dinet, A. 1976. Études quantitatives du méiobenthos dans le secteur nord de la Mer
Égée. Acta Adriatica **18**:83–88.

Dinet, A., and M.-H. Vivier. 1977. Le méiobenthos abyssal du golfe de Gascogne I:
Considérations sur les données quantitatives. Cahiers de Biologie Marine
18:85–97.

Dinet, A., and M.-H. Vivier. 1979. Le méiobenthos abyssal du Golfe de Gascogne:
Les peuplements de nématodes et leur diversité spécifique. Cahiers de Biologie
Marine **20**:109–123.

Doebeli, M., and U. Dieckmann. 2003. Speciation along environmental gradients.
Nature **421**:259–264.

Doebeli, M., U. Dieckmann, J. A. J. Metz, and D. Tautz. 2005. What we have also
learned: Adaptive speciation is theoretically plausible. Evolution **59**:691–695.

Drazen, J. C., R. J. Baldwin, and K. L. Smith Jr. 1998. Sediment community response
to a temporally varying food supply at an abyssal station in the NE Pacific.
Deep-Sea Research II **45**:893–913.

Ducklow, H. W., D. K. Steinberg, and K. O. Buesseler. 2001. Upper ocean carbon export and the biological pump. Oceanography **14**:50–58.

Dugdale, R. C., and F. P. Wilkerson. 1988. Nutrient sources and primary production
in the Eastern Mediterranean. Oceanologica Acta **9**:179–184.

Duineveld, G. C. A., P. A. W. J. de Wilde, E. M. Berghuis, A. Kok, T. Tahey, and
J. Kromkamp. 1997. Benthic respiration and standing stock on two contrasting
continental margins in the western Indian Ocean: The Yemen-Somali upwelling region and the margin off Kenya. Deep-Sea Research II **44**:1293–1317.

Duineveld, G. C. A., A. Tselepides, R. Witbaard, R. P. M. Bak, E. M. Berghuis,
G. Nieuwland, J. van der Weele, and A. Kok. 2000. Benthic-pelagic coupling in
the oligotrophic Cretan Sea. Progress in Oceanography **46**:457–480.

Eckman, J. E., and D. Thistle. 1988. Small-scale spatial pattern in meiobenthos in the
San Diego Trough. Deep-Sea Research **35**:1565–1578.

Eldrett, J. S., I. C. Harding, P. A. Wilson, E. Butler, and A. P. Roberts. 2007. Continental ice in Greenland during the Eocene and Oligocene. Nature **446**:176–
179.

Engle, V. D., and J. K. Summers. 1999. Latitudinal gradients in benthic community
composition in Western Atlantic estuaries. Journal of Biogeography **26**:1007–
1023.

Ernst, S., and B. van der Zwaan. 2004. Effects of experimentally induced raised levels of organic flux and oxygen depletion on a continental slope benthic foraminiferal community. Deep-Sea Research I **51**:1709–1739.

Escobar, E., M. López, L. A. Soto, and M. Signoret. 1997. Density and biomass of the meiofauna of the upper continental slope in two regions of the Gulf of Mexico. Ciencias Marinas **23**:463–489.

Etter, R. J., and H. Caswell. 1994. The advantages of dispersal in a patchy environment: Effects of disturbance in a cellular automaton model. Pp. 284–305, *in* C. M. Young and K. J. Eckelbarger, eds. Reproduction, Larval Biology, and Recruitment of the Deep-Sea Benthos. Columbia University Press, New York.

Etter, R. J., and J. F. Grassle. 1992. Patterns of species diversity in the deep sea as a function of sediment particle size diversity. Nature **360**:576–578.

Etter, R. J., and L. S. Mullineaux. 2001. Deep-sea communities. Pp. 367–393, *in* M. D. Bertness, S. D. Gaines, and M. E. Hay, eds. Marine Community Ecology. Sinauer, Sunderland, MA.

Etter, R. J., and M. A. Rex. 1990. Population differentiation decreases with depth in deep-sea gastropods. Deep-Sea Research **37**:1251–1261.

Etter, R. J., M. A. Rex, M. C. Chase, and J. M. Quattro. 1999. A genetic dimension to deep-sea biodiversity. Deep-Sea Research I **46**:1095–1099.

Etter, R. J., M. A. Rex, M. R. Chase, and J. M. Quattro. 2005. Population differentiation decreases with depth in deep-sea bivalves. Evolution **59**:1479–1491.

Evans, K. L., J. J. D. Greenwood, and K. J. Gaston. 2005. Dissecting the species-energy relationship. Proceedings of the Royal Society of London B: Biological Sciences **272**:2155–2163.

Fabiano, M., and R. Danovaro. 1999. Meiofauna distribution and mesoscale variability in two sites of the Ross Sea (Antarctica) with contrasting food supply. Polar Biology **22**:115–123.

Falkowski, P. G., and J. A. Raven. 2006. Aquatic Photosynthesis. 2nd Edition. Princeton University Press, Princeton, NJ.

Felsenstein, J. 1981. Skepticism towards Santa Rosalia, or Why are there so few kinds of animals? Evolution **35**:124–138.

Fenchel, T., and L. H. Kofoed. 1976. Evidence of exploitative interspecific competition in mud snails (Hydrobiidae). Oikos **27**:19–32.

Filin, I., R. D. Holt, and M. Barfield. 2008. The relation of density regulation to habitat specialization, evolution of a species' range, and the dynamics of biological invasions. American Naturalist **172**:233–247.

Fischer, G., V. Ratmeyer, and G. Wefer. 2000. Organic carbon fluxes in the Atlantic and the Southern Ocean: Relationship to primary production compiled from satellite radiometer data. Deep-Sea Research II **47**:1961–1997.

Flach, E., and W. de Bruin. 1999. Diversity patterns in macrobenthos across a continental slope in the NE Atlantic. Journal of Sea Research **42**:303–323.

Flach, E., and C. Heip. 1996. Vertical distribution of macrozoobenthos within the sediment on the continental slope of the Goban Spur area (NE Atlantic). Marine Ecology Progress Series **141**:55–66.

Flach, E., A. Muthumbi, and C. Heip. 2002. Meiofauna and macrofauna community structure in relation to sediment composition at the Iberian margin compared to the Goban Spur (NE Atlantic). Progress in Oceanography **52**:433–457.

Flessa, K. W., and D. Jablonski. 1996. The geography of evolutionary turnover: A global analysis of extant bivalves. Pp. 376–397, *in* D. Jablonski, D. H. Erwin, and J. H. Lipps, eds. Evolutionary Paleobiology. University of Chicago Press, Chicago.

Flint, R. W., and N. N. Rabalais, eds. 1981. Environmental Studies of a Marine Ecosystem: South Texas Outer Continental Shelf. University of Texas Press, Austin, TX.

Fonseca, G., and T. Soltwedel. 2007. Deep-sea meiobenthic communities underneath the marginal ice zone off Eastern Greenland. Polar Biology **30**:607–618.

Forbes, E. 1844. Report on the Mollusca and Radiata of the Aegean Sea, and on their distribution, considered as bearing on geology. Report (1843) to the 13th meeting of the British Association for the Advancement of Science, pp. 30–193.

Forster, J. R. 1778. Observations Made during a Voyage Round the World on Physical, Geography, Natural History and Ethic Philosophy. G. Robinson, London.

Fowler, S. W., and G. A. Knauer. 1986. Role of large particles in the transport of elements and organic compounds through the oceanic water column. Progress in Oceanography **16**:147–194.

France, S. C. 1994. Genetic population structure and gene flow among deep-sea amphipods, *Abyssorchomene* spp., from six California continental Borderland basins. Marine Biology **118**:67–77.

France, S. C., and T. D. Kocher. 1996a. Geographic and bathymetric patterns of mitochondrial 16s r RNA sequence divergence among deep-sea amphipods, *Eurythenes gryllus.* Marine Biology **126**:633–644.

France, S. C., and T. D. Kocher. 1996b. DNA sequencing of formalin-fixed crustaceans from archival research collections. Molecular Marine Biology and Biotechnology **5**:304–313.

Frankenberg, D., and R. J. Menzies. 1968. Some quantitative analyses of deep-sea benthos off Peru. Deep-Sea Research **15**:623–626.

Gage, J. D. 1975. A comparison of the deep-sea epibenthic sledge and anchor-box dredge samplers with the van Veen grab and hand coring by diver. Deep-Sea Research **22**:693–702.

Gage, J. D. 1977. Structure of the abyssal macrobenthic community in the Rockall Trough. European Symposium on Marine Biology **11**:247–260.

Gage, J. D. 1979. Macrobenthic community structure in the Rockall Trough. Ambio Special Report **6**:43–46.

Gage, J. D. 1994. Recruitment ecology and age structure of deep-sea invertebrate populations. Pp. 233–242, *in* C. M. Young and K. J. Eckelbarger, eds. Reproduction, Larval Biology, and Recruitment of the Deep-Sea Benthos. Columbia University Press, New York.

Gage, J. D. 1996. Why are there so many species in deep-sea sediments? Journal of Experimental Marine Biology and Ecology **200**:257–286.

Gage, J. D. 1997. High benthic species diversity in deep-sea sediments: The importance of hydrodynamics. Pp. 148–177, *in* R. F. G. Ormond, J. D. Gage, and M. V. Angel, eds. Marine Biodiversity. Cambridge University Press, Cambridge, UK.

Gage, J. D., and B. J. Bett. 2005. Deep-sea benthic sampling. Pp. 273–325, *in* A. E. Eleftheriou and A. McIntyre, eds. Methods for the Study of Marine Benthos. Blackwell, Oxford, UK.

Gage, J. D., D. J. Hughes, and J. L. Gonzalez Vecino. 2002. Sieve size influence in estimating biomass, abundance and diversity in samples of deep-sea macrobenthos. Marine Ecology Progress Series **225**:97–107.

Gage, J. D., P. J. D. Lambshead, J. D. D. Bishop, C. T. Stuart, and N. S. Jones. 2004. Large-scale biodiversity pattern of Cumacea (Peracarida: Crustacea) in the deep Atlantic. Marine Ecology Progress Series **277**:181–196.

Gage, J. D., P. A. Lamont, K. Kroeger, L. J. Paterson, and J. L. G. Vecino. 2000. Patterns in deep-sea macrobenthos at the continental margin: Standing crop, diversity and faunal change on the continental slope off Scotland. Hydrobiologia **440**: 261–271.

Gage, J. D., P. A. Lamont, and P. A. Tyler. 1995. Deep-sea Macrobenthic communities at contrasting sites off Portugal, preliminary results I: Introduction and diversity comparisons. Internationale Revue der gasamten Hydrobiologie **80**: 235–250.

Gage, J. D., and R. M. May. 1993. A dip into the deep seas. Nature **365**:609–610.

Gage, J. D., and P. A. Tyler. 1981a. Non-viable seasonal settlement of larvae of the upper bathyal brittlestar *Ophiocten gracilis* in the Rockall Trough abyssal. Marine Biology **64**:153–161.

Gage, J. D., and P. A. Tyler. 1981b. Reappraisal of the age composition, growth and survivorship of the deep-sea brittlestar *Ophiura ljungmani* from size structure in a sample time series from the Rockall Trough. Marine Biology **64**:163–172.

Gage, J. D., and P. A. Tyler. 1991. Deep-Sea Biology: A Natural History of Organisms at the Deep-Sea Floor. Cambridge University Press, Cambridge, UK.

Galéron, J., M. Sibuet, M.-L. Mahaut, and A. Dinet. 2000. Variation in structure and biomass of the benthic communities at three contrasting sites in the tropical Northeast Atlantic. Marine Ecology Progress Series **197**:121–137.

Galéron, J., M. Sibuet, A. Vanreusel, K. Mackenzie, A. J. Gooday, A. Dinet, and G. A. Wolff. 2001. Temporal patterns among meiofauna and macrofauna taxa related to changes in sediment geochemistry at an abyssal NE Atlantic site. Progress in Oceanography **50**:303–324.

Gambi, C., A. Vanreusel, and R. Danovaro. 2003. Biodiversity of nematode assemblages from deep-sea sediments of the Atacama Slope and Trench (South Pacific Ocean). Deep-Sea Research I **50**:103–117.

Gardner, W. D. 1989a. Periodic resuspension in Baltimore Canyon by focusing of internal waves. Journal of Geophysical Research **94**:18185–18194.

Gardner, W. D. 1989b. Baltimore Canyon as a modern conduit of sediment to the deep sea. Deep-Sea Research **36**:323–358.

Gardner, W. D., and L. G. Sullivan. 1981. Benthic storms: Temporal variability in a deep-ocean nepheloid. Science **213**:329–331.

Gaston, K. J. 1994. Rarity. Chapman and Hall, New York.

Gavrilets, S., and D. Waxman. 2002. Sympatric speciation by sexual conflict. Proceedings of the National Academy of Sciences, USA **99**:10533–10538.

Gerdes, D., M. Klages, W. E. Arntz, R. L. Herman, J. Galéron, and S. Hain. 1992. Quantitative investigations on macrobenthos communities of the southeastern Weddell Sea shelf based on multibox corer samples. Polar Biology **12**:291–301.

Gersonde, R., F. T. Kyte, U. Bleil, B. Diekmann, J. A. Flores, G. Grahl, R. Hagen, G. Kuhn, F. J. Sierro, D. Völker, A. Abelmann, and J. A. Bostwick. 1997. Geological record and reconstruction of the late Pliocene impact of the Eltanin asteroid in the Southern Ocean. Nature **390**:357–363.

Geyer, L. B., and S. R. Palumbi. 2003. Reproductive character displacement and the genetics of gamete recognition in tropical sea urchins. Evolution **57**:1049–1060.

Gibson, T. G., and M. A. Buzas. 1973. Species diversity: Patterns in modern and Miocene foraminifera of the eastern margin of North America. Geological Society of America Bulletin **84**:217–238.

Gilg, M. R., and T. J. Hilbish. 2003. The geography of marine larval dispersal: Coupling genetics with fine-scale physical oceanography. Ecology **84**:2989–2998.

Gillooly, J. F., and A. P. Allen. 2007. Linking global patterns in biodiversity to evolutionary dynamics using metabolic theory. Ecology **88**:1890–1894.

Gillooly, J. F., A. P. Allen, G. B. West, and J. H. Brown. 2005. The rate of DNA evolution: Effects of body size and temperature on the molecular clock. Proceedings of the National Academy of Sciences, USA **102**:140–145.

Ginger, M. L., D. S. M. Billett, K. L. Mackenzie, K. Kiriakoulakis, R. R. Neto, D. K. Boardman, V. L. C. S. Santos, I. M. Horsfall, and G. A. Wolff. 2001. Organic matter assimilation and selective feeding by holothurians in the deep sea: Some observations and comments. Progress in Oceanography **50**:407–421.

Glover, A. G., C. R. Smith, G. L. J. Paterson, G. D. F. Wilson, L. Hawkins, and M. S. Sheader. 2002. Polychaete species diversity in the central Pacific abyss: Local and regional patterns, and relationships with productivity. Marine Ecology Progress Series **240:**157–170.

Gonzalez, A., and R. D. Holt. 2002. The inflationary effect of environmental fluctuations in source-sink systems. Proceedings of the National Academy of Sciences, USA **99:**14872–14877.

Gooday, A. J. 1986. Meiofaunal foraminiferans from the bathyal Porcupine Seabight (north-east Atlantic): Size structure, standing stock, taxonomic composition, species diversity and vertical distribution in the sediment. Deep-Sea Research **33:**1345–1373.

Gooday, A. J. 1988. A response by benthic Foraminifera to the deposition of phytodetritus in the deep sea. Nature **332:**70–73.

Gooday, A. J. 1996. Epifaunal and shallow infaunal foraminiferal communities at three abyssal NE Atlantic sites subject to differing phytodetritus input regimes. Deep-Sea Research I **43:**1395–1421.

Gooday, A. J. 1999. Biodiversity of foraminifera and other protists in the deep sea: Scales and patterns. Belgian Journal of Zoology **129:**61–80.

Gooday, A. J. 2002. Organic-walled allogromiids: Aspects of their occurrence, diversity and ecology in marine habitats. Journal of Foraminiferal Research **32:**384–399.

Gooday, A. J. 2003. Benthic Foraminifera (Protista) as tools in deep-water palaeoceanography: Environmental influences on faunal characteristics. Advances in Marine Biology **46:**1–90.

Gooday, A. J., J. M. Bernhard, L. A. Levin, and S. B. Suhr. 2000. Foraminifera in the Arabian Sea oxygen minimum zone and other oxygen-deficient settings: Taxonomic composition, diversity, and relation to metazoan faunas. Deep-Sea Research II **47:**25–54.

Gooday, A. J., B. J. Bett, R. Shires, and P. J. D. Lambshead. 1998. Deep-sea benthic foraminiferal species diversity in the NE Atlantic and NW Arabian Sea: A synthesis. Deep-Sea Research II **45:**165–201.

Gooday, A. J., S. Hori, Y. Todo, T. Okamoto, H. Kitazato, and A. Sabbatini. 2004. Soft-walled, monothalamous benthic foraminiferans in the Pacific, Indian and Atlantic Oceans: Aspects of biodiversity and biogeography. Deep-Sea Research I **51:**33–53.

Gooday, A. J., and J. A. Hughes. 2002. Foraminifera associated with phytodetritus deposits at a bathyal site in the northern Rockall Trough (NE Atlantic): Seasonal contrasts and a comparison of stained and dead assemblages. Marine Micropaleontology **46:**83–110.

Gooday, A. J., J. A. Hughes, and L. A. Levin. 2001. The foraminiferan macrofauna from three North Carolina (U.S.A.) slope sites with contrasting carbon flux: A comparison with the metazoan macrofauna. Deep-Sea Research I **48:**1709–1739.

Gooday, A. J., L. A. Levin, P. Linke, and T. Heeger. 1992. The role of benthic foraminifera in deep-sea food webs and carbon cycling. Pp. 63–91, *in* G. T. Rowe and V. Pariente, eds. Deep-Sea Food Chains and the Global Carbon Cycle. Kluwer, Dordrecht, The Netherlands.

Gooday, A. J., O. Pfannkuche, and P. J. D. Lambshead. 1996. An apparent lack of response by metazoan meiofauna to phytodetritus deposition in the bathyal north-eastern Atlantic. Journal of the Marine Biological Association of the United Kingdom **76**:297–310.

Gosselin, M., M. Levasseur, P. A. Wheeler, R. A. Horner, and B. C. Booth. 1997. New measurements of phytoplankton and ice algal production in the Arctic Ocean. Deep-Sea Research II **44**:1623–1644.

Graf, G. 1989. Benthic-pelagic coupling in a deep-sea benthic community. Nature **341**:437–439.

Grassle, J. F. 1977. Slow recolonisation of deep-sea sediment. Nature **265**:618–619.

Grassle, J. F. 1989. Species diversity in deep-sea communities. Trends in Ecology and Evolution **4**:12–15.

Grassle, J. F., and J. P. Grassle. 1994. Notes from the abyss: The effects of a patchy supply of organic material and larvae on soft-sediment benthic communities. Pp. 499–515, *in* P. S. Giller, A. D. Hildrew, and D. G. Raffaelli, eds. Aquatic Ecology: Scale, Pattern and Process. Blackwell, Oxford, UK

Grassle, J. F., and N. J. Maciolek. 1992. Deep-sea species richness: Regional and local diversity estimates from quantitative bottom samples. American Naturalist **193**:313–341.

Grassle, J. F., and L. S. Morse-Porteous. 1987. Macrofaunal colonization of disturbed deep-sea environments and the structure of deep-sea benthic communities. Deep-Sea Research **34**:1911–1950.

Grassle, J. F., and H. L. Sanders. 1973. Life histories and the role of disturbance. Deep-Sea Research **20**:643–659.

Grassle, J. F., H. L. Sanders, R. R. Hessler, G. T. Rowe, and T. McLellan. 1975. Pattern and zonation: A study of the bathyal megafauna using the research submersible *Alvin*. Deep-Sea Research **22**:457–481.

Grassle, J. F., H. L. Sanders, and W. K. Smith. 1979. Faunal changes with depth in the deep-sea benthos. Ambio Special Report **6**:47–50.

Gray, J. S. 1994. Is deep-sea species diversity really so high? Species diversity of the Norwegian continental shelf. Marine Ecology Progress Series **112**:205–209.

Gray, J. S. 2001. Antarctic marine benthic biodiversity in a world-wide latitudinal context. Polar Biology **24**:633–641.

Gray, J. S., G. C. B. Poore, K. I. Ugland, R. S. Wilson, F. Olsgard, and Ø. Johannessen. 1997. Coastal and deep-sea benthic diversities compared. Marine Ecology Progress Series **159**:97–103.

Griggs, G. B., A. G. Carey Jr., and L. D. Kulm. 1969. Deep-sea sedimentation and sediment-fauna interaction in Cascadia Channel and on Cascadia Abyssal Plain. Deep-Sea Research **16**:157–170.

Grosberg, R. K., and J. R. Cunningham. 2001. Genetic structure in the sea: From populations to communities. Pp. 61–84, *in* M. Bertness, S. Gaines, and M. E. Hay, eds. Marine Community Ecology. Sinauer, Sunderland, MA.

Gross, K., and B. J. Cardinale. 2007. Does species richness drive community production or vice versa? Reconciling historical and contemporary paradigms in competitive communities. American Naturalist **170**:207–220.

Grove, S. L., P. K. Probert, K. Berkenbusch, and S. D. Nodder. 2006. Distribution of bathyal meiofauna in the region of the Subtropical Front, Chatham Rise, southwest Pacific. Journal of Experimental Marine Biology and Ecology **330**:342–355.

Haedrich, R. L., J. A. Devine, and V. J. Kendall. 2008. Predictors of species richness in the deep-benthic fauna of the northern Gulf of Mexico. Deep-Sea Research II **55**:2650–2656.

Haedrich, R. L., and N. R. Merrett. 1988. Summary atlas of deep-living demersal fishes in the North Atlantic Basin. Journal of Natural History **22**:1325–1362.

Haedrich, R. L., and N. R. Merrett. 1990. Little evidence for faunal zonation of communities in deep-sea demersal fish faunas. Progress in Oceanography **24**:239–250.

Haedrich, R. L., and G. T. Rowe. 1977. Megafaunal biomass in the deep sea. Nature **269**:141–142.

Haedrich, R. L., G. T. Rowe, and P. T. Polloni. 1975. Zonation and faunal composition of epibenthic populations on the continental slope south of New England. Journal of Marine Research **33**:191–212.

Haedrich, R. L., G. T. Rowe, and P. T. Polloni. 1980. The megabenthic fauna in the deep sea south of New England, USA. Marine Biology **57**:165–179.

Hare, M. P., C. Guenther, and W. F. Fagan. 2005. Nonrandom larval dispersal can steepen marine clines. Evolution **59**:2509–2517.

Harrison, S., and J. B. Grace. 2007. Biogeographic affinity helps explain productivity-richness relationships at regional and local scales. American Naturalist Supplement **170**:5–15.

Harrold, C., K. Light, and S. Lisin. 1998. Organic enrichment of submarine-canyon and continental-shelf benthic communities by macroalgal drift imported from nearshore kelp forests. Limnology and Oceanography **43**:669–678.

Hartman, O. 1965. Deep-water benthic polychaetous annelids off New England to Bermuda and other North Atlantic areas. Occasional Papers of the Allan Hancock Foundation **28**:1–378.

Hasegawa, K. 2005. A preliminary list of deep-sea gastropods collected from the Nansei Islands, southwestern Japan. Pp. 137–190, *in* K. Hasegawa, G. Shinohara, and M. Takeda, eds. Deep-Sea Fauna and Pollutants in Nansei Islands. National Science Museum Monographs, Number 29, Tokyo.

Hawkins, B. A., F. S. Albuquerque, M. B. Araujo, J. Beck, L. M. Bini, F. J. Cabrero-Sanudo, I. Castro-Parga, A. F. Diniz, D. Ferrer-Castan, R. Field, J. F. Gomez, J. T. Hortal, J. T. Kerr, I. J. Kitching, J. L. Leon-Cortes, J. M. Lobo, D. Montoya, J. C. Moreno, M. A. Olalla-Tarraga, J. G. Pausas, H. Qian, C. Rahbek, M. A. Rodriguez, N. J. Sanders, and P. Williams. 2007a. A global evaluation of metabolic theory as an explanation for terrestrial species richness gradients. Ecology **88**:1877–1888.

Hawkins, B. A., J. A. F. Diniz-Filho, C. A. Jaramillo, and S. A. Soeller. 2007b. Climate, niche conservatism and the global bird diversity gradient. American Naturalist **170**:S16–S27.

Hawkins, B. A., J. A. F. Diniz-Filho, and A. E. Weis. 2005. The mid-domain effect and diversity gradients: Is there anything to learn? American Naturalist **166**:140–143.

Hawkins, B. A., R. Field, H. V. Cornell, D. J. Currie, J.-F. Guégan, D. M. Kaufman, J. T. Kerr, G. G. Mittelbach, T. Oberdorff, E. M. O'Brien, E. E. Porter, and J. R. G. Turner. 2003a. Energy, water, and broad-scale geographic patterns of species richness. Ecology **84**:3105–3117.

Hawkins, B. A., E. E. Porter, and J. A. F. Diniz-Filho. 2003b. Productivity and history as predictors of the latitudinal diversity gradient for terrestrial birds. Ecology **84**:1608–1623.

Hayek, L. C., M. A. Buzas, and L. E. Osterman. 2007. Community structure of foraminiferal communities within temporal biozones from the western Arctic Ocean. Journal of Foraminiferal Research **37**:33–40.

Hayward, B. W. 2001. Global deep-sea extinctions during the Pleistocene ice ages. Geology **29**:599–602.

Hayward, B. W., S. Kawagata, H. R. Grenfell, A. T. Sabaa, and T. O'Neill. 2007. Last global extinction in the deep sea during the mid-Pleistocene climate transition. Paleoceanography **22**:1–14.

Hecker, B. 1990a. Photographic evidence for the rapid flux of particles to the sea floor and their transport down the continental slope. Deep-Sea Research **37**:1773–1782.

Hecker, B. 1990b. Variation in megafaunal assemblages on the continental margin south of New England. Deep-Sea Research **37**:37–57.

Hecker, B. 1994. Unusual megafaunal assemblages on the continental slope off Cape Hatteras. Deep-Sea Research II **41**:809–834.

Hecker, B., D. T. Logan, F. E. Gandarillas, and P. R. Gibson. 1983. Megafaunal assemblages in Lydonia Canyon, Baltimore Canyon and selected slope areas. Final report submitted to the U. S. Department of the Interior, Minerals Management Service Contact 14-12-001-29178.

Hecker, B., and A. Z. Paul. 1979. Abyssal community structure of the benthic infauna of the eastern Equatorial Pacific: Domes sites A, B, and C. Pp. 287–308, *in* D. Z. Piper, ed. Marine Geology and Oceanography of the Pacific Manganese Nodule Province. Plenum, Palisades, NY.

Hedgecock, D., P. H. Barber, and S. Edmands. 2007. Genetic approaches to measuring connectivity. Oceanography **20**:70–79.

Heinz, P., C. Hemleben, and H. Kitazato. 2002. Time-response of cultured deep-sea benthic foraminifera to different algal diets. Deep-Sea Research I **49**:517–537.

Heip, C. H. R., G. Duineveld, E. Flach, G. Graf, W. Helder, P. M. J. Herman, M. Lavaleye, J. J. Middelburg, O. Pfannkuche, K. Soetaert, T. Soltwedel, H. de Stigter, L. Thomsen, J. Vanaverbeke, and P. de Wilde. 2001. The role of the benthic biota in sedimentary metabolism and sediment-water exchange processes in the Goban Spur area (NE Atlantic). Deep-Sea Research II **48**:3223–3243.

Held, C. 2000. Phylogeny and biogeography of serolid isopods (Crustacea, Isopoda, Serolidae) and the use of ribosomal expansion segments in molecular systematics. Molecular Phylogenetics and Evolution **15**:165–178.

Hellberg, M. E., R. S. Burton, J. E. Neigel, and S. R. Palumbi. 2002. Genetic assessment of connectivity among marine populations. Bulletin of Marine Science **70**:273–290.

Helly, J. J., and L. A. Levin. 2004. Global distribution of naturally occurring marine hypoxia on continental margins. Deep-Sea Research I **51**:1159–1168.

Herman, P. M. J., K. Soetaert, J. J. Middelburg, C. Heip, L. Lohse, E. Epping, W. Helder, A. N. Antia, and R. Peinert. 2001. The seafloor as the ultimate sediment trap: Using sediment properties to constrain benthic-pelagic exchange processes at the Goban Spur. Deep-Sea Research II **48**:3245–3264.

Hermann, R. L., and H.-U. Dahms. 1992. Meiofauna communities along a depth transect off Halley Bay (Weddell Sea, Antarctica). Polar Biology **12**:313–320.

Herring, P. 2002. The Biology of the Deep Ocean. Oxford University Press, Oxford, UK.

Hess, S., and W. Kuhnt. 1996. Deep-sea benthic foraminiferal recolonization of the 1991 Mt. Pinatubo ash layer in the South China Sea. Marine Micropaleontology **28**:171–197.

Hessler, R. R., and P. A. Jumars. 1974. Abyssal community analysis from replicate box cores in the central North Pacific. Deep-Sea Research **21**:185–209.

Hessler, R. R., and H. L. Sanders. 1967. Faunal diversity in the deep-sea. Deep-Sea Research **14**:65–78.

Hessler, R. R., and D. Thistle. 1975. On the place of origin of deep-sea isopods. Marine Biology **32**:155–165.

Hessler, R. R., G. D. Wilson, and D. Thistle. 1979. The deep-sea isopods: A biogeographic and phylogenetic overview. Sarsia **64**:67–75.

Hilbish, T. J., and R. K. Koehn. 1985. The physiological basis of natural selection at the lap locus. Evolution **39**:1302–1317.

Hillebrand, H. 2004a. On the generality of the latitudinal diversity gradient. American Naturalist **163**:192–211.

Hillebrand, H. 2004b. Strength, slope and variability of marine latitudinal gradients. Marine Ecology Progress Series **273**:251–267.

Hillebrand, H., and A. I. Azovsky. 2001. Body size determines the strength of the latitudinal diversity gradient. Ecography **24**:251–256.

Hochachka, P. W., and G. N. Somero. 2002. Biochemical Adaptation: Mechanisms and Progress in Physiological Evolution. Oxford University Press, New York.

Hoegh-Guldberg, O., J. R. Welborn, and D. T. Manahan. 1991. Metabolic requirements of Antarctic and temperate asteroid larvae. Antarctic Journal **26**:163–165.

Hogg, N. G., and W. B. Owens. 1999. Direct measurement of the deep circulation within the Brazil Basin. Deep-Sea Research II **46**:335–353.

Hollister, C. D., and I. N. McCave. 1984. Sedimentation under deep-sea storms. Nature **309**:220–225.

Hollister, C. D., A. R. M. Nowell, and P. A. Jumars. 1984. The dynamic abyss. Scientific American **250**:42–53.

Holt, R. D. 1985. Population dynamics in two-patch environments: Some anomalous consequences of an optimal habitat distribution. Theoretical Population Biology **28**:181–208.

Honjo, S., S. J. Manganini, R. A. Krishfield, and R. Francois. 2008. Particulate organic carbon fluxes to the ocean interior and factors controlling the biological pump: A synthesis of global sediment trap programs since 1983. Progress in Oceanography **76**:217–285.

Horne, D. J. 1999. Ocean circulation modes of the Phanerozoic: Implications for the antiquity of deep-sea benthonic invertebrates. Crustaceana **72**:999–1018.

Houston, K. A., and R. L. Haedrich. 1984. Abundance and biomass of macrobenthos in the vicinity of Carson Submarine Canyon, northwest Atlantic Ocean. Marine Biology **82**:301–305.

Howell, K. L., D. S. M. Billett, and P. A. Tyler. 2002. Depth-related distribution and abundance of seastars (Echinodermata: Asteroidea) in the Porcupine Seabight and Porcupine Abyssal Plain, N.E. Atlantic. Deep-Sea Research I **49**:1901–1920.

Howell, K. L., A. D. Rogers, P. A. Tyler, and D. S. M. Billett. 2004. Reproductive isolation among morphotypes of the Atlantic seastar species *Zoroaster fulgens* (Asteroidea: Echinodermata). Marine Biology **144**:977–984.

Hubbell, S. P. 2001. The Unified Neutral Theory of Biodiversity and Biogeography. Princeton University Press, Princeton, NJ.

Hubbell, S. P. 2005. Neutral theory in community ecology and the hypothesis of functional equivalence. Functional Ecology **19**:166-172.

Hughes, D. J., L. Brown, G. T. Cook, G. Cowie, J. D. Gage, E. Good, H. Kennedy, A. B. MacKenzie, S. Papadimitriou, G. B. Shimmield, J. Thomson, and M. Williams. 2005. The effects of megafaunal burrows on radiotracer profiles and organic composition in deep-sea sediments: Preliminary results from two sites in the bathyal north-east Atlantic. Deep-Sea Research I **52**:1–13.

Hughes, D. J., and J. D. Gage. 2004. Benthic metazoan biomass, community structure and bioturbation at three contrasting deep-water sites on the northwest European continental margin. Progress in Oceanography **63**:29–55.

Huitema, B. E. 1980. The Analysis of Covariance and Alternatives. Wiley, New York.

Hunt, G., T. M. Cronin, and K. Roy. 2005. Species-energy relationship in the deep sea: A test using the Quaternary fossil record. Ecology Letters **8**:739–747.

Hunt, G., and K. Roy. 2006. Climate change, body size evolution, and Cope's rule in deep-sea ostracodes. Proceedings of the National Academy of Sciences, USA **103**:1347–1352.

Hurlbert, S. H. 1971. The nonconcept of species diversity: A critique and alternative parameter. Ecology **52**:577–586.

Huston, M. A. 1979. A general hypothesis of species diversity. American Naturalist **113**:81–101.

Huston, M. A. 1994. Biological Diversity: The Coexistence of Species on Changing Landscapes. Cambridge University Press, Cambridge, UK.

Hutchinson, G. E. 1959. Homage to Santa Rosalia, or Why are there so many kinds of animals? American Naturalist **93**:145–159.

Hyland, J., E. Baptiste, J. Campbell, J. Kennedy, R. Kropp, and S. Williams. 1991. Macroinfaunal communities of the Santa Maria Basin on the California outer continental shelf and slope. Marine Ecology Progress Series **78**:147–161.

Iguchi, A., H. Ito, M. Ueno, T. Maeda, T. Minami, and I. Hayashi. 2007a. Molecular phylogeny of the deep-sea *Buccinum* species (Gastropoda: Buccinidae) around Japan: Inter- and intraspecific relationships inferred from mitochondrial 16SrRNA sequences. Molecular Phylogenetics and Evolution **44**:1342–1345.

Iguchi, A., H. Ito, M. Ueno, T. Maeda, T. Minami, and I. Hayashi. 2007b. Comparative analysis on the genetic population structures of the deep-sea whelks *Buccinum tsubai* and *Neptunea constricta* in the Sea of Japan. Marine Biology **151**:31–39.

Iken, K., T. Brey, U. Wand, J. Voigt, and P. Junghans. 2001. Food web structure of the benthic community at the Porcupine Abyssal Plain (NE Atlantic): A stable isotope analysis. Progress in Oceanography **50**:383–405.

Irwin, D. E. 2002. Phylogeographic breaks without geographic barriers to gene flow. Evolution **56**:2383–2394.

Irwin, D. E., S. Bensch, J. H. Irwin, and T. D. Price. 2005. Speciation by distance in a ring species. Science **307**:414–416.

Isaacs, J. D. 1969. The nature of oceanic life. Scientific American **221**:146–162.

Isozaki, Y. 1997. Permo-Triassic boundary superanoxia and stratified superocean: Records from lost deep sea. Science **276**:235–238.

Jablonski, D. 1993. The tropics as a source of evolutionary novelty through geological time. Nature **364**:142–144.

Jablonski, D., and D. J. Bottjer. 1990. The ecology of evolutionary innovation: The fossil record. Pp. 253–288, *in* M. H. Nitecki, ed. Evolutionary Innovations. University of Chicago Press, Chicago.

Jablonski, D., and D. J. Bottjer. 1991. Environmental patterns in the origins of higher taxa: The post-Paleozoic fossil record. Science **252**:1831–1833.

Jablonski, D., K. Roy, and J. W. Valentine. 2006. Out of the tropics: Evolutionary dynamics of the latitudinal diversity gradient. Science **314**:102–106.

Jablonski, D., J. J. Sepkoski Jr., D. J. Bottjer, and P. M. Sheehan. 1983. Onshore-offshore patterns in the evolution of Phanerozoic shelf communities. Science **222**:1123–1125.

Jacobs, D. K., and D. R. Lindberg. 1998. Oxygen and evolutionary patterns in the sea: Onshore/offshore trends and recent recruitment of deep-sea faunas. Proceedings of the National Academy of Sciences, USA **95**:9396–9401.

Jaeckle, W. B., and D. T. Manahan. 1989. Feeding by a "nonfeeding" larva: Uptake of dissolved amino acids from seawater by lecithotrophic larvae of the gastropod *Haliotis rufescens.* Marine Biology **103**:87–94.

Janzen, D. H. 1967. Why mountain passes are higher in the tropics. American Naturalist **101**:233–249.

Jaźdźewski, K., W. Jurasz, W. Kittel, E. Presler, P. Presler, and J. Sicinski. 1986. Abundance and biomass estimates of the benthic fauna in Admiralty Bay, King George Island, South Shetland Islands. Polar Biology **6**:5–16.

Jennings, R. M., T. M. Shank, L. S. Mullineaux, and K. M. Halanych. 2009. Assessment of the Cape Cod phylogeographic break using the bamboo worm *Clymenella torquata* (Annelida: Maldanidae). Heredity **100**:86–96.

Jensen, P. 1988. Nematode assemblages in the deep-sea benthos of the Norwegian Sea. Deep-Sea Research **35**:1173–1184.

Jensen, P., J. Rumohr, and G. Graf. 1992. Sedimentological and biological differences across a deep-sea ridge exposed to advection and accumulation of fine-grained particles. Oceanologica Acta **15**:287–296.

Johnson, N. A., J. W. Campbell, T. S. Moore, M. A. Rex, R. J. Etter, C. R. McClain, and M. D. Dowell. 2007. The relationship between the standing stock of deep-sea macrobenthos and surface production in the western North Atlantic. Deep-Sea Research I **54**:1350–1360.

Jones, D. O. B., B. J. Bett, and P. A. Tyler. 2007. Megabenthic ecology of the deep Faroe-Shetland Channel: A photographic study. Deep-Sea Research I **54**:1111–1128.

Jones, N. S., and H. L. Sanders. 1972. Distribution of Cumacea in the deep Atlantic. Deep-Sea Research **19**:737–745.

Jorissen, F. J., H. C. de Stigter, and J. G. V. Widmark. 1995. A conceptual model explaining benthic foraminiferal microhabitats. Marine Micropaleontology **26**: 3–15.

Jumars, P. A. 1975a. Methods for measurement of community structure in deep-sea macrobenthos. Marine Biology **30**:245–252.

Jumars, P. A. 1975b. Environmental grain and polychaete species' diversity in a bathyal benthic community. Marine Biology **30**:253–266.

Jumars, P. A. 1976. Deep-sea species diversity: Does it have a characteristic scale? Journal of Marine Research **34**:217–246.

Jumars, P. A. 1978. Spatial autocorrelation with RUM (Remote Underwater Manipulator): Vertical and horizontal structure of a bathyal benthic community. Deep-Sea Research **25**:589–604.

Jumars, P. A. 1981. Limits in predicting and detecting benthic community responses to manganese module mining. Marine Mining **3**:213–229.

Jumars, P. A., and J. E. Eckman. 1983. Spatial structure within deep-sea benthic communities. Pp. 399–451, *in* G. T. Rowe, ed. The Sea, Vol. 8: Deep-Sea Biology. Wiley, New York.

Jumars, P. A., and K. Fauchald. 1977. Between-community contrasts in successful polychaete feeding strategies. Pp. 1–20, *in* B. C. Coull, ed. The Ecology of Marine Benthos. University of South Carolina Press, Columbia, SC.

Jumars, P. A., and R. R. Hessler. 1976. Hadal community structure: Implications from the Aleutian Trench. Journal of Marine Research **34**:547–560.

Jumars, P. A., L. M. Mayer, J. W. Deming, J. A. Baross, and R. A. Wheatcroft. 1990. Deep-sea deposit-feeding strategies suggested by environmental and feeding constraints. Philosophical Transactions of the Royal Society of London A **331**: 85–101.

Jumars, P. A., D. Thistle, and M. L. Jones. 1977. Detecting two-dimensional spatial structure in biological data. Oecologia **28**:109–123.

Kaiho, K. 1994. Planktonic and benthic foraminiferal extinction events during the last 100 m.y. Palaeogeography, Palaeobiology, Palaeoecology **111**:45–71.

Kallianiotis, A., K. Sophronidis, P. Vidoris, and A. Tselepides. 2000. Demersal fish and megafaunal assemblages on the Cretan continental shelf and slope (NE Mediterranean): Seasonal variation in species density, biomass and diversity. Progress in Oceanography **46**:429–455.

Karl, D. M., J. R. Christian, J. E. Dore, D. V. Hebel, R. M. Letelier, L. M. Tupas, and C. D. Winn. 1996. Seasonal and interannual variability in primary production and particle flux at station ALOHA. Deep-Sea Research II **43**:539–568.

Karlsen, F., M. Kalantari, M. Chitemerere, B. Johansson, and B. Hagmar. 1994. Modifications of human and viral deoxyribonucleic acid by formaldehyde fixation. Laboratory Investigation **71**:604–611.

Karlson, R. H., H. V. Cornell, and T. P. Hughes. 2004. Coral communities are regionally enriched along an oceanic biodiversity gradient. Nature **429**:867–870.

Kaufmann, R. S., W. W. Wakefield, and A. Genin. 1989. Distribution of epibenthic megafauna and lebensspuren on two central North Pacific seamounts. Deep-Sea Research **36**:1863–1896.

Kelly, C. K., M. G. Bowler, O. Pybus, and P. H. Harvey. 2008. Phylogeny, niches and relative abundance in natural communities. Ecology **89**:962–970.

Kelly, R. P., and D. J. Eernisse. 2007. Southern hospitality: A latitudinal gradient in gene flow in the marine environment. Evolution **61**:700–707.

Kendall, V. J., and R. L. Haedrich. 2006. Species richness in Atlantic deep-sea fishes assessed in terms of the mid-domain effect and Rapoport's rule. Deep-Sea Research I **53**:506–515.

Kennett, J. P., and L. D. Stott. 1991. Abrupt deep-sea warming, palaeoceanographic changes and benthic extinctions at the end of the Palaeocene. Nature **353**:225–229.

Khripounoff, A., J.-C. Caprais, P. Crassous, and J. Etoubleau. 2006. Geochemical and biological recovery of the disturbed seafloor in polymetallic nodule fields of the Clipperton-Clarion Fracture Zone (CCFZ) at 5,000-m depth. Limnology and Oceanography **51**:2033–2041.

Khripounoff, A., D. Desbruyères, and P. Chardy. 1980. Les peuplements benthiques de la faille Vema: données quantitatives et bilan d'énergie en milieu abyssal. Oceanologica Acta **3**:187–198.

Kiel, S., and L. A. Little. 2006. Cold-seep mollusks are older than the general marine mollusk fauna. Science **313**:1429–1431.

Killingley, J. S., and M. A. Rex. 1985. Mode of larval development in some deep-sea gastropods indicated by oxygen-18 values of their carbonate shells. Deep-Sea Research **32**:809–818.

Knoll, A. H., R. K. Bambach, D. E. Canfield, and J. P. Grotzinger. 1996. Comparative earth history and the Late Permian mass extinction. Science **273**:452–457.

Knowlton, N., and L. A. Weigt. 1998. New dates and new rates for divergence across the Isthmus of Panama. Proceedings of the Royal Society of London B: Biological Sciences **265**:2257–2263.

Knowlton, N., L. A. Weigt, L. A. Solorzano, D. K. Mills, and E. Bermingham. 1993. Divergence in proteins, mitochondrial DNA, and reproductive compatibility across the Isthmus of Panama. Science **260**:1629–1632.

Kojima, S., R. Segawa, I. Hayashi, and M. Okiyama. 2001. Phylogeography of a deep-sea demersal fish, *Bothrocara hollandi*, in the Japan Sea. Marine Ecology Progress Series **217**:135–143.

Kondoh, M. 2001. Unifying the relationships of species richness to productivity and disturbance. Proceedings of the Royal Society of London B **268**:269–271.

Koslow, T. 2007. The Silent Deep: The Discovery, Ecology and Conservation of the Deep Sea. University of Chicago Press, Chicago.

Kröncke, I. 1994. Macrobenthos composition, abundance and biomass in the Arctic Ocean along a transect between Svalbard and the Makarov Basin. Polar Biology **14**:519–529.

Kröncke, I. 1998. Macrofauna communities in the Amundsen Basin, at the Morris Jesup Rise and at the Yermak Plateau (Eurasian Arctic Ocean). Polar Biology **19**:383–392.

Kröncke, I., T. L. Tan, and R. Stein. 1994. High benthic bacteria standing stock in deep Arctic basins. Polar Biology **14**:423–428.

Kröncke, I., and M. Türkay. 2003. Structural and functional aspects of the benthic communities in the deep Angola Basin. Marine Ecology Progress Series **260**: 43–53.

Kröncke, I., M. Türkay, and D. Fiege. 2003. Macrofauna communities in the eastern Mediterranean deep sea. Marine Ecology **24**:193–216.

Kröncke, I., A. Vanreusel, M. Vincx, J. Wollenburg, A. Mackensen, G. Liebezeit, and B. Behrends. 2000. Different benthic size-compartments and their relationship to sediment chemistry in the deep Eurasian Arctic Ocean. Marine Ecology Progress Series **199**:31–41.

Kuhnt, W. 1992. Abyssal recolonization by benthic foraminifera after the Cenomanian/Turonian boundary anoxic event in the North Atlantic. Marine Micropaleontology **19**:257–274.

Kukert, H., and C. R. Smith. 1992. Disturbance, colonization and succession in a deep-sea sediment community: Artificial-mound experiments. Deep-Sea Research **39**:1349–1371.

Kurihara, K., and J. P. Kennett. 1988. Bathymetric migration of deep-sea benthic foraminifera in the southwest Pacific during the Neogene. Journal of Foraminiferal Research **18**:75–83.

Kussakin, O. G. 1973. Peculiarities of the geographical and vertical distribution of marine isopods and the problem of deep-sea fauna origin. Marine Biology **23**:19–34.

Lagoe, M. B. 1976. Species diversity of deep-sea benthic Foraminifera from the central Arctic Ocean. Geological Society of America Bulletin **87**:1678–1683.

Lambeck, K., and J. Chappell. 2001. Sea level change through the last glacial cycle. Science **292**:679–686.

Lambshead, P. J. D., and G. Boucher. 2003. Marine nematode deep-sea biodiversity-hyperdiverse or hype? Journal of Biogeography **30**:475–485.

Lambshead, P. J. D., C. J. Brown, T. J. Ferrero, N. J. Mitchell, C. R. Smith, L. E. Hawkins, and J. Tietjen. 2002. Latitudinal diversity patterns of deep-sea marine nematodes and organic fluxes: A test from the central equatorial Pacific. Marine Ecology Progress Series **236:**129–135.

Lambshead, P. J. D., J. Tietjen, T. Ferrero, and P. Jensen. 2000. Latitudinal diversity gradients in the deep sea with special reference to North Atlantic nematodes. Marine Ecology Progress Series **194:**159–167.

Lambshead, P. J. D., J. Tietjen, A. Glover, T. Ferrero, D. Thistle, and A. J. Gooday. 2001a. Impact of large-scale natural physical disturbance on the diversity of deep-sea North Atlantic nematodes. Marine Ecology Progress Series **214:**121–126.

Lambshead, P. J. D., J. Tietjen, C. B. Moncrieff, and T. J. Ferrero. 2001b. North Atlantic latitudinal diversity patterns in deep-sea marine nematode data: A reply to Rex et al. Marine Ecology Progress Series **210:**299–301.

Lamont, P. A., J. D. Gage, and P. A. Tyler. 1995. Deep-sea macrobenthic communities at contrasting sites off Portugal, preliminary results II: Spatial dispersion. Internationale Revue der gasamten Hydrobiologie **80:**251–265.

Lampadariou, N., and A. Tselepides. 2006. Spatial variability of meiofaunal communities at areas of contrasting depth and productivity in the Aegean Sea (NE Mediterranean). Progress in Oceanography **69:**19–36.

Lampitt, R. S., and A. N. Antia. 1997. Particle flux in deep seas: Regional characteristics and temporal variability. Deep-Sea Research I **44:**1377–1403.

Lampitt, R. S., B. J. Bett, K. Kiriakoulakis, E. E. Popova, O. Ragueneau, A. Vangriesheim, and G. A. Wolff. 2001. Material supply to the abyssal seafloor in the Northeast Atlantic. Progress in Oceanography **50:**27–63.

Lampitt, R. S., D. S. M. Billett, and A. L. Rice. 1986. Biomass of the invertebrate megabenthos from 500 to 4100 m in the northeast Atlantic Ocean. Marine Biology **93:**69–81.

Lampitt, R. S., R. C. T. Raine, D. S. M. Billett, and A. L. Rice. 1995. Material supply to the European continental slope: A budget based on benthic oxygen demand and organic supply. Deep-Sea Research I **42:**1865–1880.

Laubier, L., and M. Sibuet. 1979. Ecology of the benthic communities of the deep North East Atlantic. Ambio Special Report **6:**37–42.

Lauerman, L. M. L., and R. S. Kaufmann. 1998. Deep-sea epibenthic echinoderms and a temporally varying food supply: Results from a one year time series in the N.E. Pacific. Deep-Sea Research II **45:**817–842.

Lavender, K. L., R. E. Davis, and W. B. Owens. 2000. Mid-depth recirculation observed in the interior Labrador and Irminger seas by direct velocity measurements. Nature **407:**66–69.

Lavender, K. L., W. B. Owens, and R. E. Davis. 2005. The mid-depth circulation of the subpolar North Atlantic Ocean as measured by subsurface floats. Deep-Sea Research I **52:**767–785.

Laws, E. A. 2004. New production in the equatorial Pacific: A comparison of field data with estimates derived from empirical and theoretical models. Deep-Sea Research I **51:**205–211.

Lawton, J. 1996. Patterns in ecology. Oikos **75**:145–147.

Le Danois, E. 1948. Les profondeurs de la mer. Payot, Paris.

Le Goff-Vitry, M. C., O. G. Pybus, and A. D. Rogers. 2004. Genetic structure of the deep-sea coral *Lophelia pertusa* in the northeast Atlantic revealed by microsatellites and internal transcribed spacer sequences. Molecular Ecology **13**:537–549.

Leimar, O., M. Doebieli, and U. Dieckmann. 2008. Evolution of phenotypic clusters through competition and local adaptation along an environmental gradient. Evolution **62**:807–822.

Levin, L. A. 2003. Oxygen minimum zone benthos: Adaptation and community response to hypoxia. Oceanography and Marine Biology: An Annual Review **41**: 1–45.

Levin, L. A. 2005. Ecology of cold seep sediments: Interactions of fauna with flow, chemistry and microbes. Oceanography and Marine Biology: An Annual Review **43**:1–46.

Levin, L. A., N. Blair, D. DeMaster, G. Plaia, W. Fornes, C. Martin, and C. Thomas. 1997. Rapid subduction of organic matter by maldanid polychaetes on the North Carolina slope. Journal of Marine Research **55**:595–611.

Levin, L. A., S. E. Childers, and C. R. Smith. 1991a. Epibenthic, agglutinating foraminiferans in the Santa Catalina Basin and their response to disturbance. Deep-Sea Research **38**:465–483.

Levin, L. A., D. J. DeMaster, L. D. McCann, and C. L. Thomas. 1986. Effects of giant protozoans (Class: Xenophyophorea) on deep-seamount benthos. Marine Ecology Progress Series **29**:99–104.

Levin, L. A., R. J. Etter, M. A. Rex, A. J. Gooday, C. R. Smith, J. Pineda, C. T. Stuart, R. R. Hessler, and D. Pawson. 2001. Environmental influences on regional deep-sea species diversity. Annual Review of Ecology and Systematics **32**:51–93.

Levin, L. A., and J. D. Gage. 1998. Relationships between oxygen, organic matter and the diversity of bathyal macrofauna. Deep-Sea Research II **45**:129–163.

Levin, L. A., J. D. Gage, C. Martin, and P. A. Lamont. 2000. Macrobenthic community structure within and beneath the oxygen minimum zone, NW Arabian Sea. Deep-Sea Research II **47**:189–226.

Levin, L. A., and A. J. Gooday. 2003. The deep Atlantic Ocean. Pp. 111–178, *in* P. A. Tyler, ed. Ecosystems of the World 28: Ecosystems of the Deep Oceans. Elsevier, Amsterdam.

Levin, L. A., C. L. Huggett, and K. F. Wishner. 1991b. Control of deep-sea benthic community structure by oxygen and organic-matter gradients in the eastern Pacific Ocean. Journal of Marine Research **49**:763–800.

Levin, L. A., G. R. Plaia, and C. L. Huggett. 1994. The influence of natural organic enhancement on life histories and community structure of bathyal polychaetes. Pp. 261–283, *in* C. M. Young and K. J. Eckelbarger, eds. Reproduction, Larval Biology, and Recruitment of the Deep-Sea Benthos. Columbia University Press, New York.

Levin, L. A., and C. R. Smith. 1984. Response of background fauna to disturbance and enrichment in the deep sea: A sediment tray experiment. Deep-Sea Research **31**:1277–1285.

Levin, L. A., and C. L. Thomas. 1989. The influence of hydrodynamic regime on infaunal assemblages inhabiting carbonate sediments on central Pacific seamounts. Deep-Sea Research **36**:1897–1915.

Levin, S. A., and R. T. Paine. 1974. Disturbance, patch formation, and community structure. Proceedings of the National Academy of Sciences, USA **71**:2744–2747.

Levinton, J. S. 2001. Marine Biology: Function, Biodiversity, Ecology. Oxford University Press, Oxford, UK.

Lindner, A., S. D. Cairns, and C. W. Cunningham. 2008. From offshore to onshore: Multiple origins of shallow-water corals from deep-sea ancestors. PloS One **3**:e2429.

Linke, P. 1992. Metabolic adaptations of deep-sea benthic foraminifera to seasonally varying food input. Marine Ecological Progress Series **81**:51–63.

Lochte, K., and C. M. Turley. 1988. Bacteria and cyanobacteria associated with phytodetritus in the deep sea. Nature **333**:67–69.

Longhurst, A. R. 2007. Ecological Geography of the Sea. Academic, New York.

MacArthur, R. H., and R. Levins. 1967. The limiting similarity, convergence, and divergence of coexisting species. American Naturalist **101**:377.

MacAvoy, S. E., R. S. Carney, C. R. Fisher, and S. A. Macko. 2002. Use of chemosynthetic biomass by large, mobile, benthic predators in the Gulf of Mexico. Marine Ecology Progress Series **225**:65–78.

Macdonald, A. G. 1975. Physiological Aspects of Deep-Sea Biology. Cambridge University Press, Cambridge, UK.

MacIlvaine, J. C., and D. A. Ross. 1979. Sedimentary processes on the continental slope of New England. Journal of Sedimentary Petrology **49**:563–574.

Maciolek, N. J., and J. F. Grassle. 1987. Variability of the benthic fauna, II: The seasonal variation, 1981–1982. Pp. 303–309, *in* R. H. Backus, ed. Georges Bank, MIT Press, Cambridge, MA.

Maciolek, N., J. F. Grassle, B. Hecker, P. D. Boehm, B. Brown, B. Dade, W. G. Steinhauer, E. Baptiste, R. E. Ruff, and R. Petrecca. 1987b. Study of biological processes on the U.S. Mid-Atlantic slope and rise. Final Report Prepared for U.S. Department of the Interior Minerals Management Service, Washington, DC, 310 and appendices A–M.

Maciolek, N. J., J. F. Grassle, B. Hecker, B. Brown, J. A. Blake, P. D. Boehm, R. Petrecca, S. Duffy, E. Baptiste, and R. E. Ruff. 1987a. Study of biological processes on the U.S. North Atlantic slope and rise. Final Report Prepared for U.S. Department of the Interior, Minerals Management Service, Washington, DC, 362 and appendices A–L.

Maciolek-Blake, N. J., J. F. Grassle, J. A. Blake, and J. M. Neff. 1985. Georges Bank infauna monitoring program: Final report for the third year of sampling. Prepared for the U.S. Department of Interior, Minerals Management Service. Washington, DC.

Mackensen, A., H. P. Sejrup, and E. Jansen. 1985. The distribution of living benthic foraminifera on the continental slope and rise off southwest Norway. Marine Micropaleontology **9**:275–306.

Macpherson, E. 2002. Large-scale species-richness gradients in the Atlantic Ocean. Proceedings of the Royal Society of London B **269**:1715–1720.

Madurell, T., and J. E. Cartes. 2006. Trophic relationships and food consumption of slope dwelling macrourids from the bathyal Ionian Sea (eastern Mediterranean). Marine Biology **148**:1325–1338.

Magurran, A. E. 2004. Measuring biological diversity. Blackwell, Oxford, UK.

Marsh, A. G., P. K. K. Leong, and D. T. Manahan. 1999. Energy metabolism during embryonic development and larval growth of an Antarctic sea urchin. Journal of Experimental Biology **202**:2041–2050.

Martin, A. P., and S. R. Palumbi. 1993. Body size, metabolic-rate, generation time, and the molecular clock. Proceedings of the National Academy of Sciences, USA **90**:4087–4091.

Martin, P. R., and J. K. McKay. 2004. Latitudinal variation in genetic divergence of populations and the potential for future speciation. Evolution **58**:938–943.

Martin, W. R., and F. L. Sayles. 2004. Organic matter cycling in sediments of the continental margin in the northwest Atlantic Ocean. Deep-Sea Research I **51**:457–489.

May, R. M. 1988. How many species are there on Earth? Science **241**:1441–1449.

May, R. M. 1992. Bottoms up for the oceans. Nature **357**:278–279.

McCain, C. M. 2007. Area and mammalian elevational diversity. Ecology **88**:76–86.

McClain, C. R. 2004. Connecting species richness, abundance and body size in deep-sea gastropods. Global Ecology and Biogeography **13**:327–334.

McClain, C. R., and R. J. Etter. 2005. Mid-domain models as predictors of species diversity patterns: Bathymetric diversity gradients in the deep sea. Oikos **109**: 555–566.

McClain, C. R., N. A. Johnson, and M. A. Rex. 2004. Morphological disparity as a biodiversity metric in lower bathyal and abyssal gastropod assemblages. Evolution **58**:338–348.

McClain, C. R., M. A. Rex, and R. J. Etter. 2009. Patterns in deep-sea macroecology. *In* J. D. Witman and K. Roy, eds. Marine Macroecology. University of Chicago Press, Chicago (in press).

McClain, C. R., M. A. Rex, and R. Jabbour. 2005. Deconstructing bathymetric body size patterns in deep-sea gastropods. Marine Ecology Progress Series **297**:181–187.

McClain, C. R., E. P. White, and A. H. Hurlbert. 2007. Challenges in the application of geometric constraint models. Global Ecology and Biogeography **16**:257–264.

McGillicuddy Jr., D. J., A. R. Robinson, D. A. Siegel, H. W. Jannasch, R. Johnson, T. D. Dickey, J. McNeil, A. F. Michaels, and A. H. Knap. 1998. Influence of mesoscale eddies on new production in the Sargasso Sea. Nature **394**:263–266.

McPeek, M. A. 2007. The macroevolutionary consequences of ecological differences among species. Palaeontology **50**:111–129.

McPeek, M. A., and J. M. Brown. 2007. Clade age and not diversification rate explains species richness among animal taxa. American Naturalist **169**:E97–E106.

Mellor, C. A., and C. K. Paull. 1994. Sea Beam bathymetry of the Manteo 467 Lease Block off Cape Hatteras, North Carolina. Deep-Sea Research II **41**:711–718.

Menge, B. A., and J. P. Sutherland. 1976. Species diversity gradients: Synthesis of the roles of predation, competition, and temporal heterogeneity. American Naturalist **110**:351–369.

Menzies, R. J., R. Y. George, and G. T. Rowe. 1973. Abyssal Environment and Ecology of the World Oceans. Wiley, New York.

Metz, E. C., and S. R. Palumbi. 1996. Positive selection and sequence rearrangements generate extensive polymorphism in the gamete recognition protein bindin. Molecular Biology and Evolution **13**:397–406.

Miller, R. J., C. R. Smith, D. J. DeMaster, and W. L. Fornes. 2000. Feeding selectivity and rapid particle processing by deep-sea megafaunal deposit feeders: A ^{234}Th tracer approach. Journal of Marine Research **58**:653–673.

Mittelbach, G. G., D. W. Schemske, H. V. Cornell, A. P. Allen, J. M. Brown, M. B. Bush, S. P. Harrison, A. H. Hurlbert, N. Knowlton, H. A. Lessios, C. M. McCain, A. R. McCune, L. A. McDade, M. A. McPeek, T. J. Near, T. D. Price, R. E. Ricklefs, K. Roy, D. F. Sax, D. Schluter, J. M. Sobel, and M. Turelli. 2007. Evolution and the latitudinal diversity gradient: Speciation, extinction and biogeography. Ecology Letters **10**:315–331.

Mittelbach, G. G., C. F. Steiner, S. M. Scheiner, K. L. Gross, H. L. Reynolds, R. B. Waide, M. R. Willig, S. I. Dodson, and L. Gough. 2001. What is the observed relationship between species richness and productivity? Ecology **82**:2381–2396.

Mokievsky, V., and A. Azovsky. 2002. Re-evaluation of species diversity patterns of free-living marine nematodes. Marine Ecology Progress Series **238**:101–108.

Monniot, C., and F. Monniot. 1978. Recent work on the deep-sea tunicates. Oceanography and Marine Biology: An Annual Review **16**:181–228.

Moodley, L., J. J. Middelburg, H. T. S. Boschker, G. C. A. Duineveld, R. Pel, P. M. J. Herman, and C. H. R. Heip. 2002. Bacteria and foraminifera: Key players in a short-term deep-sea benthic response to phytodetritus. Marine Ecology Progress Series **236**:23–29.

Moran, A. L., and D. T. Manahan. 2004. Physiological recovery from prolonged "starvation" in larvae of the Pacific oyster *Crassostrea gigas*. Journal of Experimental Marine Biology and Ecology **306**:17–36.

Morel, A., and J.-M. André. 1991. Pigment distribution and primary production in the western Mediterranean as derived and modeled from Coastal Zone Color Scanner observations. Journal of Geophysical Research **96**:12685–12698.

Moreno, R. A., M. M. Rivadeneira, C. E. Hernández, S. Sampértegui, and N. Rozbaczylo. 2008. Bathymetric gradient of polychaete richness on the southeastern Pacific coast of Chile. Global Ecology and Biogeography **17**:415–423.

Moseley, H. N. 1880. Deep-sea dredging and life in the deep sea. Nature **21**:543–547.

Müller-Karger, F. E., C. R. McClain, and P. L. Richardson. 1988. The dispersal of the Amazon's water. Nature **333**:56–59.

Müller-Karger, F. E., J. J. Walsh, R. H. E. Evans, and M. B. Meyers. 1991. On the seasonal phytoplankton concentration and sea surface temperature cycles of the Gulf of Mexico as determined by satellites. Journal of Geophysical Research **96**:12645–12665.

Mullineaux, L. S., K. G. Speer, A. M. Thurnherr, M. E. Maltrud, and A. Vangriesheim. 2002. Implications of cross-axis flow for larval dispersal along mid-ocean ridges. Cahiers de Biologie Marine **43**:281–284.

Murray, J., and J. Hjort. 1912. The Depths of the Ocean. Macmillan, London.

Naeem, S. 2002. Ecosystem consequences of biodiversity loss: The evolution of a paradigm. Ecology **83**:1537–1552.

Narayanaswamy, B. E., B. J. Bett, and J. D. Gage. 2005. Ecology of bathyal polychaete fauna at an Arctic-Atlantic boundary (Faroe-Shetland Channel, North-east Atlantic). Marine Biology Research **1**:20–32.

Netto, S. A., F. Gallucci, and G. F. C. Fonseca. 2005. Meiofauna communities of continental slope and deep-sea sites off SE Brazil. Deep-Sea Research I **52**:845–859.

Nichols, J., and G. T. Rowe. 1977. Infaunal macrobenthos off Cap Blanc, Spanish Sahara. Journal of Marine Research **35**:525–536.

Nixon, S. W. 1995. Coastal marine eutrophication: A definition, social causes, and future concerns. Ophelia **41**:199–219.

Nodder, S. D., C. A. Pilditch, P. K. Probert, and J. A. Hall. 2003. Variability in benthic biomass and activity beneath the Subtropical Front, Chatham Rise, SW Pacific Ocean. Deep-Sea Research I **50**:959–985.

Nouvian, C. 2007. The Deep: The Extraordinary Creatures of the Abyss. University of Chicago Press, Chicago.

Nozawa, F., H. Kitazato, M. Tsuchiya, and A. J. Gooday. 2006. Live benthic foraminifera at an abyssal site in the equatorial Pacific nodule province: Abundance, diversity and taxonomic composition. Deep-Sea Research I **53**:1406–1422.

Ohta, S. 1983. Photographic census of large-sized benthic organisms in the bathyal zone of Suruga Bay, Central Japan. Bulletin of the Ocean Research Institute, University of Tokyo **15**:1–244.

Ohta, S. 1984. Star-shaped feeding traces produced by echiuran worms on the deep-sea floor of the Bay of Bengal. Deep-Sea Research **31**:1415–1432.

Okey, T. A. 2003. Macrobenthic colonist guilds and renegades in Monterey Canyon (USA) drift algae: Partitioning multidimensions. Ecological Monographs **73**: 415–440.

Olabarria, C. 2005. Patterns of bathymetric zonation of bivalves in the Porcupine Seabight and adjacent abyssal plain, NE Atlantic. Deep-Sea Research I **52**:15–31.

Ollitrault, M., and A. C. de Verdiere. 2002. SOFAR floats reveal midlatitude intermediate North Atlantic general circulation I: A Lagrangian descriptive view. Journal of Physical Oceanography **32**:2020–2033.

O'Neill, T. A., B. W. Hayward, S. Kawagata, A. T. Sabaa, and H. R. Grenfell. 2007. Pleistocene extinctions of deep-sea benthic foraminifera: The south Atlantic record. Palaeontology **50**:1073–1102.

Oppo, D. W., J. F. McManus, and J. L. Cullen. 1998. Abrupt climate events 500,000 to 340,000 years ago: Evidence from subpolar North Atlantic sediments. Science **279**:1335–1338.

Oschlies, A., and V. Garçon. 1998. Eddy-induced enhancement of primary production in a model of the North Atlantic Ocean. Nature **394**:266–269.

Osman, R. W., and R. B. Whitlatch. 1978. Patterns of species diversity: Fact or artifact? Paleobiology **4**:41–54.

Paine, R. T. 1966. Food web complexity and species diversity. American Naturalist **100**:65–75.

Palumbi, S. 2004. Marine reserves and ocean neighborhoods: The spatial scale of marine populations and their management. Annual Review of Environmental Resources **29**:31–68.

Palumbi, S. R. 1992. Marine speciation on a small planet. Trends in Ecology and Evolution **7**:114–118.

Palumbi, S. R. 1994. Genetic divergence, reproductive isolation and marine speciation. Annual Review of Ecology and Systematics **25**:547–572.

Palumbi, S. R. 1999. All males are not created equal: Fertility differences depend on gamete recognition polymorphisms in sea urchins. Proceedings of the National Academy of Sciences, USA **96**:12632–12637.

Palumbi, S. R., and E. C. Metz. 1991. Strong reproductive isolation between closely related tropical sea urchins (Genus Echinometra). Molecular Biology and Evolution **8**:227–239.

Parkes, R. J., G. Webster, B. A. Cragg, A. J. Weightman, C. J. Newberry, T. G. Ferdelman, J. Kallmeyer, B. B. Jørgensen, I. W. Aiello, and J. C. Fry. 2005. Deep subseafloor prokaryotes stimulated at interfaces over geological time. Nature **436**:390–394.

Parulekar, A. H., S. N. Harkantra, Z. A. Ansari, and S. G. P. Matondkar. 1982. Abyssal benthos of the central Indian Ocean. Deep-Sea Research **29**: 1531–1537.

Patarnello, T., F. Volckaert, and R. Castilho. 2007. Pillars of Hercules: Is the Atlantic-Mediterranean transition a phylogeographical break? Molecular Ecology **16**: 4426–4444.

Paterson, G. L. J., J. D. Gage, P. Lamont, B. J. Bett, and M. H. Thurston. 1994. Patterns of abundance and diversity from the abyss-polychaetes from northeastern Atlantic abyssal plains. Mémoires du Museum national d'Histoire naturelle **162**:503–511.

Paterson, G. L. J., and P. J. D. Lambshead. 1995. Bathymetric patterns of polychaete diversity in the Rockall Trough, northeast Atlantic. Deep-Sea Research I **42**: 1199–1214.

Paul, A. Z., and R. J. Menzies. 1974. Benthic ecology of the high Arctic deep sea. Marine Biology **27**:251–262.

Pawlowski, J., J. Fahrni, B. Lecroq, D. Longet, N. Cornelius, L. Excoffier, T. Cedhagen, and A. J. Gooday. 2007. Bipolar gene flow in deep-sea benthic foraminifera. Molecular Ecology **16**:4089–4096.

Pawlowski, J., M. Holzmann, J. Fahrni, and S. Richardson. 2003. Small subunit ribosomal DNA suggests that the xenophyophorean *Syringammina corbicula* is a foraminiferan. Journal of Eukaryotic Microbiology **50**:483–487.

Pearcy, W. G., D. L. Stein, and R. S. Carney. 1982. The deep-sea benthic fish fauna of the northeastern Pacific Ocean on Cascadia and Tufts Abyssal Plains and adjoining continental slopes. Biological Oceanography **1**:375–428.

Pequegnat, W. E., B. J. Gallaway, and L. H. Pequegnat. 1990. Aspects of the ecology of the deep-water fauna of the Gulf of Mexico. American Zoologist **30**:45–64.

Pfannkuche, O. 1985. The deep-sea meiofauna of the Porcupine Seabight and abyssal plain (NE Atlantic): Population structure, distribution, standing stocks. Oceanologica Acta **8**:343–353.

Pfannkuche, O. 1992. Organic carbon flux through the benthic community in the temperate abyssal Northeast Atlantic. Pp. 183–198, *in* G. T. Rowe and V. Pariente, eds. Deep-Sea Food Chains and the Global Carbon Cycle. Kluwer, Dordrecht, The Netherlands.

Pfannkuche, O. 1993. Benthic response to the sedimentation of particulate organic matter at the BIOTRANS station, 47° N, 20° W. Deep-Sea Research II **40**: 135–149.

Pfannkuche, O., A. Boetius, K. Lochte, U. Lundgreen, and H. Thiel. 1999. Responses of deep-sea benthos to sedimentation patterns in the North-East Atlantic in 1992. Deep-Sea Research I **46**:573–596.

Pfannkuche, O., and K. Lochte. 1993. Open ocean pelago-benthic coupling: Cyanobacteria as tracers of sedimenting salp faeces. Deep-Sea Research I **40**:727–737.

Pfannkuche, O., and T. Soltwedel. 1998. Small benthic size classes along the western European continental margin: Spatial and temporal variability in activity and biomass. Progress in Oceanography **42:**189–207.

Pfannkuche, O., R. Theeg, and H. Thiel. 1983. Benthos activity, abundance and biomass under an area of low upwelling off Morocco, Northwest Africa. Meteor Forschungsergebnisse, Reihe D Biologie **36:**85–96.

Pfannkuche, O., and H. Thiel. 1987. Meiobenthic stocks and benthic activity on the NE-Svalbard Shelf and in the Nansen Basin. Polar Biology **7:**253–266.

Phillimore, A. B., and T. D. Price. 2008. Density-dependent cladogenesis in birds. PLoS Biology **6:**483–489.

Piepenburg, D. 2005. Recent research on Arctic benthos: Common notions need to be revised. Polar Biology **28:**733–755.

Pineda, J. 1993. Boundary effects on the vertical ranges of deep-sea benthic species. Deep-Sea Research I **40:**2179–2192.

Pineda, J., and H. Caswell. 1998. Bathymetric species-diversity patterns and boundary constraints on vertical range distribution. Deep-Sea Research II **45:**83–101.

Pineda, J., J. Hare, and S. Sponaugle. 2007. Larval transport and dispersal and consequences for population connectivity. Oceanography **20:**22–39.

Podolsky, R. D. 1994. Temperature and water viscosity: Physiological versus mechanical effects on suspension feeding. Science **265:**100–103.

Polechova, J., and N. H. Barton. 2005. Speciation through competition: A critical review. Evolution **59:**1194–1210.

Poore, G. B. C., and G. D. F. Wilson. 1993. Marine species richness. Nature **361:**597–598.

Pope, R. H., D. J. DeMaster, C. R. Smith, and H. Seltmann Jr. 1996. Rapid bioturbation in equatorial Pacific sediments: Evidence from excess ^{234}Th measurements. Deep-Sea Research II **43:**1339–1364.

Potter, E., and M. A. Rex. 1992. Parallel development-depth trends in deep-sea turrid snails from the eastern and western North Atlantic. The Nautilus **106:**72–75.

Powell, S. M., R. L. Haedrich, and J. D. McEachran. 2003. The deep-sea demersal fish fauna of the northern Gulf of Mexico. Journal of Northwest Atlantic Fisheries Science **31:**19–33.

Priede, I. G., P. M. Bagley, A. Smith, S. Creasey, and N. R. Merrett. 1994. Scavenging deep demersal fishes of the Porcupine Seabight, north-east Atlantic: Observations by baited camera, trap and trawl. Journal of the Marine Biological Association of the United Kingdom **74:**481–498.

Pringle, J. M., and J. P. Wares. 2007. The maintenance of alongshore variation in allele frequency in a coastal ocean. Marine Ecology Progress Series **335:**69–84.

Psarra, S., A. Tselepides, and L. Ignatiades. 2000. Primary productivity in the oligo-trophic Cretan Sea (NE Mediterranean): Seasonal and interannual variability. Progress in Oceanography **46**:187–204.

Pulliam, H. R. 1988. Sources, sinks and population regulation. American Naturalist **132**:652–661.

Quattro, J. M., M. R. Chase, M. A. Rex, T. W. Greig, and R. J. Etter. 2001. Extreme mitochondrial DNA divergence within populations of the deep-sea gastropod *Frigidoalvania brychia*. Marine Biology **139**:1107–1113.

Rabinowitz, D. 1981. Seven forms of rarity. Pp. 205–217, *in* H. Synge, ed. The Bio-logical Aspects of Rare Plant Conservation. Wiley, New York.

Raes, M., A. Vanreusel, and W. Decraemer. 2003. Epsilonematidae (Nematoda) from a cold-water coral environment in the Porcupine Seabight, with a discussion on the status of the genus *Metaglochinema* Gourbault & Decraemer 1986. Hydro-biologia **505**:49–72.

Rahbek, C. 1997. The relationship among area, elevation and regional species rich-ness in Neotropical birds. American Naturalist **149**:875–902.

Ramaswamy, V., M. M. Sarin, and R. Rengarajan. 2005. Enhanced export of carbon by salps during the northeast monsoon period in the northern Arabian Sea. Deep-Sea Research II **52**:1922–1929.

Rasmussen, T. L., E. Thomsen, S. R. Troelstra, A. Kuijpers, and M. A. Prins. 2002. Millennial-scale glacial variability versus Holocene stability: Changes in plank-tic and benthic foraminifera faunas and ocean circulation in the North Atlantic during the last 60,000 years. Marine Micropaleontology **47**:143–176.

Raupach, M. J., C. Held, and J.-W. Wägele. 2004. Multiple colonization of the deep sea by the Asellota (Crustacea: Peracarida; Isopoda). Deep-Sea Research II **51**:1787–1795.

Raymo, M. E., K. Ganley, S. Carter, D. W. Oppo, and J. McManus. 1998. Millennial-scale climate instability during the early Pleistocene epoch. Nature **392**:699–702.

Rea, D. K., M. W. Lyle, L. M. Liberty, S. A. Hovan, M. P. Bolyn, J. D. Gleason, I. L. Hendy, J. C. Latimer, B. M. Murphy, R. M. Owen, C. F. Paul, T. H. C. Rea, A. M. Stancin, and D. J. Thomas. 2006. Broad region of no sediment in the southwest Pacific Basin. Geology **34**:873–876.

Reichart, G. L., L. J. Lourens, and W. J. Zachariasse. 1998. Temporal variability in the northern Arabian Sea oxygen minimum zone (OMZ) during the last 225,000 years. Paleoceanography **13**:607–621.

Relexans, J.-C., J. Deming, A. Dinet, J.-F. Gaillard, and M. Sibuet. 1996. Sedimen-tary organic matter and micro-meiobenthos with relation to trophic conditions in the tropical northeast Atlantic. Deep-Sea Research I **43**:1343–1368.

Rex, M. A. 1973. Deep-sea species diversity: Decreased gastropod diversity at abyssal depths. Science **181**:1051–1053.

Rex, M. A. 1976. Biological accommodation in the deep-sea benthos: Comparative evidence on the importance of predation and productivity. Deep-Sea Research **23:**975–987.

Rex, M. A. 1977. Zonation in deep-sea gastropods: The importance of biological interactions to rates of zonation. European Symposium on Marine Biology **11:** 521–530.

Rex, M. A. 1981. Community structure in the deep-sea benthos. Annual Review of Ecology and Systematics **12:**331–353.

Rex, M. A. 1983. Geographic patterns of species diversity in the deep-sea benthos. Pp. 453–472, *in* G. T. Rowe, ed. The Sea, Vol. 8: Deep-Sea Biology. Wiley, New York.

Rex, M. A. 2002. Biogeography of the deep-sea gastropod *Palazzia planorbis* (Dall, 1927): An uncommon form of rarity. The Nautilus **116:**36–38.

Rex, M. A., A. Bond, R. J. Etter, A. C. Rex, and C. T. Stuart. 2002. Geographic variation of shell geometry in the abyssal snail *Xyloskenea naticiformis* (Jeffreys, 1883). Veliger **45:**218–223.

Rex, M. A., J. A. Crame, C. T. Stuart, and A. Clarke. 2005b. Large-scale biogeographic patterns in marine mollusks: A confluence of history and productivity? Ecology **86:**2288–2297.

Rex, M. A., and R. J. Etter. 1990. Geographic variation and population differentiation in two deep-sea snails, *Benthomangelia antonia* (Jeffreys) and *Benthonella tenella* (Dall). Deep-Sea Research **37:**1229–1249.

Rex, M. A., and R. J. Etter. 1998. Bathymetric patterns of body size: Implications for deep-sea biodiversity. Deep-Sea Research II **45:**103–127.

Rex, M. A., R. J. Etter, J. S. Morris, J. Crouse, C. R. McClain, N. A. Johnson, C. T. Stuart, J. W. Deming, R. Thies, and R. Avery. 2006. Global bathymetric patterns of standing stock and body size in the deep-sea benthos. Marine Ecology Progress Series **317:**1–8.

Rex, M. A., R. J. Etter, and C. T. Stuart. 1997. Large-scale patterns of species diversity in the deep-sea benthos. Pp. 94–121, *in* R. F. G. Ormond, J. D. Gage, and M. V. Angel, eds. Marine Biodiversity: Patterns and Processes. Cambridge University Press, Cambridge, UK.

Rex, M. A., C. R. McClain, N. A. Johnson, R. J. Etter, J. A. Allen, P. Bouchet, and A. Warén. 2005a. A source-sink hypothesis for abyssal biodiversity. American Naturalist **165:**163–178.

Rex, M. A., C. T. Stuart, and G. Coyne. 2000. Latitudinal gradients of species richness in the deep-sea benthos of the North Atlantic. Proceedings of the National Academy of Sciences, USA **97:**4082–4085.

Rex, M. A., C. T. Stuart, and R. J. Etter. 2001. Do deep-sea nematodes show a positive latitudinal gradient of species diversity? The potential role of depth. Marine Ecology Progress Series **210:**297–298.

Rex, M. A., C. T. Stuart, R. R. Hessler, J. A. Allen, H. L. Sanders, and G. D. F. Wilson. 1993. Global-scale latitudinal patterns of species diversity in the deep-sea benthos. Nature **365**:636–639.

Rex, M. A., C. A. Van Ummersen, and R. D. Turner. 1979. Reproductive pattern in the abyssal snail *Benthonella tenella* (Jeffreys). Pp. 173–188, *in* S. E. Stancyk, ed. Reproductive Ecology of Marine Invertebrates. University of South Carolina Press, Columbia, SC.

Rex, M. A., and A. Warén. 1982. Planktotrophic development in deep-sea prosobranch snails from the western North Atlantic. Deep-Sea Research **29**:171–184.

Rex, M. A., M. C. Watts, R. J. Etter, and S. O'Neill. 1988. Character variation in a complex of rissoid gastropods from the upper continental slope of the western North Atlantic. Malacologia **29**:325–339.

Rhoads, D. C., D. F. Boesch, T. Zhican, X. Fengshan, H. Liqiang, and K. J. Nilsen. 1985. Macrobenthos and sedimentary facies on the Changjiang delta platform and adjacent continental shelf, East China Sea. Continental Shelf Research **4**:189–213.

Rhoads, D. C., and B. Hecker. 1994. Processes on the continental slope off North Carolina with special reference to the Cape Hatteras region. Deep-Sea Research II **41**:965–980.

Rice, A. L., D. S. M. Billett, M. H. Thurston, and R. S. Lampitt. 1991. The Institute of Oceanographic Sciences biology programme in the Porcupine Seabight: Background and general introduction. Journal of the Marine Biological Association of the United Kingdom **71**:282–310.

Rice, A. L., and P. J. D. Lambshead. 1994. Patch dynamics in the deep-sea benthos: The role of a heterogeneous supply of organic matter. Pp. 469–498, *in* P. S. Giller, A. G. Hildrew, and D. G. Raffaelli, eds. Aquatic Ecology: Scale, Pattern and Process. Blackwell, Oxford.

Rice, A. L., M. H. Thurston, and A. L. New. 1990. Dense aggregations of a hexactinellid sponge, *Pheronema carpenteri,* in the Porcupine Seabight (northeast Atlantic Ocean), and possible causes. Progress in Oceanography **24**:179–196.

Richardson, M., and D. K. Young. 1987. Abyssal benthos of the Venezuela Basin, Caribbean Sea: Standing stock considerations. Deep-Sea Research **34**:145–164.

Richardson, M. D., K. B. Briggs, F. A. Bowles, and J. H. Tietjen. 1995. A depauperate benthic assemblage from the nutrient-poor sediments of the Puerto Rico Trench. Deep-Sea Research I **42**:351–364.

Richardson, M. D., K. B. Briggs, and D. K. Young. 1985. Effects of biological activity by abyssal benthic macroinvertebrates on a sedimentary structure in the Venezuela Basin. Marine Geology **68**:243–267.

Richardson, M. J., G. L. Weatherly, and W. D. Gardner. 1993. Benthic storms in the Argentine Basin. Deep-Sea Research II **40**:975–987.

Richardson, M. J., M. Wimbush, and L. Mayer. 1981. Exceptionally strong near-bottom flows on the continental rise of Nova Scotia. Science **213**:887–888.

Richardson, P. L. 1993. A census of eddies observed in North-Atlantic SOFAR float data. Progress in Oceanography **31**:1–50.

Richardson, P. L., and D. M. Fratantoni. 1999. Float trajectories in the deep western boundary current and deep equatorial jets of the tropical Atlantic. Deep-Sea Research II **46**:305–333.

Richerson, P., R. Armstrong, and C. R. Goldman. 1970. Contemporaneous disequilibrium, a new hypothesis to explain the 'paradox of the plankton.' Proceedings of the National Academy of Sciences, USA **67**:1710–1714.

Ricklefs, R. E. 1987. Community diversity: Relative roles of local and regional processes. Science **235**:167–171.

Ricklefs, R. E. 2004. A comprehensive framework for global patterns of biodiversity. Ecology Letters **7**:1–15.

Ricklefs, R. E. 2007. Estimating diversification rates from phylogenetic information. Trends in Ecology and Evolution **22**:601–610.

Ricklefs, R. E. 2008. Disintegration of the ecological community. American Naturalist **172**:741–750.

Ricklefs, R. E., and D. Schluter. 1993. Species diversity in ecological communities. Pp. 350–363, *in* R. E. Ricklefs and D. Schluter, eds. Diversity in Ecological Communities. University of Chicago Press, Chicago.

Riginos, C., and M. W. Nachman. 2001. Population subdivision in marine environments: The contributions of biogeography, geographical distance and discontinuous habitat to genetic differentiation in a blennioid fish, *Axoclinus nigricaudus.* Molecular Ecology **10**:1439–1453.

Riginos, C., D. Wang, and A. J. Abrams. 2006. Geographic variation and positive selection on M7 lysin, an acrosomal sperm protein in mussels (*Mytilus* spp.). Molecular Biology and Evolution **23**:1952–1965.

Robison, B. H., K. R. Reisenbichler, and R. E. Sherlock. 2005. Giant larvacean houses: Rapid carbon transport to the deep-sea floor. Science **308**:1609–1611.

Rodriguez-Lazaro, J., and T. M. Cronin. 1999. Quaternary glacial and deglacial Ostracoda in the thermocline of the Little Bahama Bank (NW Atlantic): Palaeoceanographic implications. Palaeogeography, Palaeoclimatology, Palaeoecology **152**:339–364.

Rogers, A. D. 1994. The biology of seamounts. Advances in Marine Biology **30**:305–350.

Rogers, A. D. 2000. The role of the oceanic oxygen minima in generating biodiversity in the deep sea. Deep-Sea Research II **47**:119–148.

Rogers, A. D. 2002. Molecular ecology and evolution of slope species. Pp. 323–337, *in* G. Wefer, D. S. M. Billett, D. Hebbeln, B. Jørgensen, M. Schluter, and T. C. E. van Weering, eds. Ocean Margin Systems. Springer-Verlag, Berlin/Heidelberg.

Rohde, K. 1992. Latitudinal gradients in species diversity: The search for the primary cause. Okios **65**:514–527.

Rokop, F. J. 1974. Reproductive patterns in the deep-sea benthos. Science **186**:743–745.

Rokop, F. J. 1977. Patterns of reproduction in the deep-sea benthic crustaceans: A re-evaluation. Deep-Sea Research **24**:683–691.

Rokop, F. J. 1979. Year-round reproduction in the deep-sea bivalve molluscs. Pp. 189–198, *in* S. E. Stancyk, ed. Reproductive Ecology of Marine Invertebrates. University of South Carolina Press, Columbia, SC.

Romero-Wetzel, M. B. 1987. Sipunculans as inhabitants of very deep, narrow burrows in deep-sea sediments. Marine Biology **96**:87–91.

Romero-Wetzel, M. B., and S. A. Gerlach. 1991. Abundance, biomass, size-distribution and bioturbation potential of deep-sea macrozoobenthos on the Voring Plateau (1200-1500 m, Norwegian Sea). Meeresforschung **33**:247–265.

Roques, S., J. M. Sévigny, and L. Bernatchez. 2002. Genetic structure of deep-water redfish, *Sebastes mentella,* populations across the North Atlantic. Marine Biology **140**:297–307.

Rosenzweig, M. L. 1995. Species Diversity in Space and Time. Cambridge University Press, Cambridge, UK.

Rosenzweig, M. L., and Z. Abramsky. 1993. How are diversity and productivity related? Pp. 52–65, *in* R. E. Ricklefs and D. Schluter, eds. Species Diversity in Ecological Communities: Historical and Geographical Perspectives. University of Chicago Press, Chicago.

Rothwell, R. G., J. Thomson, and G. Kähler. 1998. Low-sea-level emplacement of a very large Late Pleistocene "megaturbidite" in the western Mediterranean Sea. Nature **392**:377–380.

Roughgarden, J., S. Gaines, and H. Possingham. 1988. Recruitment dynamics in complex life cycles. Science **241**:1460–1466.

Rowe, G., M. Sibuet, J. Deming, A. Khripounoff, J. Tietjen, S. Macko, and R. Theroux. 1991. "Total" sediment biomass and preliminary estimates of organic carbon residence time in deep-sea benthos. Marine Ecology Progress Series **79**: 99–114.

Rowe, G. T. 1971a. Benthic biomass and surface productivity. Pp. 441–454, *in* J. D. Costlow Jr., ed. Fertility of the Sea. Gordon and Breach, New York.

Rowe, G. T. 1971b. Benthic biomass in the Pisco, Peru upwelling. Investigacion pesquera **35**:127–135.

Rowe, G. T. 1983. Biomass and production of the deep-sea macrobenthos. Pp. 97–122, *in* G. T. Rowe, ed. The Sea, Vol. 8: Deep-Sea Biology. Wiley, New York.

Rowe, G. T., G. S. Boland, E. G. Escobar Briones, M. E. Cruz-Kaegi, A. Newton, D. Piepenburg, I. Walsh, and J. Deming. 1997. Sediment community biomass and respiration in the Northeast Water Polynya, Greenland: A numerical simulation of benthic lander and spade core data. Journal of Marine Systems **10**:497–515.

Rowe, G. T., A. Lohse, F. Hubbard, G. S. Boland, E. Escobar Briones, and J. Deming. 2003. Preliminary trophodynamic carbon budget for the Sigsbee Deep benthos, Northern Gulf of Mexico. American Fisheries Society Symposium **36**:225–238.

Rowe, G. T., and D. W. Menzel. 1971. Quantitative benthic samples from the deep Gulf of Mexico with some comments on the measurement of deep-sea biomass. Bulletin of Marine Science **21**:556–566.

Rowe, G. T., and R. J. Menzies. 1969. Zonation of large benthic invertebrates in the deep-sea off the Carolinas. Deep-Sea Research **16**:531–537.

Rowe, G. T., P. T. Polloni, and R. L. Haedrich. 1975. Quantitative biological assessment of the benthic fauna in deep basins of the Gulf of Maine. Journal of Fisheries Research Board of Canada **32**:1805–1812.

Rowe, G. T., P. T. Polloni, and R. L. Haedrich. 1982. The deep-sea macrobenthos on the continental margin of the northwest Atlantic Ocean. Deep-Sea Research **29**:257–278.

Rowe, G. T., P. T. Polloni, and S. G. Horner. 1974. Benthic biomass estimates from the northwestern Atlantic Ocean and the northern Gulf of Mexico. Deep-Sea Research **21**:641–650.

Roy, K., and E. E. Goldberg. 2007. Origination, extinction, and dispersal: Integrative models for understanding present-day diversity gradients. American Naturalist **170**:S71–S85.

Roy, K., D. Jablonski, and J. W. Valentine. 1994. Eastern Pacific molluscan provinces and latitudinal diversity gradient: No evidence for "Rapoport's rule." Proceedings of the National Academy of Sciences, USA **91**:8871–8874.

Roy, K., D. Jablonski, J. W. Valentine, and G. Rosenberg. 1998. Marine latitudinal diversity gradients: Tests of causal hypotheses. Proceedings of the National Academy of Sciences, USA **95**:3699–3702.

Roy, K. O. L., R. von Cosel, S. Hourdez, S. L. Carney, and D. Jollivet. 2007. Amphi-Atlantic cold-seep *Bathymodiolus* species complexes across the equatorial belt. Deep-Sea Research II **54**:1890–1911.

Ruhl, H. A. 2007. Abundance and size distribution dynamics of abyssal epibenthic megafauna in the northeast Pacific. Ecology **88**:1250–1262.

Ruhl, H. A. 2008. Community change in the variable resource habitat of the abyssal northeast Pacific. Ecology **89**:991–1000.

Ruhl, H. A., and K. L. Smith Jr. 2004. Shifts in deep-sea community structure linked to climate and food supply. Science **305**:513–515.

Rutgers van der Loeff, M. M., and M. S. S. Lavaleye. 1986. Sediments, fauna, and the dispersal of radionuclides at the N. E. Atlantic dumpsite for low-level radioactive waste. Report of the Dutch DORA program. Netherlands Institute for Sea Research, 134 pp.

Ryan, J. P., J. A. Yoder, and P. C. Cornillon. 1999. Enhanced chlorophyll at the shelf-break of the Mid-Atlantic Bight and Georges Bank during the spring transition. Limnology and Oceanography **44:**1–11.

Sanders, H. L. 1968. Marine benthic diversity: A comparative study. American Naturalist **102:**243–282.

Sanders, H. L. 1969. Benthic marine diversity and the stability-time hypothesis. Brookhaven Symposia in Biology **22:**71–81.

Sanders, H. L. 1977. Evolutionary ecology and the deep-sea benthos. Pp. 223–243, *in* C. E. Goulden, ed. The Changing Scenes in Natural Sciences 1776–1976. Philadelphia Academy of Natural Sciences Special Publication.

Sanders, H. L., and R. R. Hessler. 1969. Ecology of the deep-sea benthos. Science **163:**1419–1424.

Sanders, H. L., R. R. Hessler, and G. R. Hampson. 1965. An introduction to the study of deep-sea benthic faunal assemblages along the Gay Head-Bermuda transect. Deep-Sea Research **12:**845–867.

Sardà, F., J. E. Cartes, and J. B. Company. 1994. Spatio-temporal variations in megabenthos abundance in three different habitats of the Catalan deep-sea (Western Mediterranean). Marine Biology **120:**211–219.

Sars, G. O. 1872. On some remarkable forms of animal life from great depths off the Norwegian coast I. Partly from posthumous manuscripts of the late Professor Dr. Michael Sars, Christiana.

Sathyendranath, S., A. Longhurst, C. M. Caverhill, and T. Platt. 1995. Regionally and seasonally differentiated primary production in the North Atlantic. Deep-Sea Research I **42:**1773–1802.

Sayles, F. L., W. R. Martin, and W. G. Deuser. 1994. Response of benthic oxygen demand to particulate organic carbon supply in the deep sea near Bermuda. Nature **371:**686–689.

Schaff, T., L. Levin, N. Blair, D. DeMaster, R. Pope, and S. Boehme. 1992. Spatial heterogeneity of benthos on the Carolina continental slope: Large (100 km)-scale variation. Marine Ecology Progress Series **88:**143–160.

Schaff, T. R., and L. A. Levin. 1994. Spatial heterogeneity of benthos associated with biogenic structures on the North Carolina continental slope. Deep-Sea Research II **41:**901–918.

Scheiner, S. M., and M. R. Willig. 2005. Developing unified theories in ecology as exemplified with diversity gradients. American Naturalist **166:**458–469.

Schewe, I., and T. Soltwedel. 2003. Benthic response to ice-edge-induced particle flux in the Arctic Ocean. Polar Biology **26:**610–620.

Schmidt, P. S., M. D. Bertness, and D. M. Rand. 2000. Environmental heterogeneity and balancing selection in the acorn barnacle *Semibalanus balanoides*. Proceedings for the Royal Society of London B **267:**379–384.

Schmidt, P. S., and D. M. Rand. 1999. Intertidal microhabitat and selection at Mpi: Interlocus contrasts in the northern acorn barnacle, *Semibalanus balanoides.* Evolution **53:**135–146.

Schmidt, P. S., and D. M. Rand. 2001. Adaptive maintenance of genetic polymorphism in an intertidal barnacle: Habitat- and life-stage-specific survivorship of Mpi genotypes. Evolution **55:**1336–1344.

Schmittner, A. 2005. Decline of the marine ecosystem caused by a reduction in the Atlantic overturning circulation. Nature **434:**628–633.

Schmitz, W. J., and M. S. McCartney. 1993. On the North Atlantic circulation. Review of Geophysics **31:**29–39.

Schulenberger, E., and R. R. Hessler. 1974. Scavenging abyssal benthic amphipods trapped under oligotrophic central North Pacific gyre waters. Marine Biology **28:**185–187.

Schüller, M., and B. Ebbe. 2007. Global distributional patterns of selected deep-sea Polychaeta (Annelida) from the Southern Ocean. Deep-Sea Research II **54:** 1737–1751.

Schwinghamer, P. 1985. Observations on size-structure and pelagic coupling of some shelf and abyssal benthic communities. European Marine Biology Symposium **19:**347–359.

Seibel, B. A., and J. C. Drazen. 2007. The rate of metabolism in marine animals: Environmental constraints, ecological demands and energetic opportunities. Philosophical Transactions of the Royal Society of London B **362:**2061–2078.

Seiter, K., C. Hensen, J. Schröter, and M. Zabel. 2004. Organic carbon content in surface sediments: Defining regional provinces. Deep-Sea Research I **51:**2001–2026.

Seiter, K., C. Hensen, and M. Zabel. 2005. Benthic carbon mineralization on a global scale. Global Biogeochemical Cycles **19:**1–26.

Self, R. F. L., and P. A. Jumars. 1988. Cross-phyletic patterns of particle selection by deposit feeders. Journal of Marine Research **46:**119–143.

Sepkoski Jr., J. J. 1991. A model of onshore-offshore change in faunal diversity. Paleobiology **17:**58–77.

Sepkoski Jr., J. J. 1998. Rates of speciation in the fossil record. Philosophical Transactions of the Royal Society of London B **353:**315–326.

Severinghaus, J. P., T. Sowers, E. J. Brook, R. B. Alley, and M. L. Bender. 1998. Timing of abrupt climate change at the end of the Younger Dryas interval from thermally fractionated gases in polar ice. Nature **391:**141–146.

Shilling, F. M., and D. T. Manahan. 1991. Using dissolved organic-matter to meet the metabolic needs of development in an extreme environment (Antarctica). American Zoologist **31:**A4.

Shilling, F. M., and D. T. Manahan. 1994. Energy metabolism and amino acid transport during early development of Antarctic and temperate echinoderms. Biological Bulletin **187**:398–407.

Shirayama, Y. 1983. Size structure of deep-sea meio- and macrobenthos in the Western Pacific. Internationale Revue der gesamten Hydrobiologie **68**:799–810.

Shirayama, Y. 1984. The abundance of deep-sea meiobenthos in the Western Pacific in relation to environmental factors. Oceanologica Acta **7**:113–121.

Shirayama, Y., and S. Kojima. 1994. Abundance of deep-sea meiobenthos off Sanriku Northeastern Japan. Japanese Journal of Oceanography **50**:109–117.

Shoosmith, D., P. L. Richardson, A. S. Bower, and H. T. Rossby. 2005. Discrete eddies in the northern North Atlantic as observed by looping RAFOS floats. Deep-Sea Research II **52**:627–650.

Sibuet, M. 1977. Répartition et diversité des Echinodermes (Holothurides-Astérides) en zone profonde dans le Golfe de Gascogne. Deep-Sea Research **24**:549–563.

Sibuet, M. 1979. Distribution and diversity of asteroids in Atlantic abyssal basins. Sarsia **64**:85–91.

Sibuet, M., C. E. Lambert, R. Chesselet, and L. Laubier. 1989. Density of the major size groups of benthic fauna and trophic input in deep basins of the Atlantic Ocean. Journal of Marine Research **47**:851–867.

Sibuet, M., C. Monniot, D. Desbruyères, A. Dinet, A. Khripounoff, G. Rowe, and M. Segonzac. 1984. Peuplements benthiques et caractéristiques trophiques du milieu dans la plaine abyssale de Demerara. Oceanologica Acta **7**:345–358.

Siebenaller, J., and G. N. Somero. 1978. Pressure-adaptive differences in lactate dehydrogenases of congeneric fishes living at different depths. Science **201**:255–257.

Siebenaller, J. F., and G. N. Somero. 1979. Pressure-adaptive differences in the binding and catalytic properties of muscle-type (M4) lactate-dehydrogenases of shallow-living and deep-living marine fishes. Journal of Comparative Physiology **129**:295–300.

Siebenaller, J. F., and G. N. Somero. 1982. The maintenance of different enzyme-activity levels in congeneric fishes living at difference depths. Physiological Zoology **55**:171–179.

Siebenaller, J. F., and G. N. Somero. 1984. Pressure-adaptive differences in nad-dependent dehydrogenases of congeneric marine fishes living at different depths. Journal of Comparative Physiology **154**:443–448.

Siebenaller, J. K. 1978. Genetic variation in deep-sea invertebrate populations: The bathyal gastropods *Bathybembix bairdii*. Marine Biology **47**:265–275.

Siegel, D. A., and R. A. Armstrong. 2002. Corrigendum to "Trajectories of sinking particles in the Sargasso Sea: Modeling of statistical funnels above deep ocean sediment traps." Deep-Sea Research I **49**:1115–1116.

Siegel, D. A., and W. G. Deuser. 1997. Trajectories of sinking particles in the Sargasso Sea: Modeling of statistical funnels above deep ocean sediment traps. Deep-Sea Research I **44:**1519–1541.

Siegel, D. A., E. Fields, and K. O. Buesseler. 2008. A bottom-up view of the biological pump: Modeling source funnels above ocean sediment traps. Deep-Sea Research I **55:**108–127.

Slowey, N. C., and W. B. Curry. 1995. Glacial-interglacial difference in circulation and carbon cycling within the upper western North Atlantic. Paleoceanography **10:**715–732.

Smith, A. B. 2004. Phylogeny and systematics of holasteroid echinoids and their migration into the deep-sea. Paleobiology **47:**123–150.

Smith, A. B., and B. Stockley. 2005. The geological history of deep-sea colonization by echinoids: Roles of surface productivity and deep-water ventilation. Proceedings of the Royal Society of London B: Biological Sciences **272:**865–869.

Smith, C. R. 1985. Colonization studies in the deep sea: Are results biased by experimental designs? European Marine Biological Symposium **19:**183–190.

Smith, C. R. 1986. Nekton falls, low-intensity disturbance and community structure of infaunal benthos in the deep sea. Journal of Marine Research **44:**567–600.

Smith, C. R. 1992. Factors controlling bioturbation in deep-sea sediments and their relation to models of carbon diagenesis. Pp. 375–393, *in* G. T. Rowe and V. Pariente, eds. Deep-Sea Food Chains and the Global Carbon Cycle. Kluwer, Dordrecht, The Netherlands.

Smith, C. R., and A. R. Baco. 2003. Ecology of whale falls at the deep-sea floor. Oceanography and Marine Biology: An Annual Review **41:**311–354.

Smith, C. R., W. Berelson, D. J. DeMaster, F. C. Dobbs, D. Hammond, D. J. Hoover, R. H. Pope, and M. Stephens. 1997. Latitudinal variations in benthic processes in the abyssal equatorial Pacific: Control by biogenic particle flux. Deep-Sea Research II **44:**2295–2317.

Smith, C. R., and S. J. Brumsickle. 1989. The effects of patch size and substrate isolation on colonization modes and rates in an intertidal sediment. Limnology and Oceanography **34:**1263–1277.

Smith, C. R., F. C. De Leo, A. F. Bernardino, A. K. Sweetman, and P. Martinez Arbizu. 2008. Abyssal food limitation, ecosystem structure and climate change. Trends in Ecology and Evolution **23:**518–528.

Smith, C. R., and A. W. J. Demopoulos. 2003. The deep Pacific Ocean floor. Pp. 179–218, *in* P. A. Tyler, ed. Ecosystems of the World 28: Ecosystems of the Deep Oceans. Elsevier, Amsterdam.

Smith, C. R., and S. C. Hamilton. 1983. Epibenthic megafauna of a bathyal basin off southern California: Patterns of abundance, biomass, and dispersion. Deep-Sea Research **30:**907–928.

Smith, C. R., D. J. Hoover, S. E. Doan, R. H. Pope, D. J. DeMaster, F. C. Dobbs, and M. A. Altabet. 1996. Phytodetritus at the abyssal seafloor across 10° of latitude in the central equatorial Pacific. Deep-Sea Research II **43**:1309–1338.

Smith, C. R., P. A. Jumars, and D. J. DeMaster. 1986. In situ studies of megafaunal mounds indicate rapid sediment turnover and community response at the deep-sea floor. Nature **323**:251–253.

Smith, C. R., L. A. Levin, D. J. Hoover, G. McMurtry, and J. D. Gage. 2000. Variations in bioturbation across the oxygen minimum zone in the northwest Arabian Sea. Deep-Sea Research II **47**:227–257.

Smith, C. R., H. L. Maybaum, A. R. Baco, R. H. Pope, S. D. Carpenter, P. L. Yager, S. A. Macko, and J. W. Deming. 1998. Sediment community structure around a whale skeleton in the deep Northeast Pacific: Macrofaunal, microbial and bioturbation effects. Deep-Sea Research **45**:335–364.

Smith, C. R., R. H. Pope, D. J. DeMaster, and L. Magaard. 1993. Age-dependent mixing of deep-sea sediments. Geochimica et Cosmochimica Acta **57**:1473–1488.

Smith, C. R., and C. Rabouille. 2002. What controls the mixed-layer depth in deep-sea sediments? The importance of POC flux. Limnology and Oceanography **47**:418–426.

Smith, J. N., and C. T. Schafer. 1984. Bioturbation processes in continental slope and rise sediments delineated by Pb-210, microfossil and textural indicators. Journal of Marine Research **42**:1117–1145.

Smith Jr., K. L. 1978. Benthic community respiration in the N. W. Atlantic Ocean: In situ measurements from 40 to 5200 m. Marine Biology **47**:337–347.

Smith Jr., K. L. 1987. Food energy supply and demand: A discrepancy between particulate organic carbon flux and sediment community oxygen consumption in the deep ocean. Limnology and Oceanography **32**:201–220.

Smith Jr., K. L. 1992. Benthic boundary layer communities and carbon cycling at abyssal depths in the central North Pacific. Limnology and Oceanography **37**:1034–1056.

Smith Jr., K. L., R. J. Baldwin, R. C. Glatts, R. S. Kaufmann, and E. C. Fisher. 1998. Detrital aggregates on the sea floor: Chemical composition and aerobic decomposition rates at a time-series station in the abyssal NE Pacific. Deep-Sea Research II **45**:843–880.

Smith Jr., K. L., R. J. Baldwin, D. M. Karl, and A. Boetius. 2002. Benthic community responses to pulses in pelagic food supply: North Pacific Subtropical Gyre. Deep-Sea Research I **49**:971–990.

Smith Jr., K. L., R. J. Baldwin, H. A. Ruhl, M. Kahru, B. G. Mitchell, and R. S. Kaufmann. 2006. Climate effect on food supply to depths greater than 4,000 meters in the northeast Pacific. Limnology and Oceanography **51**:166–176.

Smith Jr., K. L., and E. R. M. Druffel. 1998. Long time-series monitoring of an abyssal site in the NE Pacific: An introduction. Deep-Sea Research II **45:**573–586.

Smith Jr., K. L., and R. S. Kaufmann. 1999. Long-term discrepancy between food supply and demand in the deep-eastern North Pacific. Science **284:**1174–1177.

Smith Jr., K. L., R. S. Kaufmann, and R. J. Baldwin. 1994. Coupling of near-bottom pelagic and benthic processes at abyssal depths in the eastern North Pacific Ocean. Limnology and Oceanography **39:**1101–1118.

Smith Jr., K. L., R. S. Kaufmann, R. J. Baldwin, and A. F. Carlucci. 2001. Pelagic-benthic coupling in the abyssal eastern North Pacific: An 8-year time-series study of food supply and demand. Limnology and Oceanography **46:**543–556.

Smith, T. B., R. K. Wayne, D. J. Girman, and M. W. Bruford. 1997. A role for ecotones in generating rainforest biodiversity. Science **276:**1855–1857.

Snelgrove, P. V. R., C. A. Butman, and J. F. Grassle. 1995. Potential flow artifacts associated with benthic experimental gear: Deep-sea mudbox examples. Journal of Marine Research **53:**821–845.

Snelgrove, P. V. R., J. F. Grassle, and R. F. Petrecca. 1992. The role of food patches in maintaining high deep-sea diversity: Field experiments with hydrodynamically unbiased colonization trays. Limnology and Oceanography **37:**1543–1550.

Snelgrove, P. V. R., J. F. Grassle, and R. F. Petrecca. 1994. Macrofaunal response to artificial enrichments and depressions in a deep-sea habitat. Journal of Marine Research **52:**345–369.

Snelgrove, P. V. R., J. F. Grassle, and R. F. Petrecca. 1996. Experimental evidence for aging food patches as a factor contributing to high deep-sea macrofaunal diversity. Limnology and Oceanography **41:**605–614.

Snelgrove, P. V. R., and C. R. Smith. 2002. A riot of species in an environmental calm: The paradox of the species-rich deep-sea floor. Oceanography and Marine Biology: An Annual Review **40:**311–342.

Snider, L. J., B. R. Burnett, and R. R. Hessler. 1984. The composition and distribution of meiofauna and nanobiota in a central North Pacific deep-sea area. Deep-Sea Research **31:**1225–1249.

Soetaert, K., and C. Heip. 1989. The size structure of nematode assemblages along a Mediterranean deep-sea transect. Deep-Sea Research **36:**93–102.

Soetaert, K., and C. Heip. 1995. Nematode assemblages of deep-sea and shelf break sites in the North Atlantic and Mediterranean Sea. Marine Ecology Progress Series **125:**171–183.

Soetaert, K., C. Heip, and M. Vincx. 1991. The meiobenthos along a Mediterranean deep-sea transect off Calvi (Corsica) and in an adjacent canyon. Marine Ecology **12:**227–242.

Soetaert, K., A. Muthumbi, and C. Heip. 2002. Size and shape of ocean margin nematodes: Morphological diversity and depth-related patterns. Marine Ecology Progress Series **242:**179–193.

Soltwedel, T. 1997a. Meiobenthos distribution pattern in the tropical East Atlantic: Indication for fractionated sedimentation of organic matter to the sea floor? Marine Biology **129**:747–756.

Soltwedel, T. 1997b. Temporal variabilities in benthic activity and biomass on the western European continental margin. Oceanologica Acta **20**:871–879.

Soltwedel, T. 2000. Metazoan meiobenthos along continental margins: A review. Progress in Oceanography **46**:59–84.

Soltwedel, T., V. Mokievsky, and I. Schewe. 2000. Benthic activity and biomass on the Yermak Plateau and in adjacent deep-sea regions northwest of Svalbard. Deep-Sea Research I **47**:1761–1785.

Soltwedel, T., O. Pfannkuche, and H. Thiel. 1996. The size structure of deep-sea meiobenthos in the north-eastern Atlantic: Nematode size spectra in relation to environmental variables. Journal of the Marine Biological Association of the United Kingdom **76**:327–344.

Soltwedel, T., and H. Thiel. 1995. Biogenic sediment compounds in relation to marine meiofaunal abundances. Internationale Revue der gesamten Hydrobiologie **80**:297–311.

Soltwedel, T., and K. Vopel. 2001. Bacterial abundance and biomass in response to organism-generated habitat heterogeneity in deep-sea sediments. Marine Ecology Progress Series **219**:291–298.

Somero, G. N. 1990. Life at low-volume change: Hydrostatic pressure as a selective factor in the aquatic environment. American Zoologist **30**:123–135.

Somero, G. N. 1992. Adaptations to high hydrostatic-pressure. Annual Review of Physiology **54**:557–577.

Sommer, S., and O. Pfannkuche. 2000. Metazoan meiofauna of the deep Arabian Sea: Standing stocks, size spectra and regional variability in relation to monsoon induced enhanced sedimentation regimes of particulate organic matter. Deep-Sea Research II **47**:2957–2977.

Sparrow, M., O. Boebel, and V. Zervakis. 2002. Two circulation regimes of the Mediterranean outflow revealed by Lagrangian measurements. Journal of Physical Oceanography **32**:1322–1330.

Speer, K. G., M. E. Maltrud, and A. M. Thurnherr. 2003. A global view of dispersion on the mid-ocean ridge. Pp. 287–302, *in* P. Halbach, V. Tunnicliffe, and J. Hein, eds. Energy and Mass Transfer in Marine Hydrothermal Systems. Dahlem University Press, Berlin.

Spiess, F. N., R. Hessler, G. Wilson, and M. Weydert. 1987. Environmental effects of deep-sea dredging. SIO Reference 87-5, Final Report, NOAA Contract Number 83-SAC-00659.

Spratt, T. A. B., and E. Forbes. 1847. Travels in Lycia, Milyas, and the Cybyratis, in company with the late Rev. E. T. Daniell. Van Voorst, London.

Srivastava, D. S., and J. H. Lawton. 1998. Why more productive sites have more species: An experimental test of theory using tree-hole communities. American Naturalist **152:**510–529.

Ståhl, H., A. Tengberg, J. Brunnegård, and P. O. J. Hall. 2004. Recycling and burial of organic carbon in sediments of the Porcupine Abyssal Plain, NE Atlantic. Deep-Sea Research I **51:**777–791.

Stefanni, S., and H. Knutsen. 2007. Phylogeography and demographic history of the deep-sea fish *Aphanopus carbo* (Lowe, 1839) in the NE Atlantic: Vicariance followed by secondary contract or speciation? Molecular Phylogenetics and Evolution **42:**38–46.

Stehli, F. G., and R. G. Douglas. 1969. Generation and maintenance of gradients in taxonomic diversity. Science **164:**947–949.

Stepien, C. A. 1999. Phylogeographical structure of the Dover sole *Microstomus pacificus:* The larval retention hypothesis and genetic divergence along the deep continental slope of the northeastern Pacific Ocean. Molecular Ecology **8:**923–939.

Stepien, C. A., A. K. Dillon, and A. K. Patterson. 2000. Population genetics, phylogeography, and systematics of the thornyhead rockfishes (*Sebastolobus*) along the deep continental slopes of the North Pacific Ocean. Canadian Journal of Fisheries and Aquatic Science **57:**1701–1717.

Stevens, G. C. 1992. The elevational gradient in altitudinal range: An extension of Rapoport's latitudinal rule to altitude. American Naturalist **140:**893–911.

Stockton, W. L., and T. E. DeLaca. 1982. Food falls in the deep sea: Occurrence, quality, and significance. Deep-Sea Research **29:**157–169.

Stuart, C. T., and M. A. Rex. 1994. The relationship between development pattern and species diversity in deep-sea prosobranch snails. Pp. 119–136, *in* C. M. Young and K. J. Eckelbarger, eds. Reproduction, Larval Biology and Recruitment in the Deep-Sea Benthos. Columbia University Press, New York.

Stuart, C. T., M. A. Rex, and R. J. Etter. 2003. Large-scale spatial and temporal patterns of deep-sea benthic species diversity. Pp. 297–313, *in* P. A. Tyler, ed. Ecosystems of the World 28: Ecosystems of the Deep Oceans. Elsevier, Amsterdam.

Suchanek, T. H., S. L. Williams, J. C. Ogden, D. K. Hubbard, and I. P. Gill. 1985. Utilization of shallow-water seagrass detritus by Caribbean deep-sea macrofauna: ^{13}C evidence. Deep-Sea Research **32:**201–214.

Sun, X., B. H. Corliss, C. W. Brown, and W. J. Showers. 2006. The effect of primary productivity and seasonality on the distribution of deep-sea benthic foraminifera in the North Atlantic. Deep-Sea Research I **53:**28–47.

Svavarsson, J. 1997. Diversity of isopods (Crustacea): New data from the Arctic and Atlantic Oceans. Biodiversity and Conservation **6:**1571–1579.

Svavarsson, J., T. Brattegard, and J.-O. Strömberg. 1990. Distribution and diversity patterns of asellote isopods (Crustacea) in the deep Norwegian and Greenland Seas. Progress in Oceanography **24:**297–310.

Sweetman, A. K., and U. Witte. 2008a. Response of an abyssal macrofaunal community to a phytodetrital pulse. Marine Ecology Progress Series **355:**73–84.

Sweetman, A. K., and U. Witte. 2008b. Macrofaunal response to phytodetritus in a bathyal Norwegian fjord. Deep-Sea Research I **55:**1503–1514.

Tahey, T. M., G. C. A. Duineveld, E. M. Berghuis, and W. Helder. 1994. Relation between sediment-water fluxes of oxygen and silicate and faunal abundance at continental shelf, slope and deep-water stations in the northwest Mediterranean. Marine Ecology Progress Series **104:**119–130.

Talling, P. J., R. B. Wynn, D. G. Masson, M. Frenz, B. T. Cronin, R. Schiebel, A. M. Akhmetzhanov, S. Dallmeier-Tiessen, S. Benetti, P. P. E. Weaver, A. Georgiopoulou, C. Zühlsdorff, and L. A. Amy. 2007. Onset of submarine debris flow deposition far from original giant landslide. Nature **450:**541–544.

Tendal, O. S. 1972. A monograph of the Xenophyophoria (Rhizopodea, Protozoa). Galathea Report **12:**7–103.

Terborgh, J. 1971. Distribution on environmental gradients: Theory and a preliminary interpretation of distributional patterns in the avifauna of the Cordillera Vilcabamba, Peru. Ecology **52:**23–40.

Thiel, H. 1966. Quantitative Untersuchungen über die Meiofauna des Tiefseebodens. Veröffentlichungen des Instituts für Meeresforschung Bremerhaven Supplement **II:**131–148.

Thiel, H. 1975. The size structure of the deep-sea benthos. Internationale Revue der gesamten Hydrobiologie **60:**575–606.

Thiel, H. 1979a. First quantitative data on the deep Red Sea benthos. Marine Ecology Progress Series **1:**347–350.

Thiel, H. 1979b. Structural aspects of the deep-sea benthos. Ambio Special Report **6:**25–31.

Thiel, H. 1982. Zoobenthos of the CINECA area and other upwelling regions. Rapports et Procès-Verbaux des Réunions du Conseil International pour l'Exploration de la Mer **180:**323–324.

Thistle, D. 1978. Harpacticoid dispersion patterns: Implications for deep-sea diversity maintenance. Journal of Marine Research **36:**377–397.

Thistle, D. 1979. Harpacticoid copepods and biogenic structures: Implications for deep-sea diversity maintenance. Pp. 217–231, *in* R. J. Livingston, ed. Ecological Processes in Coastal and Marine Systems. Plenum, New York.

Thistle, D., S. C. Ertman, and K. Fauchald. 1991. The fauna of the HEBBLE site: Patterns in standing stock and sediment-dynamic effects. Marine Geology **99:**413–422.

Thistle, D., B. Hilbig, and J. E. Eckman. 1993. Are polychaetes sources of habitat heterogeneity for harpacticoid copepods in the deep sea? Deep-Sea Research I **40:**151–157.

Thistle, D., L. Sedlacek, K. R. Carman, J. W. Fleeger, and J. P. Barry. 2007. Emergence in the deep sea: Evidence from harpacticoid copepods. Deep-Sea Research I **54:**1008–1014.

Thistle, D., and G. D. F. Wilson. 1987. A hydrodynamically modified, abyssal isopod fauna. Deep-Sea Research **34:**73–87.

Thistle, D., and G. D. F. Wilson. 1996. Is the HEBBLE isopod fauna hydrodynamically modified? A second test. Deep-Sea Research I **43:**545–554.

Thistle, D., J. Y. Yingst, and K. Fauchald. 1985. A deep-sea benthic community exposed to strong near-bottom currents on the Scotian Rise (Western Atlantic). Marine Geology **66:**91–112.

Thomas, E., L. Booth, M. Maslin, and N. J. Shackleton. 1995. Northeastern Atlantic benthic foraminifera during the last 45,000 years: Productivity changes as seen from the bottom up. Paleoceanography **10:**545–562.

Thomas, E., and A. J. Gooday. 1996. Cenozoic deep-sea benthic foraminifers: Tracers for changes in oceanic productivity? Geology **24:**355–358.

Thomas, E., J. C. Zachos, and T. J. Bralower. 2000. Deep-sea environments on a warm earth: Latest Paleocene-early Eocene. Pp. 132–160, *in* K. MacLeod, ed. Warm Climates in Earth History. Cambridge University Press, Cambridge, UK.

Thorson, G. 1946. Reproduction and larval development of Danish marine bottom invertebrates. Meddelelser fra Kommissionen for Danmarks Fiskeriog Havundersøgelser **4:**1–523.

Thorson, G. 1950. Reproductive and larval ecology of marine bottom invertebrates. Biological Reviews **25:**1–45.

Thurston, M. H., B. J. Bett, A. L. Rice, and P. A. B. Jackson. 1994. Variations in the invertebrate abyssal megafauna in the North Atlantic Ocean. Deep-Sea Research I **41:**1321–1348.

Tietjen, J. H. 1971. Ecology and distribution of deep-sea meiobenthos off North Carolina. Deep-Sea Research **18:**941–957.

Tietjen, J. H. 1992. Abundance and biomass of metazoan meiobenthos in the deep sea. Pp. 45–62, *in* G. T. Rowe and V. Pariente, eds. Deep-Sea Food Chains and the Global Carbon Cycle. Kluwer, Dordrecht, The Netherlands.

Tietjen, J. H., J. W. Deming, G. T. Rowe, S. Macko, and R. J. Wilke. 1989. Meiobenthos of the Hatteras Abyssal Plain and Puerto Rico Trench: Abundance, biomass and associations with bacteria and particulate fluxes. Deep-Sea Research **36:** 1567–1577.

Todo, Y., H. Kitazato, J. Hashimoto, and A. J. Gooday. 2005. Simple foraminifera flourish at the ocean's deepest point. Science **307:**689.

Tselepides, A., and A. Eleftheriou. 1992. South Aegean (Eastern Mediterranean) continental slope benthos: Macroinfaunal-environmental relationships. Pp. 139–156, *in* G. T. Rowe and V. Pariente, eds. Deep-Sea Food Chains and the Global Carbon Cycle. Kluwer, Dordrecht, The Netherlands.

Tselepides, A., K.-N. Papadopoulou, D. Podaras, W. Plaiti, and D. Koutsoubas. 2000a. Macrobenthic community structure over the continental margin of Crete (South Aegean Sea, NE Mediterranean). Progress in Oceanography **46**:401–428.

Tselepides, A., V. Zervakis, T. Polychronaki, R. Danovaro, and G. Chronis. 2000b. Distribution of nutrients and particulate organic matter in relation to the prevailing hydrographic features of the Cretan Sea (NE Mediterranean). Progress in Oceanography **46**:113–142.

Tunnicliffe, V., S. K. Juniper, and M. Sibuet. 2003. Reducing environments in the deep oceans. Pp. 81–110, *in* P. A. Tyler, ed. Ecosystems of the World 28: Ecosystems of the Deep Oceans. Elsevier, Amsterdam.

Turner, R. D. 1973. Wood-boring bivalves, opportunistic species in the deep sea. Science **180**:1377–1379.

Turner, R. D. 1977. Wood, mollusks, and deep-sea food chains. Bulletin American Malacological Union **1977**:13–19.

Tyler, P. A., and C. M. Young. 1998. Temperature and pressure tolerances in dispersal stages of the genus *Echinus* (Echinodermata: Echinoidea): Prerequisites for deep-sea invasion and speciation. Deep-Sea Research II **45**:253–277.

Tyler, P. A., C. M. Young, and A. Clarke. 2000. Temperature and pressure tolerances of embryos and larvae of the Antarctic sea urchin *Sterechinus neumayeri* (Echinodermata: Echinoidea): Potential for deep-sea invasion from high latitudes. Marine Ecology Progress Series **192**:173–180.

Underwood, A. J., M. G. Chapman, and S. D. Connell. 2000. Observations in ecology: You can't make progress on processes without understanding the patterns. Journal of Experimental Marine Biology and Ecology **250**:97–115.

Vale, F. K., and M. A. Rex. 1988. Repaired shell damage in deep-sea prosobranch gastropods from the western North Atlantic. Malacologia **28**:65–79.

Vale, F. K., and M. A. Rex. 1989. Repaired shell damage in a complex of rissoid gastropods from the upper continental slope south of New England. Nautilus **103**:105–108.

Vanaverbeke, J., K. Soetaert, C. Heip, and A. Vanreusel. 1997. The metazoan meiobenthos along the continental slope of the Goban Spur (NE Atlantic). Journal of Sea Research **38**:93–107.

Van Dover, C. L. 2000. The Ecology of Deep-Sea Hydrothermal Vents. Princeton University Press, Princeton, NJ.

Vanhove, S., J. Wittoeck, G. Desmet, B. Van den Berghe, R. L. Herman, R. P. M. Bak, G. Nieuwland, J. H. Vosjan, A. Boldrin, S. Rabitti, and M. Vincx. 1995. Deep-sea meiofauna communities in Antarctica: Structural analysis and relation with the environment. Marine Ecology Progress Series **127**:65–76.

Vanreusel, A., N. Cosson-Sarradin, A. J. Gooday, G. L. J. Paterson, J. Galéron, M. Sibuet, and M. Vincx. 2001. Evidence for episodic recruitment in a small opheliid polychaete species from the abyssal NE Atlantic. Progress in Oceanography **50**:285–301.

Vanreusel, A., M. Vincx, D. Schram, and D. Van Gansbeke. 1995. On the vertical distribution of the metazoan meiofauna in shelf break and upper slope habitats of the NE Atlantic. Internationale Revue der gesamten Hydrobiologie **80**:313–326.

Vanreusel, A., M. Vincx, D. Van Gansbeke, and W. Gijselinck. 1992. Structural analysis of the meiobenthos communities of the shelf break area in two stations of the Gulf of Biscay (N.E. Atlantic). Belgian Journal of Zoology **122**:185–202.

van Weering, T. C. E., H. C. De Stigter, W. Balzer, E. H. G. Epping, G. Graf, I. R. Hall, W. Helder, A. Khripounoff, L. Lohse, I. N. McCave, L. Thomasen, and A. Vangriesheim. 2001. Benthic dynamics and carbon fluxes on the NW European continental margin. Deep-Sea Research II **48**:3191–3221.

Vellend, M. 2005. Species diversity and genetic diversity: Parallel processes and correlated patterns. American Naturalist **166**:199–215.

Vetter, E. W. 1994. Hotspots of benthic production. Nature **372**:47.

Vetter, E. W., and P. K. Dayton. 1998. Macrofaunal communities within and adjacent to a detritus-rich submarine canyon system. Deep-Sea Research II **45**:25–54.

Villalobos, F. B., P. A. Tyler, and C. M. Young. 2006. Temperature and pressure tolerance of embryos and larvae of the Atlantic seastars *Asterias rubens* and *Marthasterias glacialis* (Echinodermata: Asteroidea): Potential for deep-sea invasion. Marine Ecology Progress Series **314**:109–117.

Vincx, M., B. J. Bett, A. Dinet, T. Ferrero, A. J. Gooday, P. J. D. Lambshead, O. Pfannkuche, T. Soltwedel, and A. Vanreusel. 1994. Meiobenthos of the deep Northeast Atlantic. Advances in Marine Biology **30**:1–88.

Vinogradova, N. G. 1958. Vertical distribution of deep-sea bottom fauna in the ocean. Transactions of the Institute of Oceanology of the Academy of Sciences of the USSR **27**:87–127.

Vinogradova, N. G. 1962. Vertical zonation in the distribution of deep-sea benthic fauna in the ocean. Deep-Sea Research **8**:245–250.

Vivier, M. H. 1978. Influence d'un déversement industriel profond sur la nématofauna (Canyon de Cassidaigne, Méditerranée). Tethys **8**:307–321.

von Humboldt, A. 1808. Ansichten der Natur mit wissenschaftlichen Erläuterungen. J. G. Cotta, Tübingen.

Walsh, J. J., D. A. Dieterle, M. B. Meyers, and F. E. Müller-Karger. 1989. Nitrogen exchange at the continental margin: A numerical study of the Gulf of Mexico. Progress in Oceanography **23**:245–301.

Waniek, J. J., D. E. Schulz-Bull, T. Blanz, R. D. Prien, A. Oschlies, and T. J. Müller. 2005. Interannual variability of deep water particle flux in relation to production and lateral sources in the northeast Atlantic. Deep-Sea Research I **52**:33–50.

Wares, J. P. 2002. Community genetics in the Northwestern Atlantic intertidal. Molecular Ecology **11**:1131–1144.

Wares, J. P., S. D. Gaines, and C. W. Cunningham. 2001. A comparative study of asymmetric migration events across a marine biogeographic boundary. Evolution **55**:295–306.

Weaver, P. P. E., and P. J. Schultheiss. 1983. Vertical open burrows in deep-sea sediments 2 m in length. Nature **301**:329–331.

Wei, C.-L., G. T. Rowe, A. H. Scheltema, G. D. F. Wilson, M. K. Wicksten, M. Chen, Y. Soliman, and Y. Wang. 2009. The bathymetric zonation and community structure of deep-sea macrobenthos in the northern Gulf of Mexico. Marine Ecology Progress Series (in press).

Weinberg, J. R., T. G. Dahlgren, N. Trowbridge, and K. M. Halanych. 2003. Genetic differences within and between species of deep-sea crabs (*Chaceon*) from the North Atlantic Ocean. Biological Bulletin **204**:318–326.

Weisshappel., J. B. F., and J. Svavarsson. 1998. Benthic amphipods (Crustacea: Malacostraca) in Icelandic waters: Diversity in relation to faunal patterns from shallow to intermediate deep Arctic and North Atlantic Oceans. Marine Biology **131**:133–143.

Welborn, J. R., and D. T. Manahan. 1991. Biochemical responses of Antarctic Asteroid Larvae to dissolved organic-matter. American Zoologist **31**:A4.

Wheatcroft, R. A., and P. A. Jumars. 1987. Statistical reanalysis for size dependence in deep-sea mixing. Marine Geology **77**:157–163.

Wheatcroft, R. A., C. R. Smith, and P. A. Jumars. 1989. Dynamics of surficial trace assemblages in the deep sea. Deep-Sea Research **36**:71–91.

Wheeler, P. A., M. Gosselin, E. Sherr, D. Thibault, D. L. Kirchman, R. Benner, and T. E. Whitledge. 1996. Active cycling of organic carbon in the central Arctic Ocean. Nature **380**:697–699.

White, B. N. 1987. Oceanic anoxic events and allopatric speciation in the deep sea. Biological Oceanography **5**:243–259.

Whittaker, R. H. 1960. Vegetation of the Siskiyou Mountains, Oregon and California. Ecological Monographs **30**:279–338.

Wiens, J. J. 2007. Global patterns of diversification and species richness in amphibians. American Naturalist **170**:S86–S106.

Wiens, J. J., G. Parra-Olea, M. Garcia-Paris, and D. B. Wake. 2007. Phylogenetic history underlies elevational biodiversity patterns in tropical salamanders. Proceedings of the Royal Society of London B: Biological Sciences **274**:919–928.

Wigham, B. D., I. R. Hudson, D. S. M. Billett, and G. A. Wolff. 2003a. Is long-term change in the abyssal Northeast Atlantic driven by qualitative changes in export flux? Evidence from selective feeding in deep-sea holothurians. Progress in Oceanography **59**:409–441.

Wigham, B. D., P. A. Tyler, and D. S. M. Billett. 2003b. Reproductive biology of the abyssal holothurian *Amperima rosea:* An opportunistic response to variable flux of surface derived organic matter? Journal of the Marine Biology Association of the United Kingdom **83**:175–188.

Wigley, R. L., and A. D. McIntyre. 1964. Some quantitative comparisons of offshore meiobenthos and macrobenthos south of Martha's Vineyard. Limnology and Oceanography **9**:485–493.

Wilhelm, R., and T. J. Hilbish. 1998. Assessment of natural selection in a hybrid population of mussels: Evaluation of exogenous vs endogenous selection models. Marine Biology **131**:505–514.

Williams, S. T., and D. G. Reid. 2004. Speciation and diversity on tropical rocky shores: A global phylogeny of snails of the genus *Echinolittorina*. Evolution **58**: 2227–2251.

Wilson, G. D. F. 1991. Functional morphology and evolution of isopod genetalia. Pp. 228–245, *in* R. Bauer and J. Martin, eds. Crustacean Sexual Biology. Columbia University Press, New York.

Wilson, G. D. F. 1998. Historical influences on deep-sea isopod diversity in the Atlantic Ocean. Deep-Sea Research II **45**:279–301.

Wilson, G. D. F. 1999. Some of the deep-sea fauna is ancient. Crustaceana **72**:1020–1030.

Wilson, G. D. F. 2009. Local and regional species diversity of benthic isopoda (Crustacea) in the deep Gulf of Mexico. Deep-Sea Research II **55**:2634–2649.

Wilson, G. D. F., and R. R. Hessler. 1987. Speciation in the deep sea. Annual Review of Ecology and Systematics **18**:185–207.

Witbaard, R., G. C. A. Duineveld, J. Van der Weele, E. M. Berghuis, and J. P. Reyss. 2000. The benthic response to the seasonal deposition of phytopigments at the Porcupine Abyssal Plain in the North East Atlantic. Journal of Sea Research **43**:15–31.

Witman, J. D., R. J. Etter, and F. Smith. 2004. The relationship between regional and local species diversity in marine benthic communities: A global perspective. Proceedings of the National Academy of Sciences, USA **101**:15664–15669.

Witte, U. 2000. Vertical distribution of metazoan macrofauna within the sediment at four sites with contrasting food supply in the deep Arabian Sea. Deep-Sea Research II **47**:2979–2997.

Witte, U., N. Aberle, M. Sand, and F. Wenzhöfer. 2003a. Rapid response of a deep-sea benthic community to POM enrichment: An *in situ* experimental study. Marine Ecology Progress Series **251**:27–36.

Witte, U., F. Wenzhöfer, S. Sommer, A. Boetius, P. Heinz, N. Aberle, M. Sand, A. Cremer, W.-R. Abraham, B. B. Jørgensen, and O. Pfannkuche. 2003b. *In situ* experimental evidence of the fate of a phytodetritus pulse at the abyssal sea floor. Nature **424**:763–766.

Wlodarska-Kowalczuk, M., M. A. Kendall, J. M. Weslawski, M. Klages, and T. Soltwedel. 2004. Depth gradients of benthic standing stock and diversity on the continental margin at a high-latitude ice-free site (off Spitsbergen, 79°N). Deep-Sea Research I **51**:1903–1914.

Wolff, T. 1956. Isopoda from depths exceeding 6000 meters. Galathea Report **2**:85–157.

Wolff, T. 1979. Macrofaunal utilization of plant remains in the deep sea. Sarsia **64**:117–136.

Wollenburg, J. E., and W. Kuhnt. 2000. The response of benthic foraminifers to carbon flux and primary production in the Arctic Ocean. Marine Micropaleontology **40**:189–231.

Wollenburg, J. E., and A. Mackensen. 1998. Living benthic foraminifers from the central Arctic Ocean: Faunal composition, standing stock and diversity. Marine Micropaleontology **34**:153–185.

Worm, B., and J. E. Duffy. 2003. Biodiversity, productivity and stability in real food webs. Trends in Ecology and Evolution **18**:628–632.

Worm, B., H. K. Lotze, H. Hillebrand, and U. Sommer. 2002. Consumer versus resource control of species diversity and ecosystem functioning. Nature **417**:848–851.

Wright, D. H. 1983. Species-energy theory: An extension of species-area theory. Oikos **41**:496–506.

Wright, S. D., R. D. Gray, and R. C. Gardner. 2003. Energy and the rate of evolution: Inferences from plant rDNA substitution rates in the western Pacific. Evolution **57**:2893–2898.

Wyville Thomson, C. 1873. The Depths of the Sea. Macmillan, London.

Wyville Thomson, C. 1878. The Voyage of the "Challenger": The Atlantic. Harper and Brothers, New York.

Xiang, Q., H.-M. Zhang, R. E. Rickleffs, H. Qian, Z. D. Chen, J. Wen, and J. H. Li. 2004. Regional differences in rates of plant speciation and molecular evolution: A comparison between eastern Asia and eastern North America. Evolution **58**:2175–2184.

Yasuhara, M., and T. M. Cronin. 2008. Climatic influences on deep-sea ostracode (Crustacea) diversity for the last three million years. Ecology **89**:S53–S65.

Yasuhara, M., T. M. Cronin, P. B. deMenocal, H. Okahashi, and B. K. Linsley. 2008. Abrupt climate change and collapse of deep-sea ecosystems. Proceedings of the National Academy of Sciences, USA **105**:1556–1560.

Yingst, J. Y., and D. C. Rhoads. 1985. The structure of soft-bottom benthic communities in the vicinity of the Texas Flower Garden Banks, Gulf of Mexico. Estuarine, Coastal and Shelf Science **20**:569–592.

Yoder, J. A., J. K. Moore, and R. N. Swift. 2001. Putting together the big picture: Remote-sensing observations of ocean color. Oceanography **14**:33–40.

Yoder, J. A., S. E. Schollaert, and J. E. O'Reilly. 2002. Climatological phytoplankton chlorophyll and sea surface temperature patterns in continental shelf and slope waters off the northeast U.S. coast. Limnology and Oceanography **47**:672–682.

Young, C. M. 2003. Reproduction, development and life-history traits. Pp. 381–426 *in* P. A. Tyler, ed. Ecosystems of the World 28: Ecosystems of the Deep Oceans. Elsevier, Amsterdam.

Young, C. M., M. G. Devin, W. B. Jaeckle, S. U. K. Ekaratne, and S. B. George. 1996a. The potential for ontogenetic vertical migration by larvae of bathyal echinoderms. Oceanologica Acta **19**:263–271.

Young, C. M., and K. J. Eckelbarger, eds. 1994. Reproduction, Larval Biology, and Recruitment of the Deep-Sea Benthos. Columbia University Press, New York.

Young, C. M., S. U. K. Ekaratne, and J. L. Cameron. 1998. Thermal tolerances of embryos and planktotrophic larvae of *Archaeopneustes hystrix* (A. Agassiz) (Spatangoidea) and *Stylocidaris lineata* (Mortensen) (Cidaroidea), bathyal echinoids from the Bahamian Slope. Journal of Experimental Marine Biology and Ecology **223**:65–76.

Young, C. M., S. Fujio, and R. C. Vrijenhoek. 2008. Directional dispersal between mid-ocean ridges: Deep-ocean circulation and gene flow in *Ridgeia piscesae*. Molecular Ecology **17**:1718–1731.

Young, C. M., M. A. Sewell, P. A. Tyler, and A. Metaxas. 1997a. Biogeographic and bathymetric ranges of Atlantic deep-sea echinoderms and ascidians: The role of larval dispersal. Biodiversity and Conservation **6**:1507–1522.

Young, C. M., and P. A. Tyler. 1993. Embryos of the deep-sea echinoid *Echinus affinis* require high pressure for development. Limnology and Oceanography **38**:178–181.

Young, C. M., P. A. Tyler, and L. Fenaux. 1997b. Potential for deep sea invasion by Mediterranean shallow water echinoids: Pressure and temperature as stage-specific dispersal barriers. Marine Ecology Progress Series **154**:197–209.

Young, C. M., P. A. Tyler, and J. D. Gage. 1996b. Vertical distribution correlates with pressure tolerances of early embryos in the deep-sea asteroid *Plutonaster bifrons*. Journal of the Marine Biological Association of the United Kingdom **76**:749–757.

Zachos, J. C., U. Röhl, S. A. Schellenberg, A. Sluijs, D. A. Hodell, D. C. Kelly, E. Thomas, M. Nicolo, I. Raffi, L. J. Lourens, H. McCarren, and D. Kroon. 2005. Rapid acidification of the ocean during the Paleocene-Eocene thermal maximum. Science **308**:1611–1615.

Zapata, F. A., K. J. Gaston, and S. L. Chown. 2005. The mid-domain effect revisited. American Naturalist **166**:144–148.

Zar, J. H. 1984. Biostatistical Analysis. Prentice-Hall, Englewood Cliffs, NJ.

Zardus, J. D. 2002. Protobranch bivalves. Advances in Marine Biology **42**:1–65.

Zardus, J. D., R. J. Etter, M. R. Chase, M. A. Rex, and E. E. Boyle. 2006. Bathymetric and geographic population structure in the pan-Atlantic deep-sea bivalve *Deminucula atacellana* (Schenck 1939). Molecular Ecology **15**:639–651.

INDEX

Page numbers for entries occurring in figures are followed by an *f* and those for entries occurring in tables, by a *t*.